Ecology of Salmonids in Estuaries around the World

Colin D. Levings

Ecology of Salmonids in Estuaries around the World
Adaptations, Habitats, and Conservation

UBCPress · Vancouver · Toronto

© UBC Press 2016

All rights reserved. No part of this publication may be reproduced, stored in a retrieval system, or transmitted, in any form or by any means, without prior written permission of the publisher, or, in Canada, in the case of photocopying or other reprographic copying, a licence from Access Copyright, www.accesscopyright.ca.

25 24 23 22 21 20 19 18 17 16   5 4 3 2 1

Printed in Canada on FSC-certified ancient-forest-free paper (100% post-consumer recycled) that is processed chlorine- and acid-free.

Library and Archives Canada Cataloguing in Publication

Levings, C. D., author
    Ecology of salmonids in estuaries around the world : adaptations, habitats, and conservation / Colin D. Levings.

Includes bibliographical references and index.
Issued in print and electronic formats.
ISBN 978-0-7748-3173-4 (hardback). – ISBN 978-0-7748-3175-8 (pdf)

    1. Salmonidae. 2. Salmonidae – Adaptation. 3. Salmonidae – Habitat. 4. Salmonidae – Conservation. 5. Estuaries. I. Title.

QL638.S2L49 2016   597.5'5   C2016-902474-1
                                                       C2016-902475-X

UBC Press gratefully acknowledges the financial support for our publishing program of the Government of Canada (through the Canada Book Fund), the Canada Council for the Arts, the British Columbia Arts Council, and with the help of the University of British Columbia through the K.D. Srivastava Fund.

The author gratefully acknowledges the substantial support of the Pacific Salmon Foundation towards publication of this book. Author royalties will be donated to the foundation.

Printed and bound in Canada by Friesens
Set in Stone by Artegraphica Design Co. Ltd.
Copy editor and proofreader: Frank Chow
Indexer: Heather Ebbs
Cartographer: Eric Leinberger

UBC Press
The University of British Columbia
2029 West Mall
Vancouver, BC V6T 1Z2
www.ubcpress.ca

Contents

List of Figures and Tables / vii

Preface / x

Acknowledgments / xiv

1 Why a Focus on Salmonids in Estuaries? / 3

2 What Salmonids and Estuaries to Consider / 13

3 Salmonid's-Eye View of the Estuary: Physical, Chemical, and Geological Aspects / 19

4 What Habitats Are Used by Salmonids in Estuaries? / 31

5 Global Distribution of Salmonid Species and Local Salmonid Diversity in Estuaries / 52

6 How Have Salmonid Abundance and Distribution Been Assessed in Estuaries? / 63

7 How Do Salmonids Behave in Estuarine Habitat? / 72

8 Salmonid Growth in the Estuary / 81

9 Smolting and Osmoregulation / 88

10 Habitat-Based Food Webs Supporting Salmonids in the Natural Estuary / 95

11 Biotic Interactions in the Natural Estuary / 109

12 How Have Habitat and Water Properties Changed for Salmonids in Estuaries? / 128

13 Salmonid Survival in Estuaries / 137

14 Effects of Habitat and Community Change on Fitness Components for Survival in the Disrupted Estuary / 149

15 Harvesting and Production of Salmonids and Other Ecosystem Services Provided by the Estuary / 187

16 Health of Salmonids in Estuaries / 193

17 What Shapes an Estuary for Salmonids? / 199

18 Future Considerations for Conservation of Salmonids in Estuaries / 215

19 Conclusion / 246

Glossary / 255

References / 265

Index / 351

Figures and Tables

Figures

1 Generalized life history of iteroparous and semelparous anadromous salmonids / 6
2 The Yamato River estuary, Japan, from the eighteenth century / 9
3 Map showing locations of the estuaries and other water bodies mentioned in the book / 14
4 Estuary classifications according to salinity stratification / 21
5 Idealized geomorphology of selected estuary forms used by salmonids / 23
6 Seasonal fluctuations in surface temperature and salinity at the Campbell River estuary, British Columbia / 27
7 Fluctuations in river discharge into the Campbell River estuary, British Columbia / 29
8 Beach and vegetation habitat near Tvärminne, Finland, on the Baltic Sea / 38
9 Eastern sector of the outer Homathko River estuary, British Columbia / 40
10 Intertidal zone of Bute Inlet, British Columbia / 42
11 Seasonal trends in abundance of chum salmon fry and subhabitats sampled at Roberts Bank, Fraser River estuary, British Columbia / 44
12 Number of native nonsalmonids, invasive nonsalmonids, and native iteroparous and semelparous salmonid species at the Fraser River estuary, British Columbia, and the Sacramento–San Joaquin River estuary, California / 56

13 Brown trout at the Klamath River estuary, California, and sea trout at the Rio Grande River estuary, Argentina / 61
14 Antennae for PIT tags and fyke net for sampling juvenile Chinook salmon at the Salmon River estuary, Oregon / 69
15 Seasonal change in abundance of wild juvenile Chinook salmon at the Campbell River estuary, British Columbia / 76
16 Seasonal change in mean length of Chinook salmon fry at sandflats and brackish marsh, Fraser River estuary, British Columbia / 82
17 One of the first portraits of an estuarine food web supporting juvenile salmonids / 98
18 Food of whitespotted char downstream migrating young and kelts at the Bogataya River estuary, Russia / 99
19 Vertical distribution of potential salmonid prey in the stratified Fraser River estuary, British Columbia / 103
20 Predation by steelhead smolts and starry flounder in relation to pink salmon fry abundance at the riverine channel of the Utka River estuary, Russia / 117
21 Relationship between size differences of subyearling and yearling Chinook salmon and a coefficient of food similarity at the Bol'shaya River estuary, Russia / 124
22 Sea wall at Île d'Orléans, St. Lawrence River estuary, Québec / 130
23 Idealized design of reach survival tracking and experimental transfer approaches to estimate survival of Chinook salmon smolts / 142
24 Exposure to poor environmental conditions in fresh water may affect osmoregulatory performance of salmonid smolts in the estuary / 151
25 Weight of wild Chinook salmon fry at the Campbell River estuary, British Columbia, in relation to total biomass of salmonid juveniles / 179
26 A ternary plot to array fitness elements thought to support survival of juvenile salmonids / 185
27 Shift in Atlantic salmon catches from coastal to estuarine and river fisheries in Northern Ireland / 190
28 Possible multiple-stage scheme that helps explain factors that enable or restrict the dispersal of salmonids into estuaries by affecting survival and hence individual fitness / 200
29 Catch data from the Chinook salmon gillnet fisheries at the Fraser River estuary, British Columbia / 220

30 Historical and contemporary life history patterns for juvenile Chinook salmon in the Columbia River estuary, Washington-Oregon / 222

Tables

1 Idealized salmonid estuarine zones, with descriptors and general physicochemical properties / 26
2 A practical framework for salmonid estuarine habitats / 36
3 Selected examples of biological factors affecting food acquisition by salmonids in estuaries / 106
4 Summary of aspects or factors to consider when selecting methods of assessing predation on salmonids in estuaries / 120
5 Number of adults recovered from a Chinook salmon smolt transfer experiment at the Campbell River estuary, British Columbia / 142
6 Approaches for assessing the effects of habitat-related food deficits on estuarine salmonid growth / 165
7 Summary of possible effects of habitat loss or community change on predation and competition / 176
8 Summary of potential external stressors upstream and downstream of the salmonid estuary / 232

Supplemental Materials

The following web-based supplemental materials can be accessed at: http://hdl.handle.net/2429/57062.

Appendix 1. Data Tables Referred to in the Book

Appendix 2. Estuaries and Other Water Bodies Mentioned in the Book and Appendix 1, with Geographic Information and Coordinates

Appendix 3. Estuary Primer: An Overview of Salmonids and Estuarine Ecosystems for Citizen Scientists

Appendix 4. Additional Literature Consulted

Appendix 5. Colour Photos Linked to Figures in the Book

Preface

My intent in writing this book is to introduce and orient present and future researchers, students, and practitioners to the study of salmonid conservation in estuaries. I focus on some important anadromous salmonines, which for the purpose of this book I call "salmonids." Although a number of recent books and review reports have dealt with the oceanic and river phases of salmonids, to my knowledge there are no comprehensive publications that focus specifically on the role of estuaries in salmonid ecology and management. Compilation of current information on estuaries is vital to conservation efforts that aim to maintain productivity and diversity, improve survival, and develop recovery plans for threatened salmonids. Another aim is to point out some of the critical data gaps that need investigation to enable us to move ahead in this area of salmonid ecology. Besides reviewing and discussing the scientific literature, I prepared a guide to the salmonid estuary and a glossary for general readers and citizen scientists; these are found online as supplemental materials to this book. Supporting data and citations for additional papers consulted are given in Supplemental Materials, available online at: http://hdl.handle.net/2429/57062. The mention of appendices or appendix tables in the text refers to those materials. For example, "Appendix Table 1.1" refers to Table 1.1 in Appendix 1.

Salmonids are now found in estuaries on all continents except Africa and Antarctica, so a global scope is required. This book presents information from 196 estuaries around the world. It reviews, synthesizes, and summarizes the scientific information on ecological relationships between the various life stages of anadromous salmonids, the fish they coexist with, and their estuarine habitats around the world. It deals with data on estuarine ecology of four salmonid genera: *Hucho* (one species), *Oncorhynchus* (nine species), *Salmo* (two species), and *Salvelinus* (six species). In the process of writing the book, I reviewed papers and reports up 2014 and a few from 2015. The References section, together with the list of additional papers consulted (found in the online Supplemental Materials), should provide a good introduction to the literature on salmonids in estuaries. To help me understand the features of the estuaries cited, I used Google Maps or reviewed my personal observations.

The study of salmonids in estuaries is rather specialized, employing sampling approaches not necessarily used in their freshwater and marine habitats. To provide guidance for estuarine sampling, I describe the methods used around the world, as well as their limitations. Because of the inherent variability of the estuary, especially tidal range and water properties, and the range of geological structures, sampling for salmonids is a challenge. Estimating vital statistics such as growth, residency, and survival of discrete populations or cohorts is also challenging because the fish are often on a rearing migration through the mosaic of estuary habitats. New tagging and biochemical techniques are making measurements easier, however.

One of the objectives of this book is to demonstrate the wide range of estuary types used by salmonids and reveal the ability of these fish to successfully adapt to different habitats and communities of fishes. Several salmonid estuary regions around the world have received the most detailed attention in the scientific literature, notably the North Pacific and northeast Atlantic estuaries. In recent years, studies have become available on salmonids in estuaries in the Southern Hemisphere, where salmonid introductions began in the mid-1800s. This book provides information on all salmonid estuary regions.

Estuaries have a profile in the literature discussing salmonid evolution and salmonid fitness research. Relative to river and ocean habitats, however, how salmonid estuarine habitats confer fitness for survival is understudied. Assessing salmonid fitness in estuaries can require a variety of methodologies because estuaries are open systems with boundary issues – the estuary is a salmonid macrohabitat on a continuum from river to the ocean for juveniles and from the ocean to river for adults. I give an overview of our knowledge of salmonid fitness components in estuaries and discuss methods of assessing survival, the chief outcome of fitness.

Three basic ecophysiological concepts are thought to underpin the benefits of estuarine residency for the fitness components of survival for most species of salmonids: (1) successfully changing from fresh to salt water, (2) obtaining a growth advantage as more food may be found in the estuary relative to the river, and (3) biotic interactions, especially predation and provision of habitat as refuge. I discuss the evidence and species differences in these generalities. I also discuss how the three components have been altered in the postindustrial-era estuary, and consider the effects of directed harvesting, domestication, and health on salmonids. The focus is on how these changes affect salmonid fitness. Many studies of salmonids in estuaries have been conducted in partially disrupted estuaries and with overexploited populations. Expanding detailed studies of salmonid estuaries in the relatively intact regions of the Northern Hemisphere would be helpful in this regard.

Perhaps because estuaries are usually a small component of the geographic range of many salmonids, estuarine habitats are often managed with few data compared with other critical habitats, such as spawning grounds or stream rearing areas. There are some cases where estuary factors affecting survival are clear, for example, severe water quality degradation (especially dissolved oxygen) and migration barriers such as barrages and dams. However, more subtle effects, such as habitat degradation and invasive species, usually require more study before they can be definitively identified as limiting factors in the estuary; I provide examples showing how these issues might be examined. A variety of ecological indicators have been developed, and I review some of their advantages and disadvantages. I also review selected estuary management strategies on the west coast of North America and Europe, and discuss the information required for salmonid conservation. One caveat: management plans are continually evolving. Managers need to realize that estuaries are ecotones and that they need to look both upstream and downstream in their planning. Adequate downstream flow from the river and unimpeded entrance of ocean water into the estuary are critical to maintaining the ecological functions supporting salmonids. The maintenance and recovery of ecosystem properties, such as ecological integrity and resilience in estuaries around the world, are a central management theme. Briefly, ecological integrity is said to exist when an ecosystem is deemed characteristic for its natural region, while resilience can be measured by the degree of disturbance that an ecosystem can withstand without changing self-organized processes and structures. Salmonid estuary conservation and management would be improved by further efforts to measure ecological integrity and resilience. Socio-economic factors also need to be taken into account in a true landscape approach to salmonid estuary management. Adaptive management of the salmonid estuary would benefit from implementation of long-term strategies for monitoring of estuarine restoration projects. Such management is a challenge, but the ever-increasing pace of estuary development demands the timely implementation of proven approaches. As well, new, innovative applied science methods are needed to ensure the sustainability of salmonids in estuaries.

I am grateful to family members, mentors, students, and colleagues who supported and guided me along a rewarding career that led to this book. Growing up in coastal British Columbia, Canada, in a family often supported by the salmonid fishing industry, I learned the basics of the salmonid ecosystem from watching gill-netters and seine boats on the coast and from fishing with rod and reel on the sandbars of the Fraser River estuary. In 1972, I was hired by the Fisheries Research Board of Canada at the laboratory in West Vancouver, British Columbia, to study estuarine habitat. Over the next thirty-five years, I was fortunate to be able to study twenty salmonid estuaries on the British Columbia coast. I also learned about salmonid estuaries elsewhere in the world by participating in various international exchanges and advisory committees, especially in Norway and the Pacific Northwest of the United States. I am extremely grateful to all the people who helped me along the way. While preparing this book, I benefited enormously from discussions with, and information provided by, many people. Most are mentioned by name in the Acknowledgments, and I apologize for any who have been omitted. I hope that this book will in some way repay the scientific community and society in general for the opportunity to work with these iconic fish in such intriguing habitats.

Acknowledgments

I began this book while serving as a Fisheries and Oceans Canada Research Scientist on special assignment to the Pacific Science Branch in the Marine Ecosystems and Aquaculture Division at the Centre for Aquaculture and Environmental Research in West Vancouver. I am grateful to Dr. John Pringle for providing time and technical support for the project during that period. I continued working on the book as a Scientist Emeritus in the Division, with the support of Drs. Steve Macdonald and Laura Brown. Thanks are due to a number of colleagues for their indispensable help: Beth Piercey (Fisheries and Oceans Canada, Centre for Aquaculture and Environmental Research, West Vancouver) for literature collection and compilation; Gordon Miller (Library, Pacific Biological Station, Fisheries and Oceans Canada) for tracking down papers and reports; Dr. J.S. Macdonald (Fisheries and Oceans Canada, Centre for Aquaculture and Environmental Research, West Vancouver), Charles A. (Si) Simenstad (University of Washington, Seattle), Drs. Bob Wissmar and Bob Naiman (University of Washington, Seattle; retired), the late Dr. Bob Emmett (Hammond Laboratory, Hammond, Oregon, of the Northwest Fisheries Science Center), and Dr. David Jay (Portland State University, Portland, Oregon) for discussions and advice on matters estuarine and salmonid over the years; the late Dr. John Luternauer (Geological Survey of Canada) for consultation on estuarine geology; T. Sakamaki (University of British Columbia) for Japanese translations; Nils Arne Hvidsten, Arne J. Jensen, and Bjorn Ove Johnsen (Norwegian Institute for Nature Research, Trondheim) for consultation when I worked as a Visiting Scientist at the Institute, and for document review; Dr. Vladimir Karpenko (Kamchatka Institute for Fisheries and Oceanography, Petropavlovsk-Kamchatsky, Russia) for discussions and consultations during my visit to Kamchatka; and Drs. Ryusuke Hosoda (Osaka

Prefecture University) and Humitake Seki (University of Tsukuba) for discussions, contacts, and consultations during my visits to Japan. Dr. Kai Chan (Institute for Resources, Environment and Sustainability at the University of British Columbia) facilitated library work and desk space. Thanks are also due to Randy Schmidt, Senior Editor at UBC Press, for his assistance and guidance as various drafts of this book were being reviewed. I am grateful to Megan Brand, Editor at UBC Press, for her excellent help with the production process.

I thank the following for detailed personal communications: Dr. Eric Taylor (Department of Zoology, UBC); Mike Wallace (California Department of Fish and Wildlife, Arcata); Dr. Greg Ruggerone (Natural Resources Consultants, Inc., Seattle); Dr. Brian Dempson (Department of Fisheries and Oceans, St. John's, Newfoundland and Labrador); Dr. David Archer (Northumberland, England); Dr. Murray Hicks (New Zealand National Institute of Water and Atmospheric Research, Christchurch); Dr. Maxim Koval (Kamchatka Institute for Fisheries and Oceanography); Dr. Tony Farrell (Land and Food Systems, University of British Columbia); Professor Carlos Garcia De Leaniz (College of Bioscience, Swansea University, United Kingdom); Dr. K. Korsu (University of Oulu, Oulu, Finland); Dr. Peter Amiro (Fisheries and Oceans Canada, Halifax, Nova Scotia; retired); Dr. Casey Rice (Northwest Fisheries Science Center, Seattle); Dr. Ted Potter (Centre for Environment, Fisheries and Aquaculture Science, Lowestoft, United Kingdom); Dr. Sagy Cohen (University of Alabama, Tuscaloosa); Albert Kettner and Dr. James Syvitski (University of Colorado, Boulder); Dr. Ian Birtwell (Fisheries and Oceans Canada; retired); Dr. Doug Spry (Environment Canada, Gatineau, Québec); and Dr. Fred Whoriskey (Department of Biology, Dalhousie University, Halifax).

I thank the following for reading and providing comments on parts of the manuscript: Dr. Dave Levy (North Vancouver, British Columbia); Dr. Kate Myers (University of Washington, Seattle; retired); Drs. Craig Clarke and Trevor Evelyn (Fisheries and Oceans Canada, Pacific Biological Station, Nanaimo, British Columbia; retired); Dr. Bob Devlin (Fisheries and Oceans Canada, Centre for Aquaculture and Environmental Research, West Vancouver); Edith Tobe (Squamish River Watershed Conservation Society, Squamish, British Columbia); Veronica Lo and David Coates (Convention on Biological Diversity, Montreal); and Dr. Dennis Scarnecchia (University of Idaho, Moscow). Dr. Marc Trudel (Fisheries and Oceans Canada, Pacific Biological Station) and two anonymous reviewers provided excellent suggestions for improving the book. Any errors or omissions are my responsibility.

The following helped with word processing, formatting, and figure preparation in various drafts: Michelle Chan, Linda Hanson, Emilia Hurd, Serena Howlett, Lynn Melcombe, Penny Nelson, and Beth Piercey. I am also particularly grateful to Dr. Laura White for assistance in these matters, as well as for editing and suggestions for improving the flow of the text.

Financial or in-kind support for word processing, figure preparation, and editing of various drafts was provided by the Murray Newman Award for Significant Achievement in Aquatic Conservation to the author from the Vancouver Aquarium and Marine Science Centre, Vancouver; the Pacific Fisheries Resource Conservation Council, Vancouver (facilitated by Dr. Jeff Marliave, Vancouver Aquarium and Marine Science Centre); Fisheries and Oceans Canada, Science Branch, Pacific (the Marine Environment and Aquaculture Division, facilitated by Dr. Laura Brown; and the Salmon and Freshwater Ecosystems Division, facilitated by Mark Saunders and Dr. James Irvine); and the Pacific Salmon Foundation, Vancouver (facilitated by Dr. Brian Riddell). Funding for final figure preparation and indexing was provided by Port Metro Vancouver (facilitated by Carolina Eliasson). The Pacific Salmon Foundation is especially acknowledged for also partially supporting publication costs.

Ecology of Salmonids in Estuaries around the World

1
Why a Focus on Salmonids and Estuaries?

Throughout history, both estuaries and salmonids have played an important role in human civilization. River mouths linked interior and coastal peoples, especially after the dawn of maritime trade. Salmon were a link between the ocean and rivers, providing a steady supply of protein-rich food to estuary dwellers, who could catch the fish with little effort in the constricted river mouths. Salmonid carcasses in river watersheds and lakes brought marine nutrients (nitrates and phosphates) to a variety of ecosystems, including the estuaries of short rivers. Subsistence fisheries for salmonids were widespread in the Northern Hemisphere, which, until the nineteenth century, was the only place in the world where these anadromous species were found. Consequently, the history of humans in the Northern Hemisphere is closely linked with salmonids in estuaries. Salmonids were introduced to the Southern Hemisphere in the 1800s and are now sought by recreational fishers in that part of the globe.

Historically, Vikings, Inuit, Beothuk, Mi'kmaq, and Mohicans harvested Atlantic salmonids at river mouths, while Aleuts, Haida, Salish, Ainu, and numerous other First Nations gathered Pacific salmon species. Estuarine fishing weirs capable of catching salmonids have been found in 5,000-year-old sites in Alaska (Tveskov and Erlandson 2003). The Kaspi people netted salmonids in the numerous estuaries around the Caspian Sea. In medieval times, salmonid fishing in European estuaries was conducted as far south as the Douro River estuary in Portugal. As the European fishing industry developed more formal marketing and distribution methods, salmonid processing plants and canneries were built at the mouths of larger rivers (Hoffmann 2005). In the early 1900s, there were hundreds of canneries in estuaries on both the northwest and northeast Pacific coasts (Bottom 1997). In Japan, commercial harvesting of salmon began as early as the 1600s (Nagata and Kaeriyama 2003).

Salmonids as a Cultural Icon

Salmonid stories or celebrations have featured in numerous cultures around the world. For example, from Victorian England:

> In every stage of their existence, salmon are surrounded by enemies innumerable, and it is really wonderful that they do not become extinct. They are, however, so prolific, that when care is bestowed upon them, and their enemies kept under as much as possible, they increase in numbers and size in a most wonderful manner. (Buckland 1880, 289)

From Scandinavia/Iceland at the time of the Norse Sagas:

> Skallagrim also had his men go up the rivers looking for salmon, and settled Odd the Lone-dweller at the Gljufur River where he attended to the salmon-fishing. (Byock 2001, 29)

From West Coast First Nations in Canada:

> Raven and a friend were invited to dinner at the village of the fish. Two children, a boy and a girl, were sent into the water, and shortly after, Raven was given his salmon dinner. Raven ate his meal, but instead of putting all the bones on his plate, he kept a small bone from the salmon's head in his mouth. The bones were gathered up and thrown back into the water, where they changed back into little children. The boy was okay, but the little girl couldn't open her eyes. The parents knew that a bone had not been put on the plate and began to look for it. They searched all over. Suddenly, Raven pulled the bone out of his mouth and remarked, "Maybe this is the one!" They told the little girl to go back into the water and then threw the missing bone in after her. When she came out, she was whole again. That is why you must always throw the salmon bones back into the ocean. (D. Kennedy and Bouchard 1983, 26)

And from contemporary Japan:

> On the island of Hokkaido, Japan, an Ainu ceremony called the *Ashiri Cheppu Nomi* is held every year in September on the Toyohira-gawa River plain in Sapporo to invoke the new run of salmon. (City of Sapporo 2015)

Importance of Salmonids around the World

Today, as they have been for centuries, salmonids are an iconic fish, especially in the northern temperate region. For example, in the northeast Pacific region, they are an indicator of ecosystem health for 130 species of fish and wildlife, and salmonid life history characteristics relate strongly to local ecosystem processes such as productivity (Cederholm et al. 2000). For a long time, salmonids have also been of significant interest to evolutionary biologists, including Darwin (1871). They are highly prized for providing ecosystem services related to food provision, culture, recreation, and industry. However, countries around the world, especially those affected by globalization and industrial development, often have conflicting goals within management agencies for salmonids and their estuarine habitats. Many estuaries have major habitat degradation and water quality problems with nutrient enrichment from farms and cities in the catchment basins of rivers. Compared with other fishes, salmonids are vulnerable to local extinction due to a variety of life history factors, including relatively low fecundity (Powles et al. 2000); habitat sensitivity associated with the complexity of the freshwater/estuary/marine habitats used in their life cycle, and with their ocean range, which extends across national boundaries and exclusive economic zones; and susceptibility of populations to bycatch in mixed-stock fisheries.

Estuarine Profile in Salmonid Evolution

How salmonids may have evolved their complex patterns of freshwater, estuarine, and ocean habitat use has intrigued evolutionary ecologists. A brief description of the salmonid life cycle is required before we turn to an overview of their findings. The family Salmonidae, which in turn includes the four anadromous genera I discuss, includes the subfamily Salmoninae (Nelson 1994). Anadromous salmonids are hatched from eggs spawned in fresh water, move to the ocean, where they attain sexual maturity, and then move back to fresh water to reproduce. They move to and from the ocean via the estuary. Anadromous salmonids display two forms of reproduction, which determine the degree of estuarine use. *Semelparous* species migrate from the ocean to spawn in fresh water and then die. *Iteroparous* species spawn in rivers and overwinter in rivers or estuaries, but repeatedly migrate into estuaries or adjacent coastal waters during summer to feed. They eventually die in the river after a final spawning. These species obviously spend more of their lives in estuaries than semelparous species. The two basic patterns are shown schematically in Figure 1.

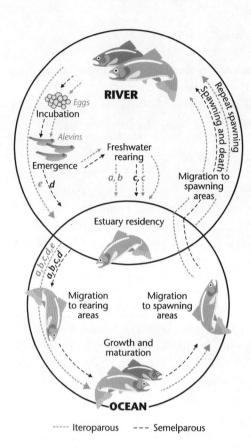

Figure 1 Generalized life history of iteroparous and semelparous anadromous salmonids, showing the central role of the estuary as a transition between the freshwater and ocean phases of the life cycle.
Notes: a – subyearling migrants, midsummer; b – subyearling migrants, autumn; c – yearling smolts; d – early spring migrants (fry or repeat migrants); e – kelts or returning veterans.
Source: Modified from Nicholas and Hankin 1988.

Most anadromous salmonids are generally considered freshwater fish that have evolved to live in the ocean, but this view is not held by all evolutionary ecologists (McPhail 2007). Most of the discussion of salmonid evolution focuses on reproductive success, which tends to emphasize the freshwater and ocean parts of the life cycle (e.g., Fleming 1998; Kinnison, Unwin, and Quinn 2008). However, the estuary has figured in the thinking about the evolution of anadromous salmonids. According to Tchernavin (1939 in Hoar 1976), sometime during the glacial period less than 1 million years ago (approximately the Pleistocene era) in the Northern Hemisphere, particularly Siberia, when "conditions of life in freshwater were unfavourable and food scarce ... some of the Salmonidae (for example, the genus *Salmo*) acquired habits of descending to the sea from the exhausted rivers to feed" (Hoar 1976, 1236). Furthermore, the type ancestor of semelparous Pacific salmonids (genus

Oncorhynchus) may have lived in estuaries as fry before they were able to live in full-strength seawater. During this period, they would have been confined to brackish estuaries until the smolt transformation evolved (Hoar 1976). Thus, the presence of salinity tolerance may also be the result of selection pressures related to the colonization of new regions during periods of changing glaciation, when fresh water from melting ice was widespread (C.C. Wilson and Hebert 1998). D.R. Montgomery (2000) proposed that the topographic variability of watersheds that occurred due to much earlier changes in geology of northwest North America during the Miocene-Pliocene times (20 to 6 million years ago) led to adaptive radiation of Pacific salmonids. They may have evolved into a variety of species using different freshwater habitats in the northeast Pacific region, and could also have taken advantage of productive ocean habitats. The freshwater changes were likely accompanied by the development of a variety of estuary types within the California-Alaska coastal reach, which the salmonids were obligated to use in their migrations between the river and the ocean. In the northwest Atlantic, these geological changes were not observed, suggesting that evolution of Atlantic salmonids in that region was not on the same trajectory (D.R. Montgomery 2000). On the New York–Newfoundland and Labrador coastline, there were only two species of endemic anadromous salmonids, *Salmo salar* (Atlantic salmon) and *Salvelinus fontinalis* (brook trout). Colonization of the region by *Salvelinus alpinus* (Arctic char) is thought to have occurred as the Pleistocene ice receded, and extant populations are considered to be vestiges of the anadromous populations that lived in the Champlain Sea and the Atlantic Ocean (i.e., glacial relict populations; Doucett et al. 1999). The study of the evolution and adaptations of salmonids to estuaries is clearly a complex and interesting challenge for the evolutionary ecologist.

Under some circumstances, salmonid evolution can occur much more quickly than profiled in the classical geological framework. The issue of short-term phenotypic plasticity in salmonids is relatively new and papers on the topic were uncommon before the 1990s (Hutching 2011). A growing body of empirical evidence shows that inheritable salmonid morphology (body size) and life history features (behaviour, growth, etc.) can shift on relatively rapid time scales when factors influencing the traits are modified (McClure et al. 2008; Hutching 2011). Domestication is a force for some of the changes. One example is the earlier return of wild coho salmon to the estuary and river after being subjected to sixty years of supplementation with early-released hatchery fish (Ford et al.

2006). Another example is the change in colour patterns of wild amago with introgression of hatchery-reared traits occurring within about thirty years (K. Kawamura et al. 2012). Short-term evolution has therefore attracted the attention of applied evolutionary ecologists, but there have been very few studies on estuarine salmonids.

In general, few researchers have explored the ability of freshwater species to establish marine populations or the ability of marine and freshwater species to establish euryhaline populations, the most likely starting point in the evolution of anadromy (Dodson, Laroche, and Lecomte 2009). The evolution of osmoregulation, which enables the removal of excess and toxic salts from the fish's body, is clearly a dominant component of the anadromous salmonid's fitness and survival suite of adaptations. However, additional selective forces in the estuary, particularly food supply and other biotic interactions, such as predation and competition, also likely conditioned the evolution of a variety of life histories, including iteroparity and the diversity of life history strategies that is now seen. As explained in following chapters, the divergence away from the conditions to which the salmonids adapted is the basis for present-day conservation concerns.

Estuarine Salmonids and Conservation Concerns
Four important genera of anadromous salmonids are currently of conservation concern, according to their listing by national or international authorities (e.g., the Red Books of the International Union for Conservation of Nature, or IUCN). Several other genera are listed as threatened or endangered by national conservation offices in some countries (see Appendix Table 1.1 at http://hdl.handle.net/2429/57062). Representatives of the salmonid genera *Salmo, Oncorhynchus, Salvelinus,* and *Hucho* are listed in conservation categories in various catchment basins over their natural range in North America, Europe, and Asia. For example, sea trout and Atlantic salmon are endemic, and now rare, in fifteen of the twenty-three estuaries summarized from the literature of northwestern Europe (M. Elliott and Hemingway 2002). In addition, forty-one of the seventy-six *Salmo, Oncorhynchus,* or *Salvelinus* taxocenes from North America listed as extinct, endangered, threatened, or vulnerable (Jelks et al. 2008) are anadromous and therefore must pass through an estuary as juveniles and adults.

Around the world, estuaries are among the most damaged or threatened habitats and deserve consideration as being critical or essential for salmonids. Compilation of current information on estuarine habitat and ecological functions is therefore vital to conservation efforts that aim

Why a Focus on Salmonids in Estuaries? 9

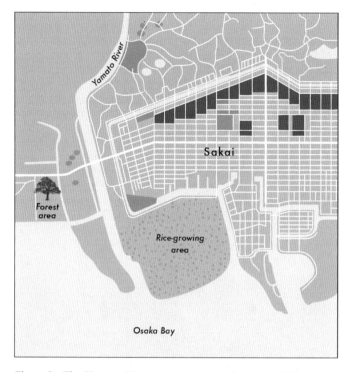

Figure 2 The Yamato River estuary, Japan, from the 18th century, showing how the mouth of the river was straightened for shipping.
Notes: Development of the port is indicated by streets and roads. This estuary is now completely developed and is part of Sakai City, near Osaka.
Source: Redrawn from Japanese Maps of the Tokugawa Era, University British Columbia, Digital Collection, with permission. Adapted from the original by Eric Leinberger. Colour photo in Appendix 5 at http://hdl.handle.net/2429/57062.

to improve survival and develop recovery plans for threatened salmonids. A number of major estuaries where one or more salmonid genera may have been once seasonally dominant organisms are now affected, in varying degrees, by overfishing and industrial or urban disruption, such as the Yamato River estuary in Japan (Luo et al. 2011) (Figure 2), the Petitcodiac River estuary in New Brunswick (Locke et al. 2003), and the Fraser River estuary in British Columbia (Levings 2004). Others have also been affected by water flow alteration, such as the Amu Darya and Syr Darya River estuaries in Uzbekistan and Kazakhstan, on the Aral Sea (Pavlovskaya 1995); by contamination, such as the Puyallup River estuary in Washington state (Stein et al. 1995); or by combinations of these factors, such as the Sacramento–San Joaquin River estuary in California

(Sommer et al. 2007) and northeast Atlantic salmonid estuaries (Limburg and Waldman 2009). In other instances, estuaries are relatively intact, such as the Stikine River estuary in Alaska (Levings 2004), or components of them are being rehabilitated, such as the Rhine River estuary in Germany (Schreiber and Diefenbach 2005) and the Columbia River estuary in Washington and Oregon (Roegner et al. 2010). Except for a few studies in Arctic or Subarctic North America – involving, for example, the Hudson Bay estuaries (Morin and Dodson 1986) and Mackenzie River estuary in the Northwest Territories (Carmack and Macdonald 2002), and the Lena River estuary in Russia (Lambelet et al. 2013) – documentation on the conditions of salmonid estuaries in the North is not extensive. This list excludes, of course, the myriad minor rivers and estuaries with small populations of salmonids that have been totally replaced by infilling to develop towns and cities around the world.

It must be stressed, however, that estuaries are only one of a series of habitats, all of which are essential for survival of wild salmonids. In the broadest sense, these habitats comprise rivers, estuaries, and the ocean (Figure 1). It is very important to realize that negative or positive changes in the estuary ecosystem are part of a continuum of events experienced by anadromous salmonids, with events occurring before (in the river) and after the estuary (in the ocean), perhaps swamping the impacts of factors in the estuary. Under some circumstances, conditions in the estuary may be more important than conditions in the river or the ocean. Most authorities recognize a continuum of ecological conditions between the three macrohabitats (e.g., Attrill and Rundle 2002).

Why a Review of Estuarine Salmonid Ecology Is Needed
One of the objectives of this book is to increase the profile of estuaries as rearing areas for juvenile salmonids and highlight the importance of maintaining estuarine ecosystems so that salmonids can successfully complete their life cycle. Some textbooks and papers state that the key importance of estuaries for salmonids is as a migratory corridor between the river and the ocean (e.g., McLusky and Elliott 2004; Lobry et al. 2008). In several review books and papers dealing with salmonids and their habitats, rivers and the coastal and open ocean receive extensive reviews of their importance for salmonids (Cushing, Cummins, and Minshall 2006; Grimes et al. 2007; Brittain et al. 2009), whereas estuaries do not. Major reviews of estuarine fish ecology conducted in areas where salmon are now relatively scarce compared with historical times (e.g., northwestern Europe) focus on nonsalmonids, even though sea trout

and Atlantic salmon are part of the estuarine ecosystem in these regions and derive food from estuarine habitats (M. Elliott et al. 2002). Finally, material on the estuarine ecology of salmonids in the Southern Hemisphere, outside of their natural range, is required to complement the papers from this part of the world (e.g., McDowall 2006).

A review and synthesis of the importance of estuaries and how salmonids around the world are adapted to them, in the context of management and conservation strategies, is overdue. Reviews of salmonid use of estuaries are provided by Simenstad and colleagues (1982) (five taxa), Healey (1982) (four taxa), Thorpe (1994) (twelve taxa), and Levings (1994a) (seven taxa), but there has been no comprehensive review of salmonids in estuaries since the early 1990s. Thorpe (1994) and Levings (1994a) dwelt on juvenile salmonids' adaptation to habitat, but did not deal with fitness, assessment and ecosystem aspects, and management problems, or suggest methodology. In this book, I deal with all life history stages of salmonids that use estuaries, including fry, parr, smolts, adults returning to spawn, and those returning to the ocean after reproducing. I provide an overview of salmonids and estuarine ecosystems for citizen scientists and conservationists, including a glossary of terms, which should be useful for these important groups of stakeholders concerned with the salmonid estuary.

The primary ecological advantages of residency for juvenile salmonids in estuaries stem from the opportunity to (1) adjust their osmoregulatory machinery; (2) take advantage of increased food availability in the estuary food web; and (3) postpone an increase in predation or competition risk such as might be encountered in the ocean, as pointed out by Moser and colleagues (1991) and the authors whom they cite. Some of these components are also applicable to adult salmonids, and I discuss this life phase in the estuary as appropriate.

My general approach in describing fitness changes in these three components affecting survival due to human activities in the salmonid estuary is as follows: (1) discuss the component from the viewpoint of mainstream ecology; (2) discuss the baseline condition in the natural estuary, recognizing that data regarding salmonid ecology may have been obtained from partially disrupted areas, and delineate where necessary any problems with methods for determining the baseline; (3) outline methods for estimating survival and point out how inference and context setting are used; (4) describe the typical and important changes that have occurred in salmonid estuaries; and (5) discuss the effects of the change on the component. This comprehensive approach should be useful for a broad audience of researchers and practitioners, and dovetail with the more focused earlier literature on salmonids in the ocean and coastal zone.

This book complements and expands on a number of previous works dealing with the marine and estuarine phases of anadromous salmonids, including Pearcy 1992; Beamish, Pearsall, and Healey 2003; Brodeur, Myers, and Helle 2003; Karpenko 1998 and 2003; and Mayama and Ishida 2003, which reviewed the ecological literature on Pacific salmonids (*Oncorhynchus* spp.) in the marine environment, including the estuary, but mainly for commercially significant species. Klemetsen and colleagues (2003) reviewed the life histories of Atlantic salmon, sea trout, and Arctic char, and sea trout were also discussed by G. Harris and Milner (2006). These authors dealt with broad aspects of the ecology of the species, as did Jonsson and Jonsson (2011), whose review paper compared the habitat ecology of Atlantic salmon and sea trout. Hansen and Quinn (1998) reviewed the marine phase of Atlantic salmon and compared this part of the life history with that of Pacific salmonids. Quinn (2005) reviewed the migrations of Pacific salmonids through estuaries. The estuarine ecology of Atlantic salmon and five species of *Oncorhynchus* were compared by Weitkamp and colleagues (2014). In their book on Atlantic salmon, Aas, Einum, Klemetsen, and Skurdal (2011a) did not specifically discuss the importance of estuaries. In summary, although there have been a number of recent comprehensive reviews on the ecology of anadromous salmonids, the material covered was wide-ranging and did not focus on the estuarine habitat of a range of species.

2
What Salmonids and Estuaries to Consider

There is sometimes confusion in discussions of salmonid taxa in the scientific literature due to a lack of consensus on the nomenclature used, especially for lesser-known species and subspecies that tend to be recognized by local names. The names I use in this book are at the species level, except for two subspecies of the genus *Oncorhynchus* from Asia. For consistency, throughout this book I mainly use the nomenclature found in FishBase (Froese and Pauly 2014), but if a replacement scientific name is mentioned in a paper, I use that name instead of attempting to explain changes in the taxonomic literature.

Salmonids use estuaries on river systems in temperate, subpolar, or polar regions in the Northern and Southern Hemispheres (Figure 3; see Appendix 2). This book includes data from estuaries in nineteen countries in the Northern Hemisphere and six countries or dependencies in the Southern Hemisphere. When I first use a river estuary name from countries other than the United States or Canada, I also give the country, plus additional regional description when helpful for clarifying context.

Salmonids That Utilize Estuaries

Based on the available literature, the eighteen salmonid taxa known to utilize estuaries are as follows: *Hucho perryi* (Sakhalin taimen), *Oncorhynchus clarkii* (cutthroat trout), *Oncorhynchus gorbuscha* (pink salmon), *Oncorhynchus keta* (chum salmon), *Oncorhynchus kisutch* (coho salmon), *Oncorhynchus masou ishikawae* (amago), *Oncorhynchus masou masou* (masu), *Oncorhynchus mykiss* (steelhead), *Oncorhynchus nerka* (sockeye salmon), *Oncorhynchus tshawytscha* (Chinook salmon), *Salmo salar* (Atlantic salmon), *Salmo trutta* (sea trout), *Salvelinus alpinus* (Arctic char), *Salvelinus confluentus* (bull trout), *Salvelinus fontinalis* (brook trout),

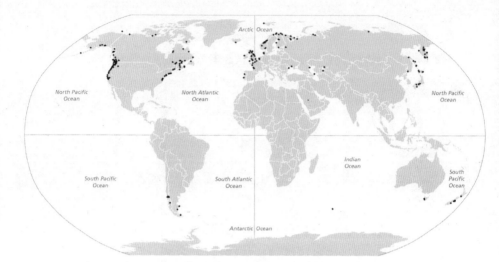

Figure 3 Map showing the world oceans and general locations (•) of the estuaries and other water bodies mentioned in the book. See Appendix 2 for names and geographic coordinates.
Source: Adapted from the original by Eric Leinberger

Salvelinus malma (Dolly Varden), *Salvelinus namaycush* (lake trout), and *Salvelinus leucomaenis* (whitespotted char). Many of these taxa are on various conservation lists, but this may be an underestimate of the number that should be conserved owing to insufficient data availability and nomenclature confusion. Because of the extensive movement of salmonids around the world, initially for stock development and more recently for aquaculture purposes, I include taxa that have successfully colonized a river-estuary system or are exploiting in some way natural estuarine habitats outside of their native range.

Given the difficulties in defining a meaningful taxonomic unit at or below the species level, conservation status data in Appendix Table 1.1 (at http://hdl.handle.net/2429/57062) should be considered provisional, giving a minimum number of taxa, and are subject to change. Species and subspecies are usually classified mainly on the basis of their morphological phenotypes – an organism's observable characteristics (e.g., body form, skeletal structure) or other traits (e.g., behaviour) that result from the expression of its genes as well as the influence of environmental factors and the interaction between the two. However, we now know that details of the genotype (the inherited instructions an organism carries within its genetic code) can be used to define meaningful ecological units for classification or conservation purposes. Identification of discrete salmonid populations below the species, subspecies, or life history type level requires complex ecological and genetic analyses. For example, in the United States, the National Marine Fisheries Service defines reproductively isolated groups of salmonids with unique evolutionary legacies (and therefore likely unique genetic backgrounds) as evolutionarily significant units under the Endangered Species Act (see Ruckelshaus et al. 2002).

The degree of estuarine dependence shown by salmonid taxa and their life histories has a genetic basis, a theme introduced here. An early example of a species with strong dependence was revealed by the genetic study by Carl and Healey (1984) for three Chinook salmon populations in the Nanaimo River, British Columbia. Results found an estuary-rearing population, a group that reared in the river for two months and then moved to the estuary, and a stream-rearing population that remained in fresh water over the winter. These populations were coined life history types, an ecological/genetic concept that I elaborate on later in this chapter. The terminology was an expansion of earlier ideas. C.H. Gilbert (1913) proposed two overarching patterns: (1) stream-type Chinook salmon that spend a year in freshwater as juveniles and then migrate to the estuary and the ocean as yearling smolts, and (2) ocean type, which move to the estuary in spring and summer, rear there, and move to the ocean as smolts. Healey (1991) adopted these patterns and proposed that the distribution of the two types were related to latitude. While Healey (1991) concluded that the subvariant life history characteristics of the overarching life history types were inherited and possibly characteristic for major groups of river systems, Brannon and colleagues (2004) concluded that the genetic differences were also mediated by temperature within a region. That is, the life history forms are "more correctly described as a continuum of forms that fall along a temporal cline related to incubation and rearing temperatures that determine spawn timing and juvenile residence patterns" (Brannon et al. 2004, 1). More detailed genetic studies support the inheritance of the life history characteristics (e.g., Rasmussen et al. 2003). Additional data suggest that the patterns may be best named by their temporal patterns of migration to the estuary (e.g., subyearlings, or fish that move to the estuary in their first year of life [Bottom et al. 2005a]) rather than type (Moran et al. 2013).

Where possible, I have restricted my data review to wild or feral fish and have attempted to separate out the characteristics of fish reared in hatcheries. Not all hatchery-reared salmonids are marked, however, so it is possible that some of the field data on salmonids in estuaries are based on hatchery fish.

Scope of Estuaries Used by Salmonids

Historically, each of the salmonid species considered in this book was probably distributed in numerous watersheds, perhaps thousands, and the same number of estuaries. For example, the Atlantic salmon was once found in about 2,600 watersheds in the North Atlantic (World Wildlife Fund 2001). Exceptions might be Sakhalin taimen and lake trout, which are probably found in fewer estuaries given their smaller geographic range (see Appendix Table 1.1 at http://hdl.handle.net/2429/57062). Salmonids are found in a wide variety of estuary sizes. There is a gross and highly variable relationship between salmonid catch (a possible proxy for freshwater productivity) and watershed area and discharge (possible proxies for available freshwater habitat) (Hindar et al. 2007). Watershed area and estuarine marsh area are positively correlated, at least in the northeast Pacific (Hood 2004a). There is a positive correlation between sediment yield and sediment load in rivers worldwide (Milliman and Syvitski 1992), and this in turn must relate to delta or estuary size, as most delta platforms are built by sediment. The situation is complex, because small pocket estuaries are sometimes nested within the freshwater influence of a larger river, as noted for the Skagit River estuary in Puget Sound, Washington (Fresh 2006). The main principles of salmonid use of estuaries can likely be drawn from a small system with a modest population of salmonids, given the complexity of studying the great salmonid estuaries of the world. However, some processes, such as productivity and species diversity, obviously need to be scaled by size. Some large brackish inland seas or lakes (e.g., Puget Sound, the Baltic Sea, and the Bras d'Or Lakes in Nova Scotia) can be classified as estuaries and, where appropriate, I include them. However, the main focus of the book is the landscape at the mouths of rivers.

Terminology for Life History Stages and Types

In this book, I have included all salmonid life history stages that occur in estuaries, but again the nomenclature used in the literature can be confusing. I consider adults to be salmonids in reproductive readiness, migrating through an estuary to a river spawning habitat or seeking a spawning site in an estuary. For iteroparous species, a *kelt* is a spawned-out fish returning to the ocean, but other terms are sometimes used for kelts, such as the *sea-run migrant* for whitespotted char (Morita 2001). The terms *virgin sea run* and *returning veterans* are sometimes used to distinguish adult fish returning to the river for the first time to spawn from those that have had several migrations already (e.g., Morita 2001 for whitespotted char). The terminology for fry, parr, and smolt is more difficult to standardize, because the terms have been used differently in the four genera mentioned in this book. I generally followed Allan and Ritter (1977) in developing operational terms that include these four genera, especially the section of their paper dealing with Atlantic salmon. *Alevins* are the stage from hatching to the end of dependence on the yolk sac as a primary source of nutrition. Alevins are sometimes found in the estuaries of rivers where spawning grounds of salmonids are a short distance above tide water. *Fry* are the stage from independence from the yolk sac as a primary source of nutrition to dispersal from the spawning nest (*redd*), which is a workable definition for the four genera, with the further complexity that some feeding *Oncorhynchus* fry are caught in estuaries with the yolk sac still present. Fry are often defined by size. For example, Chinook salmon fry were defined as fish less than 60 mm in length by Burke (2004), and this size criterion is frequently used for this species to distinguish them from larger juvenile Chinook salmon less than one year old, which are often called *subyearlings*. Definition of a *parr* is more complicated. Allan and Ritter (1977) define this stage as the form after dispersal from the redd to migration downriver as a *smolt*, although for some *Oncorhynchus* species this includes fry, as noted above. As well, the early smolt stages of some *Oncorhynchus* species have parr marks, and these pre-smolts may be found in some estuaries. The darkly pigmented blotches of skin along the lateral line of the fish are the distinguishing feature of parr, and these marks vary between species as well as between populations (Boulding et al. 2008). A smolt is a fully silvered juvenile salmonid migrating to the ocean – that is, leaving the estuary. This definition fits juveniles or first-time migrants of either semelparous and iteroparous species, but does not fit older sea-run migrants or kelts, the adult stages of iteroparous species returning to the sea. Post-smolt is the stage from departure from the estuary until onset of wide annulus formation at the end of the first winter in the ocean, according to Allan and Ritter (1977).

In essence, a variety of terms are used to describe salmonid groupings below the species and subspecies level, often regionally. They include terminology such as "life history type" and "ecotype." "Life history type" is perhaps the most widely used term in this category. In the general ecological sense, it can be thought of as the tactic that enables genetically separate groups of populations to optimize the fitness of individuals within the populations. Alternative patterns of reproduction and the degree to which these are successful in different environmental settings are a widespread demonstration of life history tactics. For example, among species of tropical freshwater fishes, ten variables or tactical features for reproduction were identified (Winemiller 1989). Among salmonids, the tactics are often named according to the region used by a specific life stage or the age of the specific life stage residing in it. For example, a subyearling Chinook salmon is a fish migrating to the estuary when less than one year old, and jacks are precocious one- or two-year-old male Chinook salmon returning to spawn, instead of the normal four- or five-year-old fish (J. Johnson, Johnson, and Copeland 2012). Sometimes the term "ecotype" is used, such as for river- and lake-rearing sockeye salmon (C.C. Wood et al. 2008). An ecotype describes fish that can be described by genetically defined morphological or physiological attributes developed by selection for specific habitats (Gregor 1944). The season that the life history stage appears in the river after ocean life is often used as a descriptor for adults. Thus, the term "fall Chinook salmon" is used to describe fish arriving in the estuary and river in the autumn and spawning within a few weeks, whereas spring Chinook salmon are those that arrive in the spring and stay in the river for several months before spawning in the autumn. Other terms are used for mature and immature estuarine salmonids, and the terminology can be confusing (e.g., "grilse" is a mature Atlantic salmon that has spent only one year in the ocean, but this term is also used to describe post-smolt coho salmon and Chinook salmon in British Columbia). Regional or international guides are often helpful for sorting out these nomenclature and ecosystem-specific terms for estuarine salmonid life history types, stages, or forms.

Conclusions

Although there is sometimes confusion about which particular taxon, stage, or life history type of anadromous salmonids are "officially" listed as a conservation problem, clearly the survival in the wild of almost all species in this group of fishes is threatened over most of their natural range and in a variety of estuary configurations. Many of the endangered salmonid taxa living within their natural range are listed in the Northern Hemisphere south of about 50°N (see Appendix Table 1.1 at http://hdl.handle.net/2429/57062), suggesting either that anthropogenic factors such as human populations are having more effects on estuaries and salmonids in this region or that salmonids in the southern end of their natural range are more susceptible to environmental change. Before discussing adaptations and how natural and human-induced factors affect survival, it is necessary to describe the salmonid estuarine environment.

3
Salmonid's-Eye View of the Estuary: Physical, Chemical, and Geological Aspects

The unique physical and chemical features of the estuary distinguish the estuarine phase of salmonid life history, as the characteristics of rivers and oceans are very different. This chapter describes some of the major oceanographic, geomorphic, and hydrologic features of estuaries. This overview helps place information on physical factors in context vis-à-vis the adaptations of salmonids to estuarine life. There are many books and journal articles that deal specifically with the physics and chemistry of estuaries (e.g., K.R. Dyer 1998; Wolanski and McLusky 2011). There are also regional works and websites that deal with particular areas, such as the British Columbia coast (Thomson 1981), the United Kingdom (Foundation for Water Research 2014), and the United States (National Oceanic and Atmospheric Administration 2014a).

I have adopted the classic estuary definition of Cameron and Pritchard (1963), as it enables a discussion of a variety of habitats between and within estuaries: "An estuary is (a) a semi-enclosed and coastal body of water, (b) with free communication to the ocean, and (c) within which ocean water is diluted by freshwater derived from land." The noun "estuary" is derived from an adaptation of the Latin *aestuarium,* a properly adjectival term meaning "tidal," hence a tidal marsh or opening. Over the centuries, the word has obviously evolved broader meaning and is often used as a synonym for delta or river mouth. Several bases can be used to classify estuaries: water balance, geomorphology, vertical structure of salinity, and hydrodynamics (Valle-Levinson 2010), as will be described next. As well, I provide a description of more broadly applicable attributes such as temperature and river discharge to provide context. In an operational sense, estuaries may be differentiated from other coastal habitats by geomorphological and various oceanographic characteristics, but their lower salinity or brackishness and tidal range are common identifying features in defining terminology.

Estuary Classification Systems and Water Properties

To place salmonid estuaries in the context of estuary ecosystems around the world, I provide an overview of some of the estuary classification systems used by various authors. I also provide comments on some of the important physicochemical properties related to salmonid ecology that are described in later chapters.

Water Balance

Positive estuaries are those where freshwater additions from river discharge, rain, and ice melting exceed freshwater losses from evaporation or freezing and establish a longitudinal water density gradient from the river to the ocean (Figure 4). There is a stronger surface outflow of brackish water relative to bottom inflow of higher-salinity water from the ocean (countercurrent flow). Low-flow estuaries, where the freshwater surface flow is weak relative to bottom inflow, include those on small river systems located in low-rainfall regions, such as Lagunitas Creek and San Antonio Creek estuaries in Tomales Bay, California (Largier, Hollibaugh, and Smith 1997). Estuaries in polar regions might be classified as low-flow estuaries during winter if river discharge is minimized because precipitation is bound up by ice and snow. Negative estuaries are those where there is almost no river discharge into the system and salinity increases towards the head of the estuary, as in estuaries found on hypersaline enclosed seas that have dams on their inflowing rivers (e.g., the Aral and Caspian Seas). Estuaries with heavy ice cover in winter can be seasonally negative estuaries when salt freezes out of the ice (e.g., Mackenzie River estuary, Northwest Territories: R.W. Macdonald 2000).

Geomorphology

The classification of estuaries using geomorphology is scale-dependent and needs to be viewed as nested within the water balance, vertical structure of salinity, and hydrodynamic bases. It is likely that most estuaries used by salmonids in the temperate zones and classified by geomorphology fall under the water balances bases of positive and low-flow estuary types. A broad and flexible definition of estuary types is important to enable discussion of salmonid estuaries formed in glaciated and unglaciated regions of the world. While there tend to be more species of endemic salmonids in recently glaciated and/or Cordilleran regions of the Northern Hemisphere (e.g., D.R. Montgomery 2000;

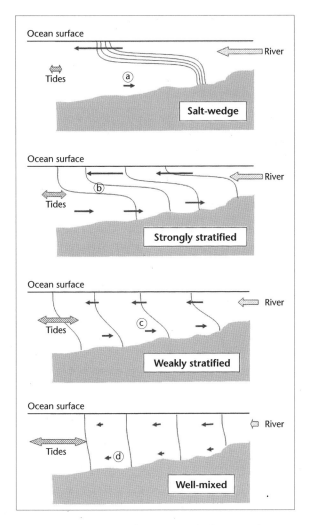

Figure 4 Estuary classifications according to salinity stratification.
Notes: (a) and (b) strong halocline; (c) weak halocline; (d) halocline absent.
Source: Modified from Valle-Levinson 2010, with permission.

McDowall 2006), salmonids are also found in lowland areas such as continental Europe and Japan. A large-scale geomorphological system suggested by Pritchard (1952) used four types:

1. coastal plain, e.g., Chesapeake Bay (includes the estuaries of the Susquehanna, Potomac, Rappahannock, York, and James Rivers), in

Delaware, Maryland, New York, Pennsylvania, Virginia, West Virginia, and the District of Columbia; and the Gironde River estuary, which receives freshwater from the Garonne and Dordogne Rivers on the Bay of Biscay, France
2. tectonic, e.g., San Francisco Bay (includes the estuaries of the Sacramento and San Joaquin Rivers), California
3. fjords, e.g., Trondheimsfjord (includes the estuaries of the Orkla and Gaula Rivers), Norway; and Howe Sound (includes the Squamish River estuary), British Columbia
4. bar-built or lagoons, e.g., the Bol'shaya River estuary, Russia; and the Scott Creek estuary, California.

Other examples are estuaries formed by the flooding of river valleys from sea level rise, such as the Columbia River estuary, Washington-Oregon (R.N. Williams 2006). To expand on Pritchard's scheme (1952), I adapted the classification system developed by MacKenzie and colleagues (2000), which takes into account these smaller systems. The scheme provides estuarine forms such as fjords, fjards, coastal plain, strands, lagoon or bar-built beaches, deltas, and fans, some of which are shown in Figure 5. There are specific lagoon types used by salmonids, such as the limans or brackish lake lagoons behind barrier beaches on the Kamchatka Peninsula, Russia (Zenkovich 1967; Bugaev 2004), that are similar to the hapua in New Zealand (Hart 2007) and the barachois in eastern Canada (M.W. Smith and Rushton 1963). Overall, the modified and idealized scheme is similar to the modern typology of river mouths described by Dürr and colleagues (2011).

The depth of an estuary is not necessarily a defining feature for salmonids, but shallowness is often mentioned as one of the main characteristics of estuaries. Coastal plain, delta, and lagoon-type estuaries are usually less than 20 m in depth, but fjord estuaries can be several hundred metres deep (Pickard and Stanton 1980).

Sediment Loading
The type and amount of sediment coming into the estuary from the river depends on a variety of factors, particularly geology of the watershed, the gradient as the river enters the estuary, and discharge (Milliman and Farnsworth 2011). Sediment loading is the product of discharge and sediment concentration, and is essential to maintaining the geomorphology of the estuary; this is a dynamic process, as in a natural estuary the river moves back and forth across the horizontal plane of the estuary.

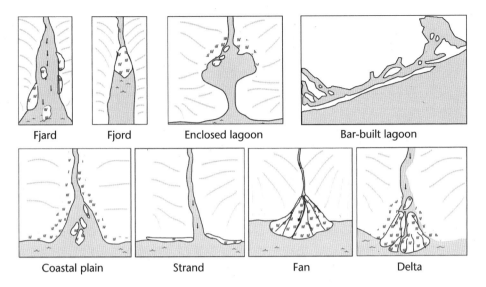

Figure 5 Idealized geomorphology of selected estuary forms used by salmonids.
Notes: Short hatching indicates the general location of vegetation (brackish marsh). The top of each diagram indicates the river, and the ocean is shown in the lower portion.
Source: Modified from MacKenzie, Remington, and Shaw 2000.

Vertical Structure of Salinity and Temperature

Large river discharge and weak tidal forcing result in salt wedge estuaries, while moderate to large river discharge and weak to moderate tidal forcing result in strongly stratified estuaries (Figure 4). In stratified estuaries, the *halocline* is the location in the water column where salinity changes with the greatest rate over depth. In deep estuaries, the stratification and halocline structure can remain strong throughout the tidal cycle; fjords are a classic example of this. Weakly stratified or partially mixed estuaries result from moderate to strong tidal forcing and weak to moderate river discharge. Many temperate-region estuaries fall into this category. Strong tidal forcing and weak river discharge result in vertically mixed estuaries where mean salinity profiles are fairly uniform over the depth of the estuary channel. It is important to realize that estuaries may change from one type to another in consecutive tidal cycles, or from month to month, from season to season, or from one location to another inside the same estuary. The Fraser River estuary in British Columbia is a good example of this phenomenon. The estuary changes from stratified flow on the flood tide to unstratified flow on the ebb tide (Kostachuk and Luternauer 2004).

If temperatures are different in the incoming salt wedge relative to the river, thermal stratification can occur along with salinity. In temperate or boreal regions, the salt wedge water is frequently warmer, and the opposite can occur in more southerly estuaries (Thomson 1981). The *thermocline* is the location in the water column where temperature changes with the greatest rate over depth. Lagoon estuaries usually lack a salt wedge or tidal prism, and are frequently characterized by fresh water or very low salinities (e.g., hapua [Hart 2007]). They may experience a tidal backwater effect where rising ocean levels limit lagoon drainage through the permeable barrier sediments. Salt inflow may be rare and limited to wave overtopping at high tide or when storms create a channel (Hart 2007).

Hydrodynamics

Data on the dynamic changes in water properties in time and space (hydrodynamics) offer good insight into the oceanographic context of salmonid estuarine ecology. However, data required for this classification system are based on complex nondimensional measurements and require specialized oceanographic observations (see Valle-Levinson 2010 for further information). The flushing or exchange rate of water in the estuary is dependent on the relative hydraulic forces of the river and the tides (Valle-Levinson 2010) and is possibly the most important hydrodynamic parameter to consider for ecological studies, as it determines the residency time of water. This in turn can affect concentration of food species for salmonids, pollutants, and passive particles such as suspended sediment in the estuary.

Tidal Range

Given the key role of tides in estuarine processes such as stratification, it is important to note the possible classification of estuaries based on tidal amplitude (McLusky and Elliott 2004). The scheme of McLusky and Elliott (2004) suggested four categories: microtidal (<2 m range), mesotidal (2–4 m range), macrotidal (4–6 m range), and hypertidal (greater than 6 m range). In areas of the world with highly convoluted coastlines, some of these tidal ranges may exist in estuaries within one region (e.g., British Columbia [Thomson 1981]). An example of a hypertidal salmonid estuary is the Mezen River estuary, Russia, with a 7 m tidal range (Dolgopolova and Isupova 2010). At the other end of the scale are the microtidal estuaries of the enclosed Aral and Caspian Seas. The water level in estuaries on these seas are affected by river discharge and evaporation; for example, the water level in the Caspian Sea decreased

2.9 m in the mid-twentieth century due to changes in flow of the Volga River, Russia (Arpe and Leroy 2007). Tidal amplitude decreases longitudinally up the estuary, and when river flow dominates, tides are not perceptible.

Other classifications related to tidal range are based on temporal changes or biological communities. The range of high and low tides can vary seasonally. Neap tides have the lowest range (amplitude, measured above a local baseline), whereas spring tides have the greatest amplitude (not necessarily in spring) (Thomson 1981). Most salmonid estuaries have two high tides and two low tides over a twenty-four-hour period (e.g., British Columbia estuaries [Thomson 1981]). Exceptions include the salmonid estuaries on the northwest coast of the Kamchatka Peninsula, Russia, where there is only one low and one high tide per day (Kowalik 2004). On most estuary shorelines, plants and invertebrate communities are vertically zoned according to tidal level and occur in predictable sequences of intertidal life (e.g., northeast Pacific estuaries [Ricketts, Calvin, and Hedgpeth 1992]).

Wave Action and Sea Level Rise

Although most estuaries are protected from open ocean waves by the sediments forming the delta at the river, some estuary types require wave action for dynamic maintenance of their features. The seasonal formation of barriers in front of estuarine lagoons and their subsequent breakdown by waves during storms is an example from the coast of California (S.A. Hayes et al. 2008). Sea level rise is predicted in world oceans due to climate change, but in some salmonid estuarine areas (e.g., northeast Pacific Ocean), recent uplifting of the land from geological forces (postglacial rise) provides somewhat of a counterbalance (Beechie, Collins, and Pess 2001).

Zonal Change along the Estuary Reaches

An estuary is an ecotone, and the physical attributes of the landforms and oceanographic characteristics (river flow and tides) of the water in estuaries change in a characteristic manner from the ocean to river (Attrill and Rundle 2002). To describe this, I recognize the zones described by McLusky and Elliott (2004), with some adaptation: tidal freshwater, upper estuary, middle estuary, lower estuary, and coastal (Table 1). Estuaries on large river systems can extend tens of kilometres into the coastal zone. R.N. Williams (2006) concluded that the landward limit of the Columbia River estuary, Washington-Oregon, was approximately 235 km from the ocean, based on the fact that tidal action (50 cm

Table 1

Idealized salmonid estuarine zones, with descriptors and general physicochemical properties

Zone	Description
Tidal freshwater	Perceptible tidal fluctuations, usually <50 cm amplitude. Salt not present. River currents flowing downstream may predominate on most tidal stages and substrates determined by bed load movement, often shifting after freshets. Sometimes called the head of the estuary.
Upper estuary	Minimal currents at high tide (can lead to turbidity maxima). Upstream currents on rising tides. River currents predominate. Mud deposition depending on wash load. Salinity <5–18 psu (Practical Salinity Units).
Middle estuary	Principally mud deposits, but sandier where currents are faster depending on bed load. Salinity 18–15 psu. Upstream currents on rising tides.
Lower estuary	Principally sand or gravel deposits, but muddier where currents weaken depending on wash load. Salinity 25–30 psu. Upstream currents on rising tides.
Coastal	Sand, mud, or gravel/cobble/rocky shores at the mouth of the estuary. Salinity similar to adjacent ocean (>30 psu), especially on fringes of the estuary. On strong tidal passages, water velocities can be >100 cm $\cdot$ s^{-1}.

Source: Modified from McLusky and Elliott 2004.

amplitude on spring tides) is observed up to the first hydroelectric dam on the river.

Conservative Water Properties: Temperature and Salinity Ranges

Conservative water properties are those that are not affected by biological processes. They are key to the physical oceanography of the estuary, affecting estuary attributes such as stratification (e.g., low-salinity water is less dense and floats above high-salinity water, developing the halocline). Temperature and salinity fall into this category. Other water properties – such as dissolved oxygen and nutrients, which are affected by biological processes such as photosynthesis and plant production as well as pollutant effects – are nonconservative; I discuss them in Chapter 4.

Temperature in the estuary is a function of the adjacent ocean, the inflowing river, and local heating. Temperature varies regionally, seasonally, and even within a twenty-four-hour period in very shallow estuaries. The summer temperature (July) of the world oceans where salmonid

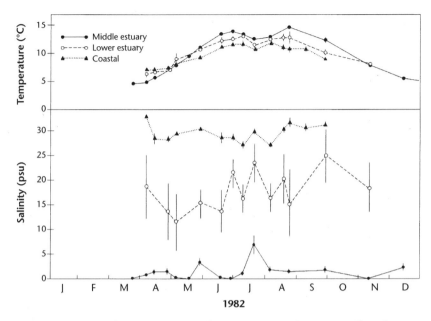

Figure 6 Seasonal fluctuations in surface temperature (upper panel) and salinity (lower panel) at three zones at the Campbell River estuary, British Columbia.
Source: Colin Levings and Steve Macdonald, unpublished data. Department of Fisheries and Oceans Canada, Centre for Aquaculture and Environmental Research, West Vancouver, BC, Canada.

estuaries are found ranges from a few degrees above zero in the polar regions (e.g., in the Arctic Ocean) to about 14–16°C in the temperate areas (e.g., northeast Pacific Ocean), as temperature is zonal and tied to latitude (Texas A&M University 2014). However, regional ocean temperatures affecting estuaries can be higher, as in the Seto Inland Sea of Japan, where coastal temperature can reach 28°C in August (Murakami, Oonishi, Kunishi 1985). The temperature in the lower reaches of salmonid rivers flowing into the estuary also needs to be considered. Temperatures in the lower river reaches affect the upper estuary and tidal freshwater zones of the estuary. The temperature of Arctic river estuaries is raised by incoming ocean water as well as local heating in the summer. Surface temperatures in August at the Mackenzie River estuary, Northwest Territories, ranged from approximately 2°C (landward) to 10°C (seaward) over a 120 km reach of the lower estuary (Emmerton, Lesack, and Vincent 2008). Temperate rivers and estuaries are milder. The middle and lower zones of the Campbell River estuary, British Columbia, show a seasonal range between approximately 5°C and 15°C (Figure 6). There may also be natural variation in temperature within a region. British Columbia rivers that are fed by glaciers have cooler estuaries relative to coastal plain estuaries (Levings and Bouillon 2008).

Salinity in the estuary originates in the adjacent ocean. Salinity levels are expressed as Practical Salinity Units (psu) in the oceanographic literature; these are the units used in this book. Salinity varies according to the concentration of mineral ions, especially sodium (Na^+) and chloride (Cl^-), but also magnesium (Mg^{2+}), calcium (Ca^{2+}), bromium (Br^-), and other ions (F.G.W. Smith 1974). High salinity levels, up to 34 psu, may be reached on the seaward portion of estuaries such as those on the warm east coast of Japan, where evaporation is high and river discharge and rainfall is low (e.g., Shinada, Shiga, and Ban 1999). Salinity is lower in oceans such as the northeast Pacific, where precipitation and runoff exceed evaporation (Pickard and Emery 1990), and can range from 34 to 30 psu.

River Discharge and Surface Salinity
River discharge is a key factor in salmonid estuaries because of its effects on salinity. Small rivers and creeks dilute the ocean water coming into the estuary to a limited extent, whereas fresh water from large rivers can dominate and create a major freshwater reach. Salinity at an estuary can range from close to 0 psu to over 30 psu at the coastal zone at the mouth (Figure 6). The river can also create a plume of fresh water that can influence the coastal ocean some distance away from the estuary, depending on tides, winds, and river discharge (Thomson 1981). The flow range of rivers discharging into salmonid estuaries around the world is huge. The Mackenzie River in the Northwest Territories and the Amur River in Russia, each with mean annual flow close to 10,000 $m^3 \cdot s^{-1}$ (see Benke and Cushing 2005 and Bogutskaya et al. 2008, respectively), are at the very high end of the range. At the low end, there are the small streams in Europe and elsewhere with mean annual flows of less than 1 $m^3 \cdot s^{-1}$ (e.g., sixteen sea trout streams in Norway listed by Jonsson et al. 2001). The salinity characteristics of an embayment may be broadly affected by the cumulative drainage from several rivers flowing into it. River discharge can affect tidal levels and can increase water elevations during maximum discharge or freshets.

Seasonal patterns of ice and/or snow melt in spring and rainfall in autumn are two river discharge features commonly found in salmonid estuaries. Flow patterns in the Campbell River estuary in British Columbia (Figure 7) and the Puelo River estuary in Chile (Lara, Villalba, and Urrutia 2008) are good examples of this dual pattern, which exists in temperate

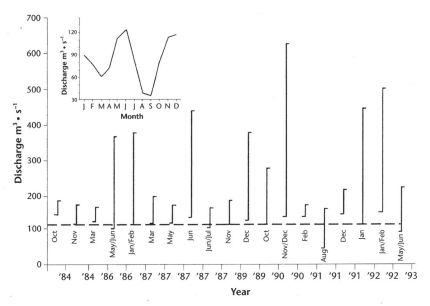

Figure 7 Fluctuations in river discharge into the Campbell River estuary, British Columbia, due to spills from a hydroelectric dam located 4 km upstream.
Notes: Vertical lines indicate spills in various months from 1984 to 1992. The dotted horizontal line is the turbine capacity of the facility. Inset: seasonal discharge pattern of the river before the dam was constructed. Note the difference in discharge scales.
Source: Modified from D. Burt and Burns 1995.

regions. In warmer regions where snow is not a factor, winter rainfall is the most important variable driving river discharge. An example is the pattern for Scott Creek in central California (S.A. Hayes et al. 2008). In watersheds affected by typhoons, such as the Razdol'naya River in Russia, there can be a major increase in discharge in summer (July and August) as weather systems bring in rainstorms from the Sea of Japan. This estuary is also one of many estuaries affected by ice. In the Razdol'naya River, estuary freeze-up occurs in October and the ice melts in March (Kolpakov and Milovankin 2010).

Conclusion

Salmonid estuaries have unique and recognizable physical and chemical features that distinguish them from the ocean and the river, the other two major habitats used by anadromous salmonids. Estuaries, like rivers, are bodies of water that are constrained by hydrological and geological forces that result in their variations in shape and size. This constraint is a feature not encountered when salmonids are in the open ocean.

Temperature and salinity are physicochemical features in the estuary that vary on a daily tidal basis as well as seasonally. These are unique features of the estuary. Perhaps the most important feature of the salmonid estuary is the continuum of water properties that develop along its axis, from the freshwater tidal zone to the mouth. The change in salinity, often abrupt in time and space, is a particular challenge along this continuum and is faced by anadromous salmonids nowhere else in their life cycle. Within a water mass, salmonid habitats in the ocean are relatively homogeneous, except for water column structure (thermoclines, haloclines, etc.), and are not conditioned by geological processes. Although river and stream salmonid habitats are, like estuaries, heterogeneous and influenced by geological processes, the water flow in their channels is unidirectional downstream. In the estuary, water flow reverses with the tide. As well, water depth and water volume change hourly on a regular basis from the tides, a feature also not found in rivers. In the following chapter, I set up a framework to describe estuarine habitats – necessary to help set the scene for later discussions of adaptations of salmonids to these distinctive water bodies.

4
What Habitats Are Used by Salmonids in Estuaries?

According to an ecological dictionary (Lincoln, Boxshall, and Clark 1987), habitat is the locality, site, and particular type of local environment occupied by an organism. This chapter describes a "practical framework" or taxonomy that is necessary to describe the estuarine habitats to which salmonids are adapted. Where applicable, the same framework is also applied to a description of how human use has changed estuaries, to effects on fitness components of survival, and to discussions on environmental management and planning.

I also give a brief overview of the likely function of the named habitats or their lower subclasses for estuarine salmonids to help provide context for more detailed discussions later. In addition, I give information on apparent differential habitat use by salmonids in estuaries; in other words, data inferring that species were more abundant in one habitat than another, perhaps preferring them or occupying them opportunistically. It is usually assumed that the differences accurately reflect the fitness benefits of using different habitats (but see Rosenfeld and Boss 2001). There are several caveats to the data I present. Because comparable sampling gear is not always used in different habitats, it is sometimes challenging to conduct analyses of abundance data to determine whether differences are statistically significant, and some authors do not apply statistical analyses. Another important caveat is that the fidelity of a species to specific habitats is usually not known. Habitat shifts in the estuary often occur on an hourly or diel time scale, for example, when young salmonids move inshore as the tide rises or as night progresses (Thedinga, Johnson, and Neff 2011). Fish are likely using a succession of interconnected habitats for short periods over a tidal cycle. In some salmonid estuaries, fish may live in surface waters (pelagic zone) at low tide and move into the intertidal zone as the tide rises, occupying low-, middle-, and high-tide zones sequentially. A good example of this movement was given by Spares and colleagues (2012) for Arctic char at the Sylvia Grinnell River estuary in Nunavut, Canada. The shift can also occur on a seasonal basis, as fish can access habitats higher on the estuary beach only during spring tides, as in June and December in British Columbia (Thomson 1981). Another issue is the question of whether the salmonids are more abundant in a particular habitat because they prefer the habitat or because the habitat type is more abundant in the area surveyed (Mathur et al. 1985). This problem is recognized in studies of freshwater habitats of salmonids (e.g., Heggenes, Northcote, and Peter 1991), but is rarely considered in salmonid estuary studies. To conclude this chapter, I also give an overview of water quality parameters that are typical for salmonids in estuaries.

How Are Habitats Described? A Practical Framework for This Book

There has been a proliferation of literature on estuarine habitat terminology and classification systems in recent years. A thorough review of the classification of estuarine and nearshore coastal ecosystems is given in Wolanski and McLusky 2011. Here, I present only an overview of this complex topic and refer specifically to habitats that relate to salmonids. Within the broad boundaries of the estuary, salmonids are found in a mosaic of habitat types, mostly defined on the basis of vegetation and/or geophysical characteristics; these are described below. A variety of estuarine habitat classification systems applicable to salmonids are in use or proposed around the world, such as the Cowardin system (Cowardin et al. 1979) and the related National Oceanic and Atmospheric Administration (NOAA) coastal and marine ecological classification standard in the United States (V.D. Engle et al. 2007). In the European Union classifications, estuaries are called transitional waters (McLusky and Elliott 2007). Under the European Union Habitat Directive, there are also specific habitat types defined under estuarine Special Areas of Conservation, as well as named "Sites of Special Scientific Interest" for their importance to salmonids. In the complementary European Union habitat classification, habitats within transitional waters are also named using vegetation and substrate terminology (D. Evans 2012; European Commission DG Environment 2013). There are also numerous regional schemes that have been developed, such as the British Columbia marine ecosystem classification (Zacharias et al. 1998), the controlling factor approach in New Zealand (Hume et al. 2007), and the classification of estuaries in Kamchatka, Russia, by Mikhailov and Gorin (2012). These are usually tailored to local management needs and governance arrangements. Some salmonid habitats in estuaries can also be described by wetland inventory systems used for waterfowl (e.g., Ramsar Convention Secretariat 2014).

Dealing first with the river-estuary-ocean continuum, it is necessary to describe where river habitats end and estuarine habitats begin. For general boundaries in the salmonid estuary, I define the upriver boundary of the freshwater tidal zone as the extent of significant tidal fluctuation (20–50 cm on spring tides) at low river discharge. For the boundary on the margins of the estuary, I use maximum mean high water tidal level to account for the importance of the land-water ecotone in the estuary and lower river (Levings and Jamieson 2001). In microtidal areas or areas lacking tides, these definitions are more difficult to apply. A boundary with the ocean is also difficult to set but is usually thought of as the edge of an estuarine water mass where salinity can be as low as 0 psu. The general issue for setting boundaries is one of scale, especially as it relates to sampling.

Habitat is often assessed at an arbitrary scale of local environment, or mesoscale, and is conditioned by size and type of gear used in the sampling. In estuaries, salmonid habitat is often sampled at a scale of between 5 and 100 m, depending on the size of the habitat patch within the estuary and the dimensions of the gear used. For example, if sand patches 10 m in diameter were found within a 50 m diameter mosaic of sand and mud, a beach seine 25 m long could not be used to sample specifically on sand. Fish caught in this situation would be described as coming from mixed mud and sand. Although use of habitats may be specific, the context of each (association/proximity to other habitats) is important (Strayer 2010). Connectivity between habitats by tidal or river water flow (e.g., tidal channels enabling the migration of fish onto a flooded beach as tide ebbs) is a basic concept in salmonid estuarine ecology (Bottom et al. 2005a). Studying habitats through a landscape approach, which by definition includes habitat connectivity (Naiman and Latterell 2005), is becoming a dominant theme for describing and managing estuarine salmonid habitat.

Habitat identification and measurement is an essential method of describing salmonid habitat in estuaries, for general ecosystem studies and especially for protocols that can be used in management and monitoring (e.g., Roegner et al. 2009). An issue is the increasing complexity of estuarine habitat classification systems themselves, and their increasing sophistication using new technology, especially remote sensing (e.g., Luck et al. 2010; vanden Borre et al. 2011).

It is important to realize that the use of simple categorical (structural) classification systems ignores the dynamic nature of habitats in estuaries. A process-oriented approach to estuarine salmonid habitat classification and management is required (Simenstad, Reed, and Ford 2006). This strategy would require assessment of physical and geological processes, such as estuary flushing time, residency time of estuary water masses, and sedimentation rate. A process-oriented approach is now being implemented in a number of river systems (e.g., Beechie et al. 2010) and is recommended for many major estuaries, such as the Columbia River estuary in Washington-Oregon (Bottom et al. 2005b) and the Sacramento–San Joaquin River estuary in California (Kimmerer et al. 2008). It is likely that dynamic strategies for habitat classification will be more widely adopted in the future.

At present, many practitioners of salmonid habitat management as well as applied researchers use structural or practical systems for estuarine habitat management that are not necessarily process-oriented. Often the focus is on mesohabitats, as defined by Pardo and Armitage (1997), for streams as visually distinct units of habitat within the stream, recognizable from the bank with apparent physical uniformity. The following is a description of a mesoscale system I use in this book that fits in with most of the structural and functional aspects of the salmonid estuary, is appropriate for most aspects of estuarine salmonid ecology, and includes terms often seen in the international literature. I am not necessarily recommending the scheme, but use it strictly as a framework for my discussions.

Naming Habitats in Relation to Their Functions: An Overview

> Geomorphology [of an estuary] is primarily concerned with the exogenous processes that mold [the sediment and surfaces at the river mouth], but the internal forces cannot be disregarded when one considers the fundamental concepts of the origin and development of land forms [within it]. (Leopold, Wolman, and Miller 1964)

This definition of estuarine geomorphology sets the stage for an explanation of the geological and hydrological processes that shape the estuarine habitats used by salmonids, and thus plays into the naming scheme. Possibly the most important external forces are river flows or currents, which in the natural flow regime are direct functions of discharge (Poff et al. 1997). Currents continuously erode and build up sediment to shape and determine the depth of channels, control the beach substrate type, and facilitate vegetation growth. In turn, vegetation can impart stability to the mud, sand, or gravel it grows in. Thus, sediments are not independent of vegetation. Currents, erosion, and local geological factors also shape the configuration of the estuary and determine whether the shoreline used by juvenile salmonids is linear (e.g., along a fjord), planar (e.g., build-up of shallow water sediment into banks in the outer estuary), or curvilinear (e.g., lagoon shorelines). Configuration determines the perimeter length of the estuary. These are some key physical landscape factors that can determine where salmonids migrate to and reside.

My goal in this section is to provide a working definition of estuarine salmonid habitats for this book. Some of the terms apply to the adjacent coastal zone because the estuary grades into the ocean. I briefly describe their usage or functional context as a prelude to more detailed discussions of behaviour, growth, feeding, and other ecological aspects later on. On-the-ground or operational methodologies often include four upper-level descriptors: beaches, channels, vegetation, and artificial or ecosystem engineer habitat; I adopt these terms. There is inevitably some overlap among the definitions and the scheme is a compromise. For example, some steep shorelines that might be called beaches are found on channels. The four upper-level descriptors are shaped by river currents, sediments, and anthropogenic activity. They include lower-level subhabitat units within them, as follows: (1) beaches: rock and boulder, gravel and cobble, sandflats, mudflats; (2) channels: riverine channels, side channels, tidal creeks, tidal swamp-forested; (3) vegetation: macroalgae, eelgrass, brackish marshes, high salt marsh, wet meadow, woody debris, riparian vegetation; and (4) artificial habitat and habitat created by invasive ecosystem engineers: docks, wharves, causeways, invasive plants, and so on.

Table 2 provides an abbreviated summary of the habitats and subhabitats and also gives information on how they are set into the general configuration of the estuary. The subhabitats are nested within the larger habitat types. Most of the habitat schemes used around the world contain nested elements or are hierarchical (e.g., Guarinello, Schumchenia, and King 2010).

Beaches

I adopted the classic definition of a beach from Bascom (1964, 14): "An accumulation of rock fragments ... and may be composed of any kind ... of rocky material, ranging from boulders to fine sand and added mudflats to this category."

Table 2

A practical framework for salmonid estuarine habitats and their general geomorphological or biological features

Habitat and subhabitats*	Shoreline configuration or growth form	Physical or biological features and examples
Beach		
Rock and boulder (8)	Linear along fjords and ice-scoured estuaries.	Sometimes solid rock; boulders >26 cm in diameter.
Gravel and cobble (8)	Deltaic or sometimes lagoonal/curved, sometimes in patches among other habitats.	Grain size 0.2–26 cm diameter.
Sandflats (9)	Typically developed into banks; seasonally developed on some high-energy shores.	Grain size 0.062–2.00 mm diameter.
Mudflats (9)	Typically developed into banks.	Grain size <0.062 mm diameter.
Channel		
Riverine channel (9)	Often sinuous; routes most fresh water through the estuary and enables salt wedge penetration upstream in stratified estuaries.	Width can range from a few metres to several kilometres; depth from <2 m to >10 m.
Side channels or sloughs (9)	Often regular but not straight; branch off the main riverine channels and typically have blind endings.	Width can range from a few metres to tens of metres; depth from <2 m to >10 m.
Tidal channels or tidal creeks (9)	Often meandering; branch off from the main riverine channels; have blind endings or connect with small streams.	Typically <5 m deep at high tide; can form pools or dewater at low tide.
Vegetation		
Macroalgae (10)	Steep or gently sloping shorelines over a range of elevation.	Vertical profile or floating length can range from a few centimetres (rockweed) to >1 m (kelp).
Eelgrass (11)	Typically gently sloping shorelines; mid to lower tidal elevation.	Vertical profile or floating length can range to about 1 m above the substrate at high tide.

Habitat and subhabitats*	Shoreline configuration or growth form	Physical or biological features and examples
Brackish or freshwater marshes (9)	Typically gently sloping or flat shorelines and along channels at mid-tidal elevation or on riverine channel banks.	Vertical profile or floating length can range to about 1 m above the substrate at high tide and stem density can be up to a few hundred stems • m^{-2}.
High salt marshes (9, 11)	Typically gently sloping or flat shorelines; higher tidal elevation.	Vertical profile or floating length can range to about 25 cm above the substrate at high tide.
Wet meadows (9)	Typically flat shorelines; extreme high tidal elevation.	Vertical profile or floating length can range to about 25 cm above the substrate when flooded.
Riparian vegetation (9, 10)	Often on edge of channels and tidal creeks; high tidal elevation in the coastal zone.	Leaves and branches of shrubs and trees overhang edges of channel shorelines or beaches.
Tidal swamp-forested (9)	Swamp vegetation (often shrub plants) and conifers are found in the freshwater tidal zone.	Swamp vegetation often overhangs tidal channels. Forests are flooded at freshet and extreme high tides.
Woody debris (10)	Often lodged on a variety of habitats and tidal elevations.	Size can range from tree trunks and stumps from the watershed to wood chips at the high-tide wrack line.
Artificial habitat and invasive ecosystem engineer habitat	Human-built structures and habitat created by invasive organisms.	Example subhabitats: Dredged sand; riprap, docks; invasive plants or animals.

* Shown in figure number given in parentheses following subhabitat name.

Figure 8 Beach and vegetation habitat near Tvärminne, Finland, on the Baltic Sea near the estuary of the Svartå River.
Note: Colour photo in Appendix 5 at http://hdl.handle.net/2429/57062.
Source: Photo by Dr. Alf Norkko, Tvärminne Zoological Station; reproduced with permission.

Rock and boulder
In recently glaciated regions, fjords, or Arctic estuaries heavily scoured by ice, solid rock is the predominant shoreline feature on beaches (e.g., north Baltic Sea coast; Figure 8). Boulders are carried down glacier-fed rivers and these large rocks can form the shoreline. Juvenile salmonids move along rocky fjord shores in schools to coastal regions; for example, sea trout in a Norwegian fjord (Aurland River estuary [Lyse, Stefansson, and Ferno 1998]) and chum salmon in a British Columbia fjord (Sechelt Inlet, estuary of the Clowholm River [Levings, Birtwell, and Piercey 2003]). Arctic char sea-run migrants were caught on rock and boulder beaches at the Freshwater Creek estuary, Nunavut [Bégout Anras et al. 1999]).

Gravel and cobble
Because of low fine sediment load in steep-sloped watersheds or rivers draining gravel-dominated basins, some estuarine beaches are characterized by gravel or cobble. Gravel and cobble are often the main shoreline substrate on permanent lagoons in glaciated areas, since these features

are usually built with coarse substrate moved from the river or adjacent coastline (Hart 2007). River currents and wave action may prevent the development of extensive vascular plants or macroalgae at these areas, but microbenthic algae can grow on gravel and cobble in more stable estuaries (e.g., Porcupine Creek estuary, Alaska [Murphy 1984]). Chinook salmon fry and smolts were abundant on gravel-cobble beaches in British Columbia deltas (e.g., Fraser River estuary [J.S. Macdonald and Chang 1993]). Arctic char at the Dieset River estuary in Svalbard, Norway, were found on gravel for about four weeks (Nilssen et al. 1997), although some of this time might also have been spent in coastal regions. Gravel and cobble in the intertidal zone are used for pink salmon spawning (e.g., Olsen Creek estuary, Alaska [Helle, Williamson, and Bailey 1964]).

Sandflats
Sandflats are dominated by material originating from rivers with a high sand bed load or from a sediment source on the adjacent coastline, on the flank of the estuary (Figure 9). Sandflats on wave-exposed shores are often eroded to a gentle 1:20 slope and tend to be in the low intertidal zone of estuaries. Steelhead abundance on sandflats at the Scott Creek estuary in California varied seasonally; parr abundance was highest when the sandbar at the mouth of the estuary was closed (S.A. Hayes et al. 2008). Chum salmon fry biomass on the sandflat in Izembek Lagoon, Alaska, was estimated at 14–16 kg • ha^{-1} in July (Tack 1970). At high tide, Chinook salmon fry and smolts were abundant over the extensive sandflats at the Fraser River estuary in British Columbia. When the water drained on extreme low tides and the sandflats dewatered, they were found in tide pools, providing low-tide refuges (Levings 1982).

Mudflats
In most meso- or macrotidal estuaries a band of mud, silt, or clay is usually observed, seaward of brackish marshes and sometimes at a slightly higher tidal elevation than sandflats (Figure 9). As with sandflats, mudflats are often eroded to a gentle slope by wave action. Mudflats can be dominant at estuaries with rivers that carry a high sediment load of mud and silt. On mudflats at the Willapa River estuary in Washington, juvenile Chinook salmon (life stage not given) tended to be more abundant in fyke net samples than juvenile coho salmon (Hosack et al. 2006). Several authors note catches of salmonids in an estuarine habitat mosaic that included mud (e.g., Atlantic salmon parr at the Western Arm Brook estuary, Newfoundland and Labrador [Cunjak, Chadwick, and Shears 1989]).

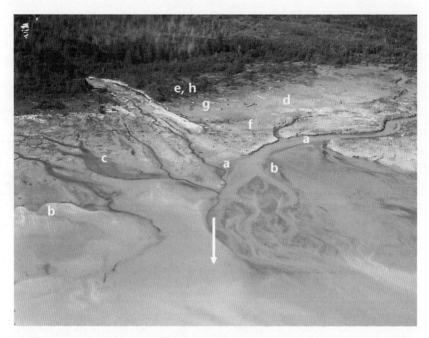

Figure 9 Eastern sector of the outer Homathko River estuary, an undisturbed estuary on fjordic Bute Inlet, British Columbia.
Notes: a = riverine channel of a branch of the main river, b = sandflats, c = mudflats, d = brackish marsh, e = riparian vegetation, f = tidal channel, g = wet meadow, h = tidal swamp-forested. Arrow in lower centre is seaward towards the coastal zone. On the left, a new channel recently created by the main river breaking through the forested flood plain can be seen. Colour photo in Appendix 5 at http://hdl.handle.net/2429/57062.
Source: Author photo.

Channels

I define a channel as a landform that directs water into the estuary from the river or the ocean, with flow direction depending on whether the tide is flooding or ebbing or whether river currents predominate.

Riverine channels

Riverine channels lead surface fresh water from the river to the ocean through the complex of intertidal islands, sandflats, and mudflats (Figure 9). Riverine channels are migration routes for adult salmonids moving into spawning rivers from the ocean. Depth is a key controlling factor that can limit the migration of adult salmonids through riverine channels and upriver to the spawning grounds. The salmonids may move upstream at high tide or hold off the river mouth until depth increases due to flow increases from rainfall. This is the typical pattern for chum salmon in small coastal streams. In Japan, for example, the main migration season of chum salmon into estuaries and upriver for spawning is from September to December, a period often characterized by high rainfall-generated spates related to typhoons (Tsuda et al. 2006). Upstream migrating sea trout in the Rhine River estuary in the Netherlands appeared to choose riverine channels according to the flow in them (bij de Vaate et al. 2003). Juvenile salmonids migrate downstream and into other estuarine habitats via riverine channels. Some sorting of the routes taken occurs, depending on life history stages (Perry et al. 2010). Curry, van de Sande, and Whoriskey (2006) found that brook trout in the Laval River estuary in Québec used deep channels at low tide and moved into shallow water at high tide, using an average water depth of 1.7 m. Smolts move to the ocean in riverine channels relatively quickly in schools and sometimes travel at the same speed as tidal and river currents, so their abundance appears as a pulse. There was a positive relationship between flow in the Sacramento River in California and Chinook salmon fry and smolt catches in the estuary (Brandes and McLain 2001). Smolts travel in deeper water than smaller fry, which tend to hug the shoreline or migrate into side channels or tidal creeks. The distribution of sockeye salmon smolts and fry in the main riverine channel of the lower Fraser River estuary in British Columbia is an example (Levings 2004).

Side channels or sloughs
Side channels or sloughs are blind or nonconnecting channels that branch off main riverine channels and often penetrate marshes. Water fills the side channels at high tide or during river freshet. Side channels usually do not dry, depending on river discharge, tidal stages, and elevation relative to the depth profile of the riverine channel they join. On large estuaries, they can be up to 10 m deep at high water. The width of side channels varies according to the size of the estuary but is usually greater than 1.0 m. Juvenile salmonids are found in side channels at high water or freshet and move back into riverine channels when the side channels dewater at low tide (e.g., Chinook salmon fry at Salmon River estuary, Oregon [Hering et al. 2010]).

Tidal creeks
Tidal creeks drain higher intertidal habitats into lower-elevation parts of the estuary. They are usually shallow (<4 m at high water depending on elevation at macrotidal estuaries). They may also branch directly off riverine channels or side channels (Figure 9). Tidal creeks that drain

uplands are flooded with estuarine water on high tides and are lotic stream habitats at low tide, although some may dry on extreme low tides. Sockeye salmon fry moved into tidal creeks at high tide in the Fraser River estuary, British Columbia (Levings, Boyle, and Whitehouse 1995). Because they are typically narrow (often <1.0 m wide), they tend to be dominated by, and sometimes named for, the vegetation on their shorelines. A tidal creek a few metres wide could have small trees and shrubs completely overhanging the water body.

Vegetation
I define vegetation as plant material in estuaries, including algae, herbaceous vascular plants, shrubs, and trees.

Macroalgae (kelp and rockweed)
Kelp is the common name for large brown algae of the order Laminariales that grow on hard substrates in the shallow brackish areas of estuaries. This includes the genus *Laminaria*. The sugar kelp, *Saccharina latissima*, is a bladed species commonly found in British Columbia estuaries

Figure 10 Intertidal zone of Bute Inlet, British Columbia, in the coastal zone further seaward from the outer Homathko River estuary shown in Figure 9.
Note: Macroalgae at the low and mid-intertidal zone grade into riparian vegetation at higher elevations. Colour photo in Appendix 5 at http://hdl.handle.net/2429/57062.
Source: Photo by Neil McDaniel, August 1976, reproduced with permission.

(Levings and McDaniel 1976) that can grow up to 2.5 m in length. Chum salmon fry were as abundant in kelp as in eelgrass habitat in southeastern Alaska (Murphy, Johnson, and Csepp 2000). Rockweed (*Fucus* spp.) is a speciose genus of low-lying algae typically less than 25 cm in length that dominates some estuarine beaches (Figure 10).

Eelgrass
Eelgrass can be the dominant vascular plant in estuaries, especially on the fringes of the estuary or in the salt wedge where freshwater effects are lessened, as the plants are not tolerant of low salinities. *Zostera marina* (e.g., Fraser River estuary, British Columbia [Gordon and Levings 1984]), *Z. japonica* (e.g., Shimanto River estuary, Japan [Ishikawa 2004]), and other subgeneric taxa are recognized. Depending on specific adaptations, the plants may be found at various elevations, but primarily from shallow subtidal to middle intertidal zones. Juvenile Chinook salmon and coho salmon (life stage not given) were about equally abundant in eelgrass and on mudflats in fyke net samples at the Willapa River estuary, Washington (Hosack et al. 2006). Chum salmon fry were more abundant in eelgrass beds in the Yaquina River estuary, Oregon, relative to Chinook salmon and coho salmon smolts (Bayer 1981). A.L. Macdonald (1984) showed that this species tended to be more abundant in eelgrass relative to sandflats at the Fraser River estuary in British Columbia (Figure 11). Abundance of juvenile salmonids in eelgrass can be highly variable and is not always higher relative to other habitats, although sampling efficiency can be low if beach seines are used to sample eelgrass. Murphy, Johnson, and Csepp (2000) found that chum salmon fry were about equally abundant in eelgrass, kelp, and filamentous algae in Alaskan estuaries.

Brackish or freshwater marshes
Brackish marshes are found in low-salinity regimes and are characterized by herbaceous vascular plants found in mud, sand, and gravel. Euryhaline genera such as sedges (e.g., *Carex* spp.), reeds (*Phragmites* spp.), and grasses (*Spartina* spp.) are some of the dominant plants in Northern Hemisphere estuaries, and there are analogous genera in the Southern Hemisphere. Brackish marshes in northeast Pacific estuaries are at the approximate midtide level (Figure 9) and thus are submerged 40–60 percent of the time on an annual basis (Levings and Nishimura 1997). Although the extensive *Spartina alterniflora* marshes in estuaries on the Atlantic coast are often called salt marshes, the biomass of the plant may be not correlated with salinity (Howes, Dacey, and Goehringer 1986). I include them as

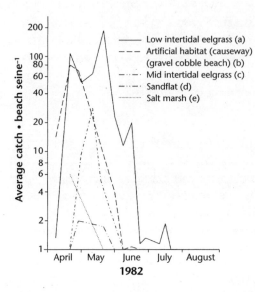

Figure 11
Upper panel: Seasonal trends in abundance of chum salmon fry at Roberts Bank, Fraser River estuary, British Columbia.
Note: The data also show a trend in availability of subhabitats with respect to tidal flooding as they are arrayed along a gradient in elevation from low intertidal eelgrass, to mid-intertidal eelgrass, to sandflats, to high salt marsh.
Source: Adapted from A.L. Macdonald 1984, with permission.

Lower panel: Low tide infrared aerial photo of Roberts Bank, 1975.
Note: Letters indicate subhabitats as in Upper Panel.
Source: Tarbottom and Harrison, 1996, with permission. Infrared photo in Appendix 5 at http://handle.net/2429/57062.

brackish marshes. Freshwater marshes are characteristic of the upper estuary or freshwater tidal zone. They are typically totally inundated during freshet. Sedges and rushes are also some of the dominant plants in this subhabitat.

There have been numerous studies of juvenile salmonids in brackish marshes, especially in estuaries of the northeast Pacific. Brackish marshes are often recognized as critical for rearing of juvenile salmonids in the region (see Brodeur, Myers, and Helle 2003; Beamish, Pearsall, and Healey 2003). Salmonid use of brackish marshes elsewhere in the world has also been documented. Brackish marshes in northeastern France may be important for Atlantic salmon and sea trout, e.g., Somme River estuary; (Ducrotoy and Dauvin 2008). In the Simojoki River estuary in Finland, reeds (probably *Phragmites australis* and *Scirpus* spp. [Dijkema 1990]) were the dominant Brackish marsh species where Atlantic salmon post-smolts were migrating (Jutila, Jokikokko, and Ikonen 2009). Brook trout use brackish marsh (*Spartina alterniflora* and *S. patens*) along tidal creeks in estuaries in Maine (Dionne, Bonebakker, and Whiting-Grant 2003), and probably other estuaries on the northwest Atlantic coast where brook trout migrate to the ocean (e.g., Massachusetts [Annett et al. 2012]). Chinook and coho salmon smolts were found in the freshwater marshes in the freshwater tidal zone of the Fraser River (R.S. Gregory and Levings 1998).

High salt marshes
High salt marshes develop on the infrequently flooded high-elevation flanks of the estuary where salinities are higher. The biomass of these plants is typically positively correlated with salinity (e.g., Hutchinson 1982). Typical vascular plant genera found in high salt marshes are glasswort (*Salicornia* spp.), saltmarsh grass (*Puccinellia* spp.), and saltgrass (*Distichilis* spp.).

Examples of salmonid use of high salt marsh are scarce, because the plants are in the highest elevations of the intertidal zone, are usually submerged for a short period of time (often only approximately an hour on extreme high tides), and have not been sampled in many studies. At the Navarro River estuary in California, steelhead smolts were abundant in high salt marshes in this estuary-lagoon system, and coho salmon smolts were also caught (Cannata 1998). Chum salmon fry used high salt marshes in the Fraser River estuary in British Columbia in April, when higher tides enabled the fry to access the habitat (A.L. Macdonald 1984; Figure 11). Sea trout were recognized as a component of high salt marsh (*Puccinellia maritima*) communities along tidal creeks in the Forth River estuary in Scotland (Mathieson et al. 2000).

Seasonal floodplains or wet meadows
Seasonal floodplains or wet meadows are found on the flat floodplain of estuaries, on the infrequently flooded high elevations between riverine or side channels on the delta. Often wet meadows are characterized by grasses (e.g., tule grass, *Scirpus acutus*) or sedges adapted to higher elevations (e.g., *Juncus* spp.), and floodplain shrubs (e.g., willows, *Salix* spp.). Toft and colleagues (2003) found juvenile Chinook salmon in wet meadows at the Sacramento–San Joaquin River estuary in California.

Woody debris
Tree trunks and stumps, usually called woody debris, are carried into the estuary by river currents. In some instances, they scar mud and sand flats, enabling colonization by rhizomes of brackish marsh plants (Maser and Sedell 1994). Woody debris can also create pools in mud or sand flats by scouring the substrate or by damming side channels or tidal creeks. The pools and woody debris in them can provide low-tide refuges for juvenile salmonids (e.g., coho salmon fry [K.V. Koski 2009]). Chum salmon fry were caught in woody debris at the wrack line at the outer Squamish River estuary in British Columbia (Romanuk and Levings 2005).

Riparian vegetation
Riparian vegetation is found along the shores of riverine and side channels and tidal creeks, usually including shrubs ("tidal shrub marsh" [Hood 2012]) and trees that overhang the channel edges. The vegetation is likely immersed in water only during river freshets and/or higher tides. In British Columbia and Washington, a variety of trees and shrubs in fresh water (especially willows, *Salix* spp., and alders, *Alnus* spp.) and cedars and spruce (*Thuja* spp. and *Picea* spp., respectively) on coastal shorelines comprise the riparian vegetation (Romanuk and Levings 2003; Figure 10). Juvenile Chinook salmon and trout (steelhead, cutthroat trout) in the Smith River estuary in California were more abundant along shorelines with riparian vegetation relative to where vegetation was absent, but the differences varied between estuarine reaches (Quiñones and Mulligan 2005). Shade provision and water cooling are likely the two functions of riparian vegetation for salmonids.

Tidal swamp-forested
Tidal swamp-forested wetlands are found in the upper estuarine zone, where higher tides and river freshet inundate the edge of the forest. Tidal pools or running water (extreme landward end of some tidal creeks) may be present at low tide, and substrates are typically mud. In addition to

coniferous trees on the floodplain or margins, plants that grow in standing water, such as skunk cabbage (*Lysichiton americanum*), may be present. Tidal swamp-forested subhabitats were rearing areas for juvenile coho salmon in a tributary of the Columbia River estuary in Washington-Oregon (B.E. Craig, Simenstad, and Bottom 2014).

Artificial Habitat and Habitat from Invasive Ecosystem Engineers

I discuss the functioning of these two habitat types in later chapters dealing with disruptions since they are not typical of the natural estuary. However, they need to be introduced here and included in the framework description. Artificial habitats are the "trademark" of the disrupted estuary. They include docks, wharves, islands or beaches made of dredged sand or gravel, rock revetments (hereafter riprap), river training walls, groins, artificial channels, and other structures. An "ecosystem engineer" is a species that controls the flow of resources in an estuary because of its ability to reorganize the ecosystem's structure and physicochemical attributes (Crooks 2002). Invasive plants are often ecosystem engineers in disrupted habitat (Toft et al. 2003).

Estuarine habitat use is also closely constrained by the quality of the water overlying the geological, biological, and physical features, as discussed next.

Water Properties and Salmonids

Salmonids have long been recognized as one of the most sensitive groups of fishes with respect to their behaviour and response to the chemical and physical properties of the water around them; literature on the topic is extensive. Here, I give an overview of some of the major water property values for salmonids in the undisrupted estuary: temperature, salinity, dissolved oxygen, pH, nutrients, micro constituents, and turbidity. The behaviour of salmonids living in water with these natural ranges should be normal and reflect their adaptations to the undisrupted condition of the estuary. Early salmonid physiological experiments showed that understanding of multiple factor effects on salmonids was required to determine water property requirements for survival (e.g., Alderdice 1972). Therefore, it should be recognized that the factors need to be considered together.

Temperature

Temperature is a classic factor influencing salmonid physiology and survival (see references in C. Groot, Clarke, and Margolis 1995), but there are few empirical examples from estuaries as most experiments have been done in fresh water. The early work by Brett (1952) with juvenile Chinook, coho, sockeye, pink, and chum salmon (fry 4–5 cm) showed that upper lethal temperature ranged from 23.8 to 25.1°C. The optimal temperature varies within a relatively narrow range for salmonids, but results from the literature are often not comparable because of differences in acclimation temperatures in experiments (J.M. Elliott 1991). When optimal temperature data were averaged over fourteen species of estuarine salmonids, the lower optimum temperature was about 11°C and the higher optimum temperature was approximately 16°C (Barton 1996). This is a crudely calculated average, however, and specific surface temperature in shallow salmonid estuaries can reach 22°C or higher in some estuaries (e.g., outer Sacramento–San Joaquin estuary (San Francisco Bay) in California [Gewant and Bollens 2012]). Suboptimal warm temperature can be reached in small estuaries with low river flows or reduced ocean exchange, but generally temperatures are cooler in the main geographic range of salmonids. At the lower end of the range, the lethal limit may be 0°C, where ice forms in the freshwater part of the estuary, although juvenile Atlantic salmon will feed at 0.9°C, according to Higgins and Talbot (1985). There is the expected latitudinal trend involved, with estuaries in lower latitudes showing warmer average annual temperatures. For example, temperatures in June, when Atlantic salmon smolts were migrating through estuaries, ranged from 14–21°C in the Gironde River estuary, France (Burdloff et al. 2000; Lobry et al. 2003), to 9.8–18°C in the Simojoki River, Finland (Jutila, Jokikokko, and Julkunen 2005). The distribution and abundance of juvenile salmonids within the estuary are influenced by temperature. Abundance of young masu at the Kushiro River estuary in Japan was higher where water temperature was 10–12°C than in cooler regions (8–9°C) (Sano and Abe 1967). Other examples for Pacific salmonids were given by C. Groot and Margolis (1991).

Adult salmonids returning to a river must position themselves at the estuary when they are physiologically prepared to resume life in fresh water; temperature is an important component of this preparation. Sockeye salmon returning to the Fraser River, British Columbia, are known to show preparatory physiological changes for passing through the Fraser River estuary and into fresh water while still several hundred kilometres northward. Gill Na^+-K^+–ATPase activity, an index of osmoregulatory preparedness, is lower at 250 km than at around 700 km away, yet both locales are full-strength seawater (Hinch et al. 2006). The reduced gill Na^+-K^+–ATPase levels likely enable the fish to move freely between the shallow-water salt wedge in the deeper and sometimes cooler parts of the estuary and the overlying fresh water leaving the river, as discussed below. Temperature units are calculated from the number of days fish are exposed to various temperatures as they migrate through a range of thermal conditions. Temperature units accumulated by adult fish ultimately affect their survival once they arrive at the estuary and are required to enter the warmer river fresh water in the estuary that lies above the thermocline. Adult migratory sockeye salmon with higher accumulated temperature units can, if such temperature units are excessive, suffer higher mortality rates before spawning (Farrell et al. 2008). Exploitation of cooler temperatures below the thermocline and typically in the salt wedge could be important.

Juvenile salmonids also use different estuarine strata relative to temperature. However, even though cooler water is usually found in the salt wedge below the thermocline, in a vertical tank experiment juvenile Chinook salmon, chum salmon, and coho salmon were oriented to warmer surface water (Birtwell and Kruzynski 1989).

Salinity
Although salinity in salmonid estuaries can range from 0 to 34 psu, exposure to intermediate salinities in the estuary may be necessary for osmoregulatory change during smoltification for some, but not all, salmonids; the influence of this factor is species-specific, as discussed in Chapter 9. Returning adults often use peaks in discharge, which result in salinity decreases in the estuary, as a cue for upstream migration (e.g., pink salmon [Heard 1991]). Salinity can affect estuarine use by salmonids in both horizontal and vertical dimensions. Because estuaries are often stratified, salmonids in the upper layers of the water column are exposed to lower salinities than those occupying deeper levels (Birtwell and Kruzynski 1989).

Dissolved Oxygen
Dissolved oxygen is a very important variable for salmonid physiology, and most salmonids are considered sensitive to this water property. Pacific salmonids have a high capacity for aerobic swimming, and oxygen uptake can increase between ten and fifteen times when the fish change from resting to maximum activity (Randall and Wright 1995). In the estuary, this rapid shift in swim speed enables feeding, predator avoidance, or migratory and other behaviours (Kramer 1987). Dissolved oxygen concentrations or percent saturation values in estuarine water decrease with high salinity and temperature (Davis 1975), so interactions between these oceanographic and river hydrology variables need to be

considered. Dissolved oxygen requirements of salmonids are often given in pollution-related water quality criteria documents (see Chapter 17), but the data are also applicable to the natural estuary. In Canada, an interim dissolved requirement for aquatic life in marine and estuarine waters was established at 8.0 mg • L^{-1} (still in place: Canadian Council of Ministers of the Environment 1999), with the following caveats: (1) when ambient dissolved oxygen levels are less than 8.0 mg • L^{-1}, human activities should not cause dissolved oxygen levels to decrease by more than 10 percent of the natural concentration expected in the receiving environment at that time; (2) when the natural dissolved oxygen level is less than the recommended interim guideline, the natural concentration should become the interim guideline at that site; and (3) when ambient dissolved concentrations are greater than 8.0 mg • L^{-1}, human activities should not cause dissolved oxygen levels to decrease by more than 10 percent of the natural concentration expected at that time. Other jurisdictions have published estuarine dissolved oxygen guidelines that are lower than 8.0 mg • L^{-1}, but specify that it may be appropriate to derive site-specific dissolved oxygen criteria when a threatened or endangered species is at a site, and when data indicate that it is sensitive at concentrations above the recommended criteria (e.g., United States Environmental Protection Agency 2000). Adult migratory salmonids began to use the Severn River estuary in the United Kingdom once dissolved oxygen levels were improved to the 48–65 percent saturation level (Jones 2006).

Decomposition of natural organic material such as algae, including sinking phytoplankton, can impose a biological oxygen demand on estuarine waters, depending on the flushing rate of the particular estuary. Dissolved-oxygen depletion is often a deep-water phenomenon, because mixing at the surface aerates shallow waters. In some circumstances (e.g., upwelling of bottom water), however, low-dissolved-oxygen water will be pushed to the surface layers of salmonid estuaries (e.g., Hood Canal, Washington [Warner, Kawase, and Newton 2001]).

Turbidity

Turbidity is common in estuaries because sediments in these shallow water bodies are easily brought into suspension by wave and current action. As well, in numerous rivers and estuaries along the glaciated west coasts of North America, Europe, South America, and elsewhere, high turbidity levels are caused by sediments eroded by glaciers in the catchment basin (Pickard and Stanton 1980). Turbidity in salmonid estuaries can range from less than 1 Nephelometric Turbidity Unit (NTU) in the clear water in a freshwater tidal marsh of a tributary in the upper Fraser River estuary, British Columbia (R.S. Gregory and Levings 1998) to about 400 NTU in the riverine channels of the glacial Taku River, Alaska (Murphy et al. 1989). The estuary turbidity maximum at the bottom on the upstream face of the salt wedge is usually where the highest turbidity levels are found in stratified estuaries (e.g., Columbia River estuary, Washington-Oregon [C.A. Morgan, Cordell, and Simenstad 1997]).

Nutrients and Microconstituents
Nutrients in the salmonid estuary are an indirect requirement, because nitrates, silicates, phosphates, and others are needed for primary production at the base of the estuary food web (Lalli and Parsons 1997). Some waterborne microconstituents, such as trace metals (e.g., magnesium), may be required for fish nutrition (Shearer and Åsgård 1992).

pH
The pH of estuarine water used by salmonids is a function of the acidic or basic characteristics of the fresh water and ocean water mixing in the estuary. Nearshore ocean water pH ranges from approximately 8.0 to 8.5 (Wootton, Pfister, and Forester 2008) and river water from 6.5 to 8.5 (Hem 1986).

Other Important Water Properties
Daytime light levels in clear-water salmonid estuaries are in the range of 50–60 lux (P.S. Young, Swanson, and Cech 2004), going down to 15–30 lux at dawn and dusk (M. Koski and Johnson 2002), to zero in turbid conditions (see Helfman 1986). Estuaries undisturbed by vessel traffic or industrial activity are quiet except for marine mammal sounds, e.g., the voices of beluga whale (*Delphinapterus leucas*) at 81–87 decibels in the St. Lawrence River estuary in Québec (Scheifele et al. 2005), which is close to noise from low wave action in the ocean (Gausland 2000).

My description of the habitat and water properties of the estuary in relation to salmonids has set the scene for later discussions of fitness and salmonid adaptations to specific habitats. However, in order to place salmonids fully in an ecosystem context and to provide a setting for describing biotic interactions at both the global and local level, it is necessary to describe the fish that live within the estuary, which is the topic of the next chapter.

5
Global Distribution of Salmonid Species and Local Salmonid Diversity in Estuaries

Salmonids contribute to the overall biodiversity in the estuary, a metric that considers the total number of species of plants, invertebrates, and fish. Biodiversity is a key factor for ecosystem services maintenance and conservation planning (Maclaurin and Sterelny 2008; Perrings et al. 2011). This chapter discusses only the narrower aspect of fish biodiversity, that is, the number and type of fish species in estuaries where salmonids are found. In the glaciated areas of the Northern Hemisphere, the status of salmonid species and their life history forms in estuarine communities may be important for the maintenance of fish biodiversity, because fish species richness is low (Curry et al. 2010). There may be a relationship between estuarine area and estuary-dependent fish species richness, as found between watershed area and freshwater fish species richness (e.g., northwest Europe river systems [Billen et al. 2006]). For estuaries with extensive freshwater tidal zones, the relationship might hold for freshwater species in the upper estuary and estuarine area. Large river and estuary systems with species complexes of the family Coregonidae (whitefish) originating from the river fish community, for example, have increased species richness relative to smaller systems with fewer whitefish species.

This chapter gives an overview of the salmonid community in estuaries around the world as background and contextual information for later discussions on biotic interactions. It also describes how human intervention spread salmonids into estuaries where they had not lived before. I have adopted Hubbell's definition (2001) of a community: "a group of trophically similar, sympatric species that actually or potentially compete in a local area for the same or similar resources." A caveat here is the contemporary view held by Ricklefs (2008) that "local communities are not integral entities, and that the species assemblages are products of separate local adaptations of the individual species." Further, it should be noted that the description of a community is dependent on how the organisms are sampled (E.L. Mills 1969). I first describe the natural salmonid community with its endemic species and then discuss the community that can result if invasive fish species are part of the estuarine assemblage.

The Natural Salmonid Community

Salmonids coexist and have coevolved with a wide variety of families of endemic nonsalmonid or nonsalmonine fish in estuaries of the Northern Hemisphere. Sturgeon (Acipenseridae), minnows (Cyprinidae), cod (Gadidae), whitefish (Coregonidae), sticklebacks (Gasterosteidae), and cottids (Cottidae) were some of the most frequently recorded families from salmonid estuaries in the literature (see Appendix Table 1.2 at http://hdl.handle.net/2429/57062). Other native families less frequently found with salmonids in estuaries include wrasses (Labridae) (e.g., Numedals River estuary, Norway [Hvidsten et al. 2000]), suckers (Catostomidae) (e.g., Fraser River estuary, British Columbia [J.S. Richardson et al. 2000]), and killifish (Fundulidae) and topminnows (Atherinopsidae) in the Miramichi River estuary, New Brunswick (J.M. Hanson and Courtenay 1995). The coexisting families and species are from a wide range of lifestyles, from pelagic (e.g., topminnows) to benthic (e.g., sturgeon). Salmonids in the Arctic and near Arctic coexist with only a few species (e.g., fourteen species in the Kara and Livar-Yakha River estuaries, Russia [Semushin and Novoselov 2009]). More fish species are found in southern estuaries. The fish community at the Gironde estuary in France is representative of an estuary that includes salmonids on the Atlantic coast of southern Europe (M. Elliott and Dewailly 1995). Seventy-five fish species were found in the Gironde estuary (Lobry et al. 2003). The Atlantic salmon and sea trout from the Gironde estuary were components of an ecotrophic guild of eleven species that included European sturgeon (Acipenseridae: *Acipenser sturio*); allis shad and twaite shad (Clupeidae: *Alosa alosa* and *A. fallax*, respectively); European eel (Anguillidae: *Anguilla anguilla*); three-spined stickleback (Gasterosteidae: *Gasterosteus aculeatus*); thinlip grey mullet (Mugilidae: *Liza ramada*); smelt (Osmeridae: *Osmerus eperlanus*); and river and sea lamprey (Petromyzontidae: *Lampetra fluviatilis* and *Petromyzon marinus*, respectively) (Lobry et al. 2003). In Japan, where the estuaries are located on small rivers with low run-off, most species living with salmonids have marine affiliations – for example, ponyfishes (Leiognathidae) and cardinal fishes (Apogonidae) (e.g., Ashida River estuary, Japan [Kakuda 1973]). The native fish community in the temperate Fraser River estuary shows a typical gradient of families with varying salinity tolerance along the estuary. Fishes with fully marine affinities, such as whitespotted greenling (*Hexagrammos stelleri*), were found at sites closer to the ocean, while freshwater species such as dace (*Rhinichthys* spp.) and northern pikeminnow (*Ptychocheilus oregonensis*) were found in the upper estuary. The Neva River estuary in Russia is also characterized by a very diverse estuarine fish community that includes salmonids. Sixty species were recorded in this estuary, many from fresh water, reflecting the low salinity regime of the estuary (Telesh, Golubkov, and Alimov 2008). Sea trout using estuaries on enclosed water bodies such as the Caspian Sea and the Aral Sea also coexist with communities dominated by freshwater families such as cyprinids (e.g., Amu Darya River estuary, Uzbekistan [Shaposhnikova 1950]).

The estuarine salmonid community is dynamic, changing with location, tidal levels, and seasons. A good example of the tidal effect was provided by surface trawl data from the Columbia River estuary, Washington-Oregon, where the dominance of juvenile Chinook salmon, Pacific herring (*Clupea pallasi*), eulachon (*Thaleichthys pacificus*), northern anchovy (*Engraulis mordax*), and whitebait smelt (*Allosmerus elongatus*) varied with tidal levels (Emmett, Brodeur, and Orton 2004). Most salmonids use estuaries in a migratory pulse (Wissmar and Simenstad 1998) and seasonal dominance of juvenile salmonids can lead to identification of specific temporal patterns. Most of the community survey data presented in the literature were obtained in spring, summer, and autumn, when salmonids were present either as downstream migrating juveniles, estuary residents, or upstream migrating adults. Usually salmonids are not present in the estuary in winter but some exceptions do exist, such as sea trout in northern Norwegian estuaries (O. Berg and M. Berg 1989). Species richness and diversity were modified by the seasonal numerical abundance of juvenile salmonids in the Duwamish River estuary in Washington (Cooksey 2006). Both species richness and diversity increased in May and June when juvenile coho salmon, chum salmon, Chinook salmon, and steelhead were present. Diversity decreased later in the summer when juvenile shiner perch (*Cymatogaster aggregata*) dominated the community and the salmonids were less abundant. The dominance of species in the fish community is strongly influenced by the abundance measure used, as biomass will typically determine a different relationship than numerical counts (Tokranov 1994).

Invasive Nonsalmonid Fishes

Concern has been raised about the ecosystem effects of invasive nonsalmonid fishes in rivers and lakes (Strayer 2010). The topic has received relatively little detailed attention in salmonid estuaries, but the variety and complexity of invasive species that salmonids are now encountering in estuaries are increasing. It is important to note the difference between nonindigenous and/or invasive species (see Colautti and MacIsaac 2004). Some nonindigenous fish may not spread from their point of introduction and hence do not deserve the term invasive. Nevertheless, even if temporarily resident, or present in a restricted area, the nonindigenous species could be impacting the natural ecological niche of salmonids or other species in the estuary. As well, the term "invasive" is conditioned by societal values – a species that was once considered desirable as an introduced species may over time become undesirable, even if no drastic effects on the native ecosystem have been found (Garcia de Leaniz, Gajardo, and Consuegra 2010). For consistency with much of the literature, I use the term "invasive" in this book.

The scope of the problem of invasive nonsalmonid species impacts is illustrated in the heavily invaded Sacramento–San Joaquin River estuary in California. In one study, 39 species of fish were collected in this estuary, including 15 (38 percent) native species and 24 (62 percent) invasive species (L.R. Brown and Michniuk 2007; Figure 12). A total of 39,095 fish were captured, of which 1,409 (4 percent) were native and 37,686 (96 percent) were invasive. Chinook salmon and steelhead accounted for 0.7–0.9 percent and 0.1 percent of the catches, respectively, and some salmonids known from the estuary (e.g., coho salmon and cutthroat trout) were absent. It is likely that the proportion of salmonids was higher before invasive nonsalmonid fishes colonized the estuary. Matern, Moyle, and Pierce (2002) noted that 25 species of invasive nonsalmonids were found in a brackish marsh in the Sacramento–San Joaquin River estuary. Questions are now arising regarding the significance of invasive nonsalmonid fish in additional salmonid estuaries (e.g., Washington and Oregon estuaries [Sanderson, Barnas, and Wargo Rub 2009]).

Representative invasive nonsalmonid fish reported from estuaries with salmonids are given in Appendix Table 1.2 (at http://hdl.handle.net/2429/57062). A variety of fish families have been found and include many species that salmonids have not coevolved with. Examples include freshwater-related species such as bottom-dwelling cyprinids like carp (*Cyprinus carpio*), ictalurids (e.g., brown bullhead, *Ictalurus nebulosus*), and centrarchids (pumpkinseed, *Lepomis gibbosus*) in the

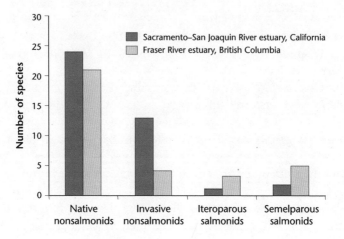

Figure 12 Number of native nonsalmonids, invasive nonsalmonids, and native iteroparous and semelparous salmonid species at the Fraser River estuary, British Columbia, and the Sacramento–San Joaquin River estuary, California.
Sources: Data from J.S. Richardson et al. 2000 and L.R. Brown and Michniuk 2007.

Fraser River estuary (J.S. Richardson et al. 2000), and unusual estuarine families such as Odontobutidae (Amur sleeper, *Perccottus glenii*) in the Neva River estuary, Russia (Telesh, Golubkov, and Alimov 2008). Additional invasive families with estuarine or marine affinities include Gobidae, such as black goby (*Gobius niger*) (Wonham et al. 2000) and Shimofuri goby (*Tridentiger bifasciatus*) in the Sacramento–San Joaquin estuary (Bennett, Kimmerer, and Burau 2002); Clupeidae, such as threadfin shad (*Dorosoma petenense*), also in the Sacramento–San Joaquin estuary (Nobriga et al. 2005), and American shad (*Alosa sapidissima*) in the Columbia River estuary in Washington-Oregon (Sanderson, Barnas, and Wargo Rub 2009); and Sciaenidae, such as Atlantic croaker (*Micropogonias undulatus*) in the Scheldt River estuary, Belgium (Stevens et al. 2004). Invasive nonsalmonid fish abundance changes seasonally and these fishes may dominate the salmonid estuary fish communities in summer, possibly because many invasive nonsalmonid fish have invaded from the south and are adapted to warmer temperatures than salmonids. In the Sacramento–San Joaquin River estuary, a spring fish community characterized by Chinook salmon and several other native fish species was replaced by a summer assemblage dominated by invasive inland silverside (*Menidia beryllina*), threadfin shad, American shad, and redear sunfish (*Lepomis microlophus*) (Nobriga et al. 2005).

Factors controlling the invasibility of estuarine communities with native salmonids by invasive nonsalmonid fishes include a number of interacting factors (Moyle 1999): (1) propagule pressure (the number of individuals, such as eggs or other life history stages, in an introduction event), and the frequency of these events; (2) environmental similarity of the estuary to the invading species' native habitat; (3) competitive interactions with native species and other invasives (biotic resistance); and (4) degree of habitat disruption or variance from the pre-invasion state of community. These factors also apply to the question of how anadromous salmonids invade river-estuary systems outside their native range.

Salmonids as Invasives

At least eleven salmonids – Atlantic salmon, cutthroat trout, pink salmon, chum salmon, coho salmon, masu, rainbow trout (anadromous form, steelhead), brook trout (anadromous form, sea-run brook trout), sockeye salmon, brown trout (anadromous form, sea trout), and Chinook salmon – have been introduced by intentional or accidental stocking outside of their native habitats (Crawford and Muir 2008; Savini et al. 2010). They have been transplanted into freshwater habitats on every continent except Antarctica, in most instances at the egg, fry, or smolt stage. Anadromous salmonids have colonized numerous river-estuary-ocean systems outside their native range; by definition, they need to be able to exploit the three interconnected ecosystems. They are coexisting with nonsalmonid species very different from those they evolved with, especially the Southern Hemisphere estuarine fishes (see Appendix Table 1.2 at http://hdl.handle.net/2429/57062). For example, sea trout, steelhead, and Chinook salmon are found in the Kakanui River estuary, New Zealand. At this estuary, eight native families (Tripterygiidae, Eleotridae, Galaxidae, Retropinnidae, Geotriidae, Leptoscopidae, Gempylidae, and Cheimarrichthyidae) were found (Jellyman et al. 1997) that do not coexist with salmonids in the Northern Hemisphere, where salmonids evolved. Here, I give profiles for selected invasive salmonids.

The polytypic brown trout and its anadromous form, the sea trout, may be the most widely distributed salmonid in rivers, lakes, estuaries, and oceans around the world. In numerous locations, and on all the continents except Africa, when a brown trout stock with a genetic profile for anadromy has been introduced to a river system with direct connections to the ocean, the species has displayed the sea trout form. Brown trout were one of the earliest salmonids moved to the Southern Hemisphere; Buckland (1880) reported the transplantation of this species to New Zealand in 1878 and the estuary form was found soon after. Colonization of Southern Hemisphere river-estuary systems has been very successful. Sea trout are now found in estuaries in New Zealand (Jellyman et al. 1997); Tasmania and Australia (Kalish 1990); Chile (Zama 1987); Argentina (Bravo et al. 2006; O'Neal and Stanford 2011); the Falkland Islands, United Kingdom Overseas Territories (McDowall, Allibone, and Chadderton 2001); and Kerguelen Islands, District of the French Southern and Antarctic Lands (Jarry, Davaine, and Beall 1998; Ayllon et al. 2006; Launey et al. 2010). Sea trout are also found in estuaries in Japan (Kitano 2004) and the Atlantic coast of Canada (Westley, Ings, and Fleming 2011).

Pink salmon are now probably the most widespread salmonid in Northern Hemisphere estuaries, occurring from Norway on the North Atlantic eastward to the Lena River, Russia, on the Arctic Ocean. Its native range in the Pacific is from Korea to Oregon, with some reports from the western Canadian Arctic Ocean (Nielsen, Ruggerone, and Zimmerman 2013). Millions of fertilized pink salmon eggs were transplanted from the Pacific coast of Russia between 1957 and 2001 to hatcheries on the White Sea (western Arctic Russia) in acclimatization experiments (Gordeeva et al. 2003). Pink salmon populations were established that spread further west to North Atlantic European estuaries.

Chinook salmon from the Sacramento River in California were successfully introduced to the south island of New Zealand in 1901 and 1907 for development of a recreational fishery (McDowall 1994), and into Chile in 1924 or earlier by aquaculturists (Iriarte, Lobos, and Jaksic 2005). Chinook salmon are found in New Zealand and Chilean estuaries (Becker, Pascual, and Basso 2007; Correa and Gross 2008) as they have established in several river systems in those countries. On the other hand, Chinook salmon transplants to Tasmanian rivers in Australia were not successful (Cadwallader and Eden 1981).

Coho salmon were introduced to rivers outside their natural range in several areas of the world, including New Brunswick (Symons and Martin 1978), several countries in northern Europe (Crawford and Muir 2008), and Chile (Iriarte, Lobos, and Jaksic 2005). In some experiments, populations from a region where coho salmon are abundant have been transplanted to another part of the world where less abundant native stocks already exist (e.g., coho salmon from Washington to Hokkaido, Japan [Umeda et al. 1981]). Establishment success of the various transplants has apparently been mixed, but definitive data are not available in the literature. The New Brunswick populations did not survive (Dr. Peter Amiro, personal communication, March 10, 2011).

Chum salmon juveniles were released into the Aysén Fjord in Chile in the 1980s (Hirakawa 1990) and were reported as an established exotic species by B.S. Dyer (2000). However, I could not find data on this species' estuarine ecology outside its natural range.

Steelhead are probably the second most widely distributed estuarine salmonid in the world and are found in numerous estuaries outside their native range. The species has been reported from eight of the twenty-three European estuaries summarized by M. Elliott and Hemingway (2002), as well as estuaries in eastern Canada (Thibault et al. 2010), Argentina (Pascual et al. 2001), and New Zealand (Jellyman et al. 1997). Some of the populations arose from transplants of rainbow trout, the nonanadromous form of steelhead. Some developed from rainbow trout transplanted into rivers (Pascual et al. 2001), while others may have originated from rainbow trout that had escaped from fish farms. In Norway, the situation is different. In 2001–09, between 7,000 and 315,000 rainbow trout per year escaped from fish farms (Ø. Jensen et al. 2010); none have colonized rivers or assumed anadromy (Arne J. Jensen, personal communication, April 10, 2015). In the United Kingdom, there is concern about the potential of introduced rainbow trout to assume anadromy and disperse from one river system to another (Fausch 2007); obviously, such spread would involve estuarine habitat.

Brook trout have dispersed from rivers along the south coast of Chile as anadromous or sea-run populations. Sometimes they are found in estuaries whose headwaters were not stocked, such as the estuaries of the Puelo and Petrohue Rivers flowing on the Relconcavi Fjord, Chile (Karas 2002). Chile is apparently the only area where brook trout are found in estuaries outside their natural ranges. Hatchery-reared brook trout are capable of moving long distances downstream (up to 80 km) in the Kemijoki River in Finland (Korsu and Huusko 2009). A few individuals were caught in the ocean about 10–20 km from the river mouth, but none were found in the Kemijoki River estuary or other estuaries nearby (Dr. Kai Korsu, personal communication, February 18, 2011).

Atlantic salmon have not become established in many North or South Pacific systems, despite several attempts at intentional introductions and unintentional escapes from aquaculture facilities (MacCrimmon and Gots 1979; Thorstad et al. 2008). There is evidence that adult Atlantic salmon ascended a river in British Columbia to spawn, and therefore used the estuary en route from the ocean (Volpe et al. 2000). Atlantic salmon are cultured in large coastal operations in southern Chile, and although many have escaped from the sea pens over the years, the species has not become established in the local rivers and estuaries (Professor

Carlos Garcia de Leaniz, Swansea University, personal communication, June 12, 2012). In contrast, Thorstad and colleagues (2008) reported that they have been found in Argentine systems. Lake Llanquihue in Chile is used for rearing Atlantic salmon smolts (Soto and Jara 2007), and escaped fish from these facilities migrate downstream via the Maullín River estuary and return as adults to the lake. I could not find data on whether or not these returning adults spawn successfully and thus form an established population. Atlantic salmon have colonized river systems on the Kerguelen Islands, District of the French Southern and Antarctic Lands (Ayllon et al. 2004). On the Faroe Islands in the North Atlantic Ocean, escaped farmed Atlantic salmon have colonized river-estuary systems where their wild counterparts are found (Thorstad et al. 2011).

Utilization of foreign estuaries by anadromous salmonids has occurred because some of the transplanted stocks have had the genes for anadromy, and short-term evolution may have facilitated adaptations to estuary features. It should be noted that the presence of invasive salmonids in the estuary does not necessarily mean that they have adopted anadromy – they may be using the upper estuary and/or freshwater tidal zones as an extension of the river and not going to the ocean. This appears to be the case for brown trout, for example, at the Adam River estuary in British Columbia (G. Wilson et al. 1999) and the Klamath River estuary in California (Figure 13). In these two instances, the sea trout form does not appear to have been realized, possibly relating to the genetics of the stocked brown trout. Below I discuss some of the additional ecological factors relating to salmonid invasions to provide context for later discussions on biotic interactions and expanded narrative on short-term evolution.

As with invasive nonsalmonid fish, a variety of ecological processes can lead to invasive salmonid colonization of estuaries. Propagule pressure from invasive salmonids can arise from juvenile fish moving downstream into the estuary, from fish moved directly to the estuary by human activity, and from adult or juvenile fish moving in from the ocean. In addition to propagule pressure and the competitive ability of native fish in estuaries, physicochemical factors will influence whether or not salmonids can use an estuary. Estuaries have been explored as saline bridges or invasion portals to permit the colonization of new watersheds by freshwater fish invaders (J.A. Brown, Scott, and Wilson 2007), but not for anadromous salmonids. In future studies, it would be logical to include estuarine considerations in dispersal of invasive anadromous salmonids. Key variables such as temperature, salinity, currents, and depth are all important as attractants in such dispersals. Another variable

Figure 13 Left panel: Brown trout, Klamath River estuary, California. Right panel: Sea trout, Rio Grande River estuary, Argentina.

Note: Colour photos in Appendix 5 at http://hdl.handle.net/2429/57062.
Source: Brown trout photo by Mike Wallace, California Fish and Wildlife, Arcata, California; reproduced with permission. Sea trout photo by Lynn Palensky, Portland, Oregon; reproduced with permission.

that might lead to the suitability of an estuary is the flood disturbance regime of the river flowing into the estuary. Floods have been identified by Fausch and colleagues (2001) as a factor influencing rainbow trout invasion success. It is possible that very large run-offs would spread brackish water along a coastline, possibility enlarging the attraction area for invading salmonids.

Various models could help predict the suitability of estuaries as portals for salmonid invasion, but they would have to be linked with submodels to estimate reproductive success in fresh water as well as ocean factors affecting survival. Three general types of models (MacIsaac et al. 2007) could be used, including vector-based gravity models, environmental suitability models, and ecological niche modelling. Of these, environmental suitability models are probably most appropriate for investigating salmonid invasions of estuaries. These models can be multivariate models, which can take into account multiple attributes of habitats under consideration. Marchetti and colleagues (2004) successfully estimated the invasion potential of freshwater fishes in California watersheds using attributes such as trophic status, size of native range, parental care, maximum adult size, physiological tolerance, distance from nearest native source, prior invasion success, and propagule pressure. The degree of habitat disruption or variance from the natural environment (e.g., increased temperature or other water quality aspects) are also likely to be important (Marchetti et al. 2004). Domestication of the invasive species is also a factor, as newly introduced hatchery-reared salmonids tend to disperse more than wild salmonids (e.g., brook trout [D.P. Peterson and Fausch 2003]). Some of these attributes are appropriate for salmonids invading river-estuary systems and could be calibrated in the models.

Conclusions

Salmonids in estuaries share habitats with very diverse groups of fishes, now including many that they have not coevolved with. In general, we have only limited knowledge of how the dynamics within the community relate to the relevant fitness components of estuarine salmonid survival. Part of the challenge in developing a global quantitative synthesis of community dynamics of estuarine salmonids is the issue of data synthesis. As discussed next, basic data such as abundance and distribution of salmonids are sometimes not comparable because of the variety of methods and sampling designs used in estuaries around the world.

6
How Have Salmonid Abundance and Distribution Been Assessed in Estuaries?

Because of the geomorphological and oceanographic characteristics of estuaries, estuarine sampling involves special techniques. Sharp and variable spatial-temporal gradients in depth, habitat type, water properties, and flow are some of the factors that challenge the salmonid estuarine ecologist. Sampling biases may exist with various techniques and can affect assessments of survival, biotic interactions, and other ecological topics that utilize abundance and distribution data.

Methods to Enumerate Salmonids in Estuaries
Methodological manuals provide general guidance for fish sampling and enumeration (e.g., for fish in a variety of aquatic habitats [Zale et al. 2012]). However, a comprehensive tabulation of techniques potentially useful for a range of salmonid species and life history stages in estuaries is not available. A selection of methods for sampling salmonids used in representative estuarine studies are shown in Appendix Table 1.3 (at http://hdl.handle.net/2429/57062). In general, the methods used are a mixture of active capture techniques, where the equipment is actively moved to enclose the fish in a net (e.g., beach seines), and passive capture techniques, where the behaviour and movements of the fish themselves result in capture (e.g., gillnets). Due to spatial and temporal changes in habitat use, the techniques employed and when they are used need to be tailored to specific salmonid life history stages and behavioural effects on catchability. For example, Methven and colleagues (2001) caught sea-run brook trout and sea trout in a Newfoundland and Labrador estuary at night by beach seining, but none were captured during daytime sampling. This finding suggests that the salmonids were more vulnerable to capture during dark hours. Gillnetting depends on the capability of the net to entangle the head and body of a salmonid as it tries to swim

through the fixed gear. The capability of a gillnet to catch juvenile or adult fish is determined by mesh size. Very small mesh will block water flow, so gillnets are typically not effective for salmonid fry. Coho salmon and Chinook salmon smolts at the Squamish River estuary in British Columbia were caught by gillnets with mesh size 2–5 cm (Levy and Levings 1978), but bigger fish (e.g., returning veteran cutthroat trout) were less vulnerable as their heads did not get entangled. Adult salmonids are captured with large-mesh gillnets in estuarine industrial fisheries around the world (e.g., a mesh size of 21 cm caught Chinook salmon in the Columbia River estuary [Olson and Quinn 1993]). Enumeration of estuarine salmonids harvested by Aboriginal peoples may involve data gathered in the past by traditional methods. For example, Arctic char were caught in Canadian Arctic estuaries using spears or stone fences in the intertidal zone (Balikci 1980). Remote sensing methods such as sonar, which depends on sound reflection from underwater targets, are applicable to larger fish. Sonar has been used to assess abundance of adult salmonids in estuarine channels (e.g., Fraser River estuary, British Columbia [Xie et al. 2005]). Thus, while a wide variety of techniques are available for salmonids in estuaries, methods have to be tailored to the objectives of the research or study program, especially the target organism's life stage.

Juvenile salmonids are often the focal life history stage in the estuary, and beach seining is frequently used for sampling salmonid fry, parr, and smolts. To be effective, beach seining requires a relatively gently sloping sand, mud, or gravel beach so that when the net is pulled from deep to shallow water, the weighted bottom of the net forms a seal on the substrate, preventing the fish from escaping the net bag. When beach seines are used on rock or cobble beaches, a tight seal is not achieved; effectiveness is low (Levings, Birtwell, and Piercey 2003) and capture data are not comparable to those from the softer sediment. Effectiveness is usually a challenge to measure, with some exceptions where a "within subhabitat" comparison may be possible. Beach seine effectiveness can be determined by comparing the total number of fish caught when a large tidal creek is drained with beach seining data from within the enclosure (Congleton et al. 1982). Effectiveness may also be measurable in the freshwater tidal zone for some species that maintain territories and are relatively stationary between repeat seine sets. The important question of fishing gear effectiveness and other statistical aspects of estimating fish density from samples has been discussed by Pope and colleagues (2010).

Other methods are appropriate for enumerating juvenile salmonids in estuaries. In deeper riverine channels, purse seines, which use rings to form a sealed bag or purse, can be used. Block nets or modified fyke nets can be used effectively in side channels and tidal creeks (Levy and Northcote 1982). Determining water column use by salmonids in deeper water is a greater challenge. Net sampling with shallow purse seines or fish trawls that sample the surface provide limited information on abundance in the vertical plane. If two boats trawl side by side and each boat samples different depths, useful data can be obtained. This was the approach taken to document juvenile Chinook salmon abundance in the water column at the mouth of the Columbia River, Washington-Oregon, by Emmett, Brodeur, and Orton (2004). However, trawls often have limited effectiveness because salmonids can escape by swimming out of the trawl net mouth. The use of baited hooks set at various depths is another possible methodology (Orsi and Wertheimer 1995). Clearly, geomorphology, depth, tidal stage, and other characteristics of estuarine juvenile salmonid habitat will determine the type of sampling gear used. To mitigate some of these sampling problems, investigators have turned to visual or photogrammetric enumeration of estuarine juvenile salmonids.

Enumeration using snorkel or scuba gear in clear water can give information on estuary use by juvenile salmonids (e.g., Toft et al. 2007). Fixed underwater cameras can also be used to enumerate migrating juvenile estuarine salmonids as well as adults. There are limitations to direct observation and camera methods, however. Because of similarities in morphology and colour, even species identification can be difficult (McMahon and Hartman 1988), and observing fish close-up has limitations in highly turbid estuaries. Generally speaking, direct observational and species identification data can be obtained more effectively in the freshwater tidal zone, where, depending on river condition, water is less turbid. As well, some juveniles salmonids may not have completed smoltification in this zone, and thus retain their characteristic parr and fry markings.

Where possible, methods used for salmonid estuary sampling should be calibrated with mark-and-recapture experiments or other techniques to provide accurate abundance data for comparisons between studies. Calibration or enumeration involves marking or tagging of individuals to provide accurate estimates of fish densities (number per square metre or number per cubic metre). Usually this can be achieved by estimating population size in a measured area. Population size can be estimated using the formula $N = MC/R$, where N is the total population size estimate, M is the total number of fish caught and marked at time 1, C is total number of fish caught at time 2, and R is the number of fish caught at time 1 that were recaptured at time 2 (Ricker 1975). This method assumes that the population of fish has remained in the study area between time 1 and time 2, an assumption not often met in the case of estuarine juvenile salmonids, which often move between habitats with tidal cycles, current changes, and other variables. If the mark-and-recapture experiments can be conducted in a constrained subhabitat, the accuracy of the density estimates are likely to be considerably improved; at least sufficient fish can be recaptured to compute population size. Estimates of juvenile salmonid densities were made in side channels in brackish marshes at the Fraser River estuary, using fyke nets as tidal traps (Levy and Northcote 1982). Fish were captured and marked at high tide and subsequently recaptured at low tide when the channels were almost drained. Mean density of Chinook salmon fry was estimated at 0.18 individuals • m^{-2}. In some instances, population sizes of the whole estuary can be estimated by mark-and-recapture experiments. Using a rotary screw or inclined plane trap in the river channel immediately above the estuary to count and mark fry or smolts moving downstream, together with a recapture program across all estuarine habitats, is an effective method for enumerating the total population entering the estuary (e.g., Chinook salmon fry in the Nanaimo River estuary, British Columbia [Healey 1980a]). The majority of estuarine juvenile salmonid beach seine studies report abundance or density in terms of catch per seine haul or estimate the individuals per square metre from the area covered in a selected beach subhabitat. Beach seining effectiveness is usually not accounted for.

Important international, regional, or local forums have been established in attempts to standardize salmonid sampling in the estuary and to provide guidance for improving data accuracy and comparability. Examples are the Pacific Northwest Aquatic Monitoring Partnership (2014) and initiatives for standardizing fish sampling in European estuaries (e.g., Champ et al. 2009; Cowx et al. 2009). Sampling method manuals for specific management problems are also available (e.g., estuarine salmonid habitat restoration [C.A. Rice et al. 2005]). In some areas, a multimethod strategy has been recommended to assess the whole fish community. If this strategy is used, limitations of the various methods for salmonid sampling should be recognized; for example, three methods suggested for sampling in the Thames River estuary in England are beach seining, hand net, and trawl, but only seines caught salmonids (Colclough et al. 2002). Further local, regional, and global collaboration on sampling and enumeration methods would greatly expand opportunities for comparative ecological studies of salmonids in estuaries. Comparative studies also need to include consideration of the various marking methods to determine distribution of various life stages of salmonids in estuarine habitats.

Tagging and Marking Methods to Assess Abundance and Distribution

Marking or tagging individual salmonids or groups of salmonids to determine their movement in time and space has a long history in fisheries science. Using fin clips, W.H. Rich and Holmes (1929) conducted some of the earliest marking studies with juvenile salmonids in estuaries using Chinook salmon in the Columbia River and estuary, Washington-Oregon. A wide variety of techniques are now available (McKenzie et al. 2012), including physical and chemical marking, telemetry using radio- and hydroacoustic tags, and genetic methods. The efficacy of physical methods for marking salmonid fry was reviewed by Skalski and colleagues (2009). Below, I give an overview of the range of techniques available for studying distribution of salmonids in the estuary and provide some comments on their suitability.

Physical or Chemical Marks

Marks of various kinds have been used in tagging programs to assess salmonid distribution in estuaries, including tags inserted into musculature, fin clips, cold branding, fluorescent grit, immersion dyes, and elastomer dye (see Appendix Table 1.4 at http://hdl.handle.net/2429/57062). There are advantages and disadvantages of each tag type, which usually has to be tailored to the specific stage of interest and the study design. Dispersal of Chinook salmon fry into the Campbell River estuary in British Columbia was assessed with coded wire nose-tagged fish that could be identified by a missing adipose fin (J.S. Macdonald et al. 1988), but this method is very expensive and requires euthanizing fish to recover the tag and decode it in the laboratory. This is likely not allowed when working with endangered populations. In the Campbell River project, the tags had a dual purpose as the fish were recovered as adults in a survivorship study. Effects of marking on the health of the fish are a concern with the more invasive methods. Marking may injure or affect the health of estuarine salmonids, which may be sensitive to handling and disfigurement if they are already stressed from smoltification.

Natural chemical signatures or marks have been used or induced to assess salmonid distributions. When a salmonid entered into the estuary can be detected by a change in microchemistry at various locations across the otolith surface. Sometimes the ratio of strontium to calcium ions is used, as this ratio is very different in estuarine or ocean water compared with fresh water (e.g., Chinook salmon [Volk et al. 2010]). However, this method is not without problems as temperature and other elements, such as barium, can complicate the deposition of strontium and calcium (J.A. Miller 2011). Thermal marking of otolith (ear bones) rings is a benign technique for mass-marking unique groups of hatchery-reared juvenile salmonids (Volk, Schroder, and Grimm 1999) and can also be used in distribution and survival studies. In this method, temperature is varied while the fish are growing. Distinctive ring patterns can be established as the width of the ring spacing is proportional to growth rate, which in turn is related to temperature, controllable in the hatchery. Otolith thermal marking enables identification of salmonids originating from different hatcheries as they return to the estuary as adults and are caught in fisheries.

Electronic Tagging
Electronic tagging has significantly advanced our knowledge of estuarine salmonid distribution patterns and other aspects of estuarine salmonid ecology (Drenner et al. 2012). The method has been used with a variety of species and a range of estuaries (see Appendix Table 1.4 at http://hdl.handle.net/2429/57062). Electronic tagging to determine distribution of individual fish can involve several stages: capturing the salmonid, implanting a battery-powered tag that generates a signal, then releasing the fish and tracking its movement with a fixed or mobile receiver. Both the release location and the detection site need to be accurately mapped, typically with a global positioning system (GPS). Radio tagging can be used for work in the freshwater regions of the estuary, but radio tag signals cannot fully penetrate saline ocean water and are therefore of limited use in the main estuarine zones. Hydroacoustic tags, which transmit a signal that can penetrate salt water, are required. Stasko (1975) provided one of the earliest examples, and used hydroacoustic tags to track migratory patterns of adult Atlantic salmon in the Miramichi River estuary in New Brunswick. Hydroacoustic tags were also used to investigate the migration patterns of Atlantic salmon kelts in the La Have River estuary in Nova Scotia (Hubley et al. 2008). Sonobuoys (a type of fixed receiver) tracked returning Atlantic salmon adults with hydroacoustic tags in the Fowey River estuary in England (Solomon and Potter 1988). Hydroacoustic tags have been used to investigate the migration and distribution of several species of juvenile salmonids in estuaries. In early studies, tag weight limited application to larger life stages, but now hydroacoustic tags that weigh less than 1 g are available, enabling tracking of smolts and larger fry or subyearling salmonids (Drenner et al. 2012).

Figure 14 Antennae for PIT tags and fyke net used to assess migrations and residency of juvenile Chinook salmon at a tidal channel in the Salmon River estuary, Oregon.
Note: Colour photo in Appendix 5 at http://hdl.handle.net/2429/57062.
Source: Reproduced from Hering et al. 2010, copyright 2008 Canadian Science Publishing or its licensors. Reproduced with permission.

Passive Integrated Transponder (PIT) tags are small radiofrequency devices that do not require a battery. They can also be used with small life history stages of salmonids and as well as with larger fish. When hit by a signal sent by a device, the tag responds by transmitting a unique code that is picked up by a fixed or mobile antenna and transmitted to a digital recorder, so the fish has to pass by a detector or antenna. There are limitations to using PIT tags in estuaries as they are less effective in high-salinity water, but a number of studies have used them on salmonids in freshwater or brackish areas of the estuary. PIT tags were used to track migrations of sea-run brook trout in the Quashnet River estuary in Massachusetts (Winders 2013) and cutthroat trout in the Salmon River in Oregon (Krentz 2007). They have also been used to document juvenile Chinook salmon residency in tidal channels of the Salmon River estuary in Oregon (Hering et al. 2010; Figure 14). Data storage tags, sometimes called archival tags or data logging tags, use small computers that contain a real-time clock, various sensors, and internal memory for storing data on parameters such as depth, temperature, and salinity. Archival tags were used to document the depth preferences of Arctic char in a Norwegian estuary (Rikardsen et al. 2007).

The development of electronic tags of various types has clearly been a very important addition to the researcher's toolbox when studying the ecology of salmonids in estuaries. The currently available hydroacoustic and PIT tags are very useful for investigating the broad-scale distributional ecology and survival of subyearlings, smolts, and larger life history stages. However, tag weight and battery life do limit the application of hydroacoustic tags for studies of salmonid fry.

Parasitological Methods
The use of parasites as tags to study the distribution of fish populations, especially for stock management purposes, has a long history and is a method that requires careful application due to the complex life history of parasites (Lester and MacKenzie 2009). Although not widely explored in estuaries, this method also shows some promise in helping to identify the habitat use of juvenile salmonids. Parasites acquired by juvenile Chinook salmon at the Columbia River estuary, Washington-California, were different in fish that had been feeding in brackish marsh compared with those feeding on sandflats, and also varied between the lower and upper estuary (Claxton et al. 2013).

Genetic Marks
Genetic methods are rapidly increasing our knowledge of salmonid ecology in estuaries. Besides explaining adaption and fitness, they can help explore distribution patterns in the estuary. Microsatellite size is a unique marker for genetically different fish populations (Chistiakov et al. 2006). Microsatellite analysis involves the electrophoretic separation of DNA fragments on genes. It is a powerful technique for determining the distribution of salmonid populations in estuaries, where there is typically a complex mixture of fish populations originating from multiple tributaries in the watershed. Samples of microsatellites are collected from adult spawners in a tributary, and then from juveniles or other life history stages in the estuary. By matching the juvenile and adult microsatellite signatures, the origin of the estuarine salmonids can be identified. Microsatellite analysis has been used to study the distribution of endangered salmonid populations relative to seal predation sites (Wright et al. 2007), to contaminated habitats (Sloan et al. 2010), and to habitat restoration locations (Teel et al. 2009). The distribution of various life

history types can also be explored with microsatellite data (e.g., separating anadromous and freshwater resident subpopulations of brook trout while both are at the juvenile stage in fresh water [Boula et al. 2002]). Further exploration of genetic methods may help us understand the distribution of salmonids in estuaries at finer scales, perhaps even at the mesohabitat scale.

Conclusion

Determining the abundance and distribution of salmonids in estuaries presents special challenges relative to the river and the ocean. Because of estuarine complexity and variable physicochemical conditions, methods for assessing abundance have to be tailored to specific habitats or subhabitats. Because local or regional methods are sometimes not standardized or assessed for accuracy, caution is required when data are compared. It is likely that the common beach seine will remain the most common assessment tool for juvenile salmonids in estuaries. Fixed or roving video cameras do show promise for obtaining data on juvenile salmonid abundance in some estuaries, and there is scope for greater use of scuba and snorkel methods. New marking technology is becoming available to obtain distribution and movement patterns of salmonids in estuaries. Electronic tags and sophisticated biological methods, such as genetic and parasite analyses, are rapidly improving our knowledge of how salmonids exploit the estuarine environment. Further exploration and interpretation of abundance and distribution patterns may help us understand several key ecological questions, including the adaptive behaviour of salmonids in estuaries, to which we turn next.

7
How Do Salmonids Behave in Estuarine Habitat?

Behaviour is an important component of salmonid fitness that has been studied extensively in freshwater and marine habitats but much less so in estuaries. Fish behaviour is now considered to be more complex than previously thought, and fishes "exhibit a rich array of sophisticated behaviour and ... learning plays a pivotal role in (their) behavioural development" (C. Brown, Laland, and Krause 2011). I give an overview of some of the common behaviours of salmonids in estuarine habitats to provide background for later discussions of how these various behaviour patterns interrelate with topics such as feeding, predation, and competition.

Migration
Migration is possibly the most important behaviour of salmonids in estuaries. The concept of estuarine importance to salmonids rests on factors influencing movements into, within, and out of the estuary. There is extensive literature on factors influencing migration from fresh water into the estuary and many papers relate to smoltification (discussed in Chapter 9). The anadromous punctuated migration (Simenstad et al. 2002), exhibited as some species of juvenile salmonids move through the estuary seeking habitat and food, can also be called a rearing migration. The fish grow as they move downstream from one zone to another, sometimes reversing direction on flood tides. Factors relating to movement out of the estuary are also primarily related to osmoregulation and salinity preference (McInerney 1964), at least for some life history types, but this migration also relates to size. There are typically a variety of salmonid sizes in the estuary, and fish size or weight can determine osmoregulatory ability. Larger fish (smolts, older sea-run migrants, and kelts) may have developed the ability to move seaward into fully saline

waters, but smaller fish linger in the estuary. As a result, estuaries are "leaky" systems: small fish are arriving in the estuary from upstream as larger fish are leaving the estuary (e.g., coho salmon smolts and coho salmon parr at the Carnation Creek estuary in British Columbia [McMahon and Holtby 1992]). Migration to coastal areas of lower salinity some distance further from the immediate river mouth has been observed in some species. Sea trout in the Baltic Sea dispersed a median distance of 27 km from the mouth of their natal river (Degerman et al. 2012).

Migration to the estuary for rearing and growth, and not to the open ocean, may be a form of partial migration (see review article by Jonsson and Jonsson [2011]), a behaviour usually observed when part of an anadromous species' population remains in fresh water. A behavioural specialist in freshwater salmonid migrations noted that "the control of migration is ... determined by the exigencies of the habitat occupied" (Dingle 1980, 44). This is likely to be true in estuaries as well; however, the literature on partial migration into estuaries is not nearly as extensive.

Below I describe some of the short-, medium-, and long-term migration patterns for salmonid behaviour. It should be noted that holding behaviour and residency behaviour, described in the following subsections, are overlapping migration patterns. A halt in migration will result in one of these behaviours.

The movement of salmonids into brackish marshes from riverine channels as water floods the estuary on rising tides is one example of a short-term migration (Hering et al. 2010). This migration is on a minute or hourly scale. A medium-term migration (daily or weekly) in the estuary might be the downstream migration of smolts and kelts through the estuary (e.g., Hubley et al. 2008; Chittenden et al. 2008) and the upstream migration of adults (Olson and Quinn 1993). Atlantic salmon post-smolts migrated at a rate of 1 km • h^{-1} through riverine channels of the Penobscot River estuary in Maine (Renkawitz et al. 2012). The transit time of Atlantic salmon smolts through the York River estuary in Québec was about two days (Martin et al. 2009). The migration rate of adult Atlantic salmon varies with tidal cycles and river flows. For example, over-the-ground speed for Atlantic salmon adults at the Aberdeenshire Dee River estuary in Scotland ranged from 0.31 to 2.46 km • h^{-1}, with the fish mostly travelling at night on ebb tides (L.P. Smith and Smith 1997). Depending on the distance from the freshwater tidal zone to the mouth of the estuary, and on factors such as river discharge, tidal cycles, and temperature, these migrations could take a number of days. Tidal cycles are thought to be particularly important for Atlantic salmon smolts, although this parameter may be autocorrelated with salinity as rising tides typically bring more saline water into the estuary (Hedger et al. 2008). The rearing migration of juvenile salmonids as they move through the various estuarine habitats over a number of weeks or months is an example of a longer-term migration period (Hering et al. 2010). In very large systems, the freshwater tidal zone is important for this function (e.g., subyearling Chinook salmon in the Columbia River estuary, Washington-Oregon [Teel et al. 2014]).

Holding or Staging
Holding or staging behaviour may be thought of as the cessation of one activity before the beginning of another, usually for minutes or hours. The terms "holding" or "staging" convey a sense of inactivity during "suspended migration" (Brawn 1982). When salmonids enter an estuary from the river as juveniles or from the ocean as adults, they often temporarily stop a directed migration, possibly in response to tidal conditions. Adult salmonids in the estuary may show diel patterns of vertical movements (e.g., Chinook salmon [Candy and Quinn 1999]). This behaviour can result in significant energy consumption if the holding fish have to swim against tidal currents to maintain position (Moser et al. 1991). Juvenile holding fish may have substantial food requirements and adult holding fish may consume stored energy, with possible consequences for upstream migration (Cooke et al. 2004). The length of time spent holding in the estuary is dependent on the species and their physiological adaptations to changes in salinity and temperature. For example, migrating Chinook salmon adults held at the mouth of the Somass River estuary in British Columbia to avoid warm river temperatures (S.C. Johnson et al. 1996). Other oceanographic conditions are related to holding. Coho salmon smolts in an Oregon estuary spent 52–84 percent of their time holding in the estuary, swimming against the downstream-flowing current on ebbing tides (B.A. Miller and Sadro 2003). Holding behaviour may provide time for the fish to learn about their surroundings, search for food, or watch for predators (C. Brown, Laland, and Krause 2011).

Residency
Residency is longer-term holding behaviour in which fish move within a restricted area of their home range and do not migrate volitionally. W.H. Burt (1943) described home range as that area traversed by the individual in the course of its normal activities of food gathering, mating, and caring for young. Only the first function is relevant to salmonids in estuaries. Evidence for the existence of home ranges in salmonids

using estuaries is ambiguous. Over a fourteen-day monitoring period of hydroacoustically tagged sea trout post-smolts in two Scottish estuaries, most individual fish were shown to spend their time within the estuaries that their natal rivers drained to, even though they could have moved to other connecting estuarine water bodies (Middlemas et al. 2009). Although localized use may have been suggestive of the home range establishment, as time progressed the fish changed their use of space and favoured using new locations, thereby providing no definitive evidence for home ranges. It was unclear whether the sea trout distribution was a response to physical or biological differences between the two estuaries or reflects differences in behaviour between the two source populations (Middlemas et al. 2009). An estimate of the estuary home range of bull trout, or at the least the area occupied by the species over a circumscribed time period, was provided from the Swinomish River estuary in Washington (M.C. Hayes et al. 2011). Over a four-month period, fish tracked with hydroacoustic tags stayed within a 1,000 m^2 area, characterized by a mosaic of shallow-water (<4 m deep) habitats. Whereas home range has been scarcely studied, residency of individual fish or groups of fish in the broader estuary has been more widely investigated.

Residency data reported for salmonids in estuaries vary widely with study goals, assessment methods, species, life history stages, habitats, seasons, and locations. Juvenile salmonids that are adapted to reside in estuaries often do so during spring and summer, probably to achieve maximum growth. Certain specialized populations of some species are adapted to living in the estuary during winter for reasons related to osmoregulation; details of residency patterns of these specialized forms are discussed in Chapter 9.

An approximation of population residency can be based on a temporal sequence of maximum or average catch data, but this is not the same as individual residency (Simenstad et al. 1982). Individuals with entirely different life history patterns in terms of residency might not be identified. Conclusions regarding the adaptive significance of different residency patterns could result as the individuals at the long and the short end of the migration spectrum might have better survival than the average fish. Determining how long an individual juvenile or adult salmonid resides in an estuary requires marking, otolith microchemistry, or genetic analysis.

The temporal sequence method at the population level has been used in studies of fry or pre-smolts of semelparous species that show gradual migration through the estuary. An example with juvenile Chinook

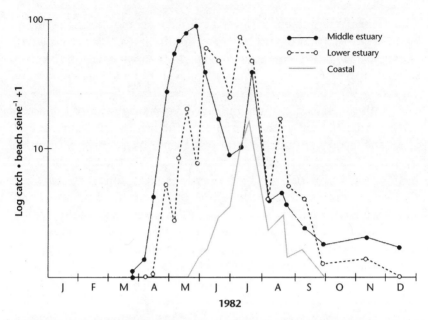

Figure 15 Seasonal change in abundance of wild juvenile Chinook salmon at three zones at the Campbell River estuary, British Columbia.
Source: Reproduced from Levings, McAllister, and Chang 1986. Copyright 2008, Canadian Science Publishing or its licensors. Reproduced with permission.

salmon at the Campbell River estuary in British Columbia is shown in Figure 15. For the middle estuary, residency of juvenile wild Chinook salmon was approximately 60 days (Levings, McAllister, and Chang 1986). Tagged fish provide a better estimate of residency for groups or individual fish. Juvenile salmonid residency in tidal channels was documented in the Fraser River estuary in British Columbia using fish marked with fluorescent grit (e.g., Chinook salmon fry residency up to 30 days, chum salmon fry up to 11 days, and pink salmon fry up to 2 days [Levy and Northcote 1982]). PIT-tagged juvenile Chinook salmon stayed in a side channel in the Salmon River, Oregon, for about three months (Hering et al. 2010). Coho salmon smolts marked with hydroacoustic tags resided in an Oregon estuary (lower estuary of Winchester Creek dominated by eelgrass habitat) for an average of about 5 days (B.A. Miller and Sadro 2003), similar to those in the Campbell River estuary in British Columbia (7 days [Chittenden et al. 2008]). Coho salmon fry marked with dye resided in the upper estuary of Winchester Creek for much longer periods, especially in early spring (up to 64 days [B.A. Miller and

Sadro 2003]). Chinook salmon smolts resided for about 40 days in the outer estuary of the Sacramento–San Joaquin River estuary and San Francisco Bay, California (MacFarlane and Norton 2002).

Residency estimates for iteroparous species also need to be discussed in the context of their estuarine life history. In their classic spring migration, wild Atlantic salmon smolts marked with ultrasonic tags in riverine channels of the Esk River estuary in England moved through the estuary quickly and resided 3 days (Tytler et al. 1978). This is a typical residency estimate for Atlantic salmon smolts, but there are some populations that exhibit a wide range. At the Nabisipi River estuary in Québec, most smolts resided in the estuary for about a week, but some stayed for 4–50 days (Power and Shooner 1966). Atlantic salmon parr residency in estuaries was studied in the Western Arm Brook estuary in Newfoundland and Labrador, where Cunjak (1992) found that marked parr stayed a minimum of two weeks in the inner estuary. Using hydroacoustic tags in spring and summer, Bégout Anras and colleagues (1999) found that Arctic char post-smolts and returning veterans resided for up to 10 days in the middle estuary of Freshwater Creek, Nunavut. Specialized populations that show residency in the estuary in winter, outside of the expected spring and summer seasons, have also been found (e.g., older age groups of sea-run migrant Arctic char and sea trout in the Skibotn River estuary in Norway [J.L.A. Jensen and Rikardsen 2008]). At the Scott Creek estuary in California, steelhead parr or pre-smolts resided in the estuary for one to ten months (S.A. Hayes et al. 2008). There are few data on residency of salmonid kelts. The mean residency time for Atlantic salmon kelts in the La Have River estuary in Nova Scotia was 5 days, but a few fish stayed for more than 10 days (Hubley et al. 2008). Sea trout kelts at the Fowey River estuary in England resided for only a few tidal cycles (Bendall et al. 2005). Salmonids in estuaries thus show highly variable residency patterns, ranging from hours to months. The adaptive significance of the residency patterns needs to be looked at in the context of the other important behaviours discussed in this chapter.

Cover-Seeking Behaviour

The adaptive significance of cover-seeking behaviour in estuarine habitats providing refuge is likely to be related to predator avoidance (discussed in Chapter 11). There are surprisingly few direct observations of juvenile salmonid behaviour in and around estuarine subhabitats that may provide cover, such as marshes, woody debris, and rocks. Hydroacoustically tagged Chinook salmon smolts may have shown cover-seeking behaviour in eelgrass in the enclosures used by Semmens (2008), but no direct observations were made in the study. Snorkel data showed that coho salmon smolts aggregated around woody debris in the estuary of Carnation Creek, British Columbia, and were more abundant there relative to freshwater cover habitat upstream (McMahon and Holtby 1992). The few available data from brackish marshes, obtained using indirect methods (gillnets, surface trawls), suggested that Chinook salmon fry in brackish marsh do not penetrate the marsh but stay in the shallows near the marsh edge (Simenstad et al. 2002; Northcote et al. 2007). Although behavioural observations near brackish marshes could be made by in situ observations in an estuary with clear water and good visibility, this has not been done.

Feeding

Some data have been obtained on in situ feeding behaviour of salmonids in estuaries. For example, using scuba at the Campbell River estuary in British Columbia, J.S. Macdonald, Birtwell, and Kruzynski (1987) observed juvenile Chinook salmon and coho salmon making vertical or horizontal dashes to feed on plankton or drift insects. Current speed affected juvenile Chinook salmon feeding, with velocities of 46–60 cm • s^{-1} impairing the fish's ability to catch prey items.

Almost all of our understanding of salmonid feeding behaviour in estuaries has been obtained by inference following examination of stomach contents and subsequent interpretation of behaviour in the prey organisms' habitat. For example, Levings, Conlin, and Raymond (1991) identified three general prey production-foraging modes important for juvenile Chinook salmon in the Fraser River estuary in British Columbia: (1) terrestrial-riparian vegetation, (2) benthic-brackish marsh, and (3) epibenthic-mud-sandflats. A few studies have used laboratory tanks or mesocosms to examine feeding behaviour and habitat preferences of juvenile salmonids in mimicked estuarine conditions. Webster and colleagues (2007) used a vertical tank to demonstrate that depth and salinity preferences for juvenile Chinook salmon can be changed when food is offered. Although they preferred deep, high-salinity water without food, when food was presented the fish moved to the fresher water at the surface of the mesocosm.

Adult salmonids on directed spawning migrations through estuaries typically do not feed, but may be attracted by bait used by sport fishers (e.g., coho salmon in the Little Susitna River estuary, Alaska [Vincent-Lang et al. 1993]). There are other exceptions. Based on stomach content data, sea trout returning veterans (J.M. Elliott 1997) and Atlantic salmon kelts (Power 1969) consumed food in the estuary, but there are no data on the behaviour involved.

Territoriality and Schooling

When territorial salmonids in fresh water migrate to estuaries, their inter- and intraspecific behaviour changes: they become schooling fish and do not show the typical territory-defending (agonistic) behaviour shown in streams and rivers (e.g., Chinook salmon [Taylor 1990]; Atlantic salmon and sea trout [Klemetsen et al. 2003]). Species such as pink salmon and chum salmon, which school in fresh water, continue this behaviour in the estuary. However, there may be some exceptions. The ability to alternate between schooling and territorial behaviour has been observed in Chinook salmon fry or parr at the Sixes River estuary in Oregon. Territoriality is likely to be linked to incomplete smolting and may be more common in the freshwater tidal zone of the estuary. Schooling behaviour may enable juvenile Chinook salmon fry or parr and young stages of other salmonids to respond to local patterns of food distribution, competition, and predation risk (Reimers 1968).

Many species of salmonids returning from the ocean to an estuary-river system as adults show classic schooling behaviour, with aggregations composed of populations that return at a specific time to a specific spawning location (e.g., Pacific salmonids [O'Malley et al. 2010]).

Homing

Homing by adult salmonids as they approach an estuary from the ocean is one of the most intensively studied aspects of salmonid behaviour (Quinn 2005). There are genetic differences between salmon populations in different rivers, and the fish's choice of an estuary is a genetic trait (Koljonen et al. 1999), as is the timing of arrival at the estuary (O'Malley et al. 2010). Homing on brackish water may be an important mechanism for maintaining metapopulations of salmon (a group of populations exchanging a limited number of migrants) (Hindar 2003), especially for estuaries on large rivers with numerous tributaries. If flow from a large river is reduced but flow on adjacent rivers is not, migrating salmon may be attracted to the latter rivers rather than their natal stream. This is known as *straying*. Salmonid pheromones emanating from estuaries of rivers that support salmon may be a mechanism for attracting adult salmon to rivers (Nordeng 2009). Along a coastline with multiple rivers, few salmon are found in the estuaries of nonsalmon rivers. This suggested to earlier researchers that homing had a genetic component (Solomon 1973).

Spawning

Almost all salmonid species are adapted to spawn only in totally freshwater systems, but there are exceptions. Some chum and pink salmon populations spawn in the weakly saline parts of estuaries, primarily areas where fresh water from seepage and groundwater upwelling bathes the eggs (E.P. Groot 1989). Examples include chum salmon spawning at Porcupine Creek estuary in Alaska (Murphy 1985) and pink salmon at estuarine beaches at the Olsen Creek estuary, also in Alaska (Helle, Williamson, and Bailey 1964).

Conclusion

Almost all of our knowledge of salmonid behaviour in estuaries has been indirect, obtained from tracking marked fish, and there is scope to improve behaviour data, based on improved tracking methods or direct observations. Electronic tags are increasingly being used to follow the movement of individual fish or groups of fish in order to determine estuarine residency. Large-scale tracking programs such as the Ocean Tracking Network (D'Or and Stokesbury 2009) have significantly increased the data banks on broad-scale migration. The large-scale tagging information could be supplemented by smaller-scale field experiments and direct observations within the estuary to improve our understanding of how various behaviours are adaptive relative to mesoscale and microscale habitats. Tanks or large-scale enclosure experiments can yield some information, but can be influenced by wall effects and artifact behaviour in enclosed spaces. Our understanding of the significance of estuaries for salmonids would be enhanced by the results of comprehensive observations of in situ behaviour, conducted in a variety of estuaries around the world. Although challenging, it would be especially important to focus on residency, feeding, and other behaviours that likely influence salmonid growth and physiological maintenance in the estuary. Growth of the younger stages of salmonids is a basic process affecting survival in the estuary, and measurement and syntheses of growth rate data in the estuary also present challenges, as discussed in the next chapter.

8
Salmonid Growth in the Estuary

This chapter provides an overview of growth rate data for estuarine salmonids, including a discussion of how growth can be measured and the limitations involved. Salmonids show indeterminate growth, which means they display rapid growth rate in juvenile stages. Iteroparous species might continue to grow after they reach reproductive age, although at a slower pace, as they spawn several times in their lives. Temperature is widely acknowledged to be a master variable affecting growth (Brett 1995). An overview of temperature effects will provide context for later detailed discussion of growth in relation to water properties. Many studies on salmonid growth have focused on the freshwater and marine phase (Weatherley and Gill 1995), but estuarine growth is also a key life history aspect. Ecological studies on estuarine salmonids have likely focused on growth rates for two reasons: (1) the assumption that increased food availability in the estuary relative to the river will lead to faster growth (e.g., Arctic char [Gulseth and Nilssen 2000], steelhead [Bond et al. 2008], and many others) and larger fish; and (2) the anticipated fitness benefit of dispersing a larger fish to the ocean, on the assumption that bigger animals will survive better (e.g., MacFarlane 2010; Irvine et al. 2013). Growth rate influences life history evolution in organisms that grow throughout their lives because of the positive influence on age-specific fecundity of the individual and hence fitness (Shine and Schwarzkopf 1992). Growth is therefore important for an understanding of numerous aspects of salmonid estuarine ecology.

Assessing Change in Length or Weight over Time
A common method for assessing growth in the estuary involves measuring the length or weight of fish at sequential time periods (days or weeks),

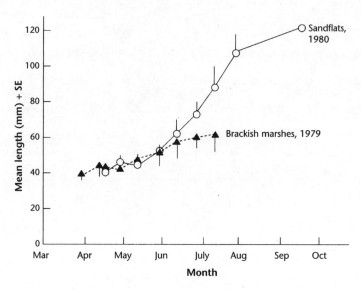

Figure 16 Seasonal change in mean length (with standard error) of Chinook salmon fry on sandflats at Sturgeon Bank and brackish marsh about 5 km upstream, Fraser River estuary, British Columbia.
Source: Reproduced from Levings 1994c, with permission.

beginning when the fry, parr, or pre-smolts first appear in the estuary. The assumption that the measurements are tracking representative data on a single cohort of fish, within which each fish in the cohort resides for the same time period, is often difficult to prove, because fish of different sizes (usually smaller) may be moving into the estuary from the river (Healey 1991), as well as possibly moving in from the ocean. An example is the change in length of Chinook salmon fry on the sandflats (Sturgeon Bank, open to the coast) and brackish marshes a few kilometres upstream at the Fraser River estuary in British Columbia (Figure 16). Salmon fry at the marsh showed a change in length of about 20 mm between April and July, whereas fish on the sandflats increased by 40 mm over the same period. Fish in the brackish marsh were likely diluted with smaller fry moving in from upstream, and those on the sandflats were possibly mixed with smolts moving inshore from the Strait of Georgia marine habitat (Levings 1994c). Size-dependent mortality may also be involved. Relative to faster-growing individuals, slower-growing juvenile salmonids may not survive as well and are therefore underrepresented in length samples (e.g., Chinook salmon in Puget Sound, Washington [Duffy and Beauchamp 2011]). Multiple age classes of iteroparous salmonids (post-smolt or virgin fish to returning veterans) use estuaries on their initial and repeat migrations to the ocean. Assessing the estuarine growth rate of these age classes is problematic, because the larger fish may use coastal habitats as well as the estuary (e.g., Arctic char at the Sylvia Grinnell River estuary, Nunavut [Spares et al. 2012]; sea trout in the Baltic Sea [Degerman et al. 2012]). This problem can be partially resolved by detailed examination of growth structures and microchemistry.

Examining Growth Structures: Otoliths and Scales
Otoliths and scales, which deposit growth rings of calcium carbonate or protein on a daily, monthly, or annual basis, can help resolve how much time a salmonid has spent in the estuary relative to freshwater and marine habitats. When a juvenile fish enters the estuary from the river might be recorded as a growth rate "check," or widening between the rings, because of increased growth relative to fresh water. Estuarine growth rate of juvenile fish has also been measured from scales of adult salmonids (e.g., Chinook salmon at the Sixes River estuary, Oregon [Reimers 1973]; masu from Naiba River estuary, Russia [Ivankov et al. 2003]), although this can be difficult without attention to scale tissue dissolution (scale resorption) (McNicol and MacLellan 2010). From field collections, estuarine growth rates of juvenile salmon can be determined using detailed growth ring data on otoliths from Chinook salmon fry and coho salmon fry (e.g., J.A. Miller and Simenstad 1997) and on scales from Atlantic salmon smolts (Kazakov 1994). Laboratory studies can be used to validate the daily ring deposition. In a study with juvenile chum salmon in tanks simulating estuarine conditions, otolith increments were produced daily for at least the first 160 days after hatching, and there was a direct relationship between mean daily otolith increment width and fish growth rate (Volk et al. 1984).

Growth of Tagged Fish
The growth of individual fish can also be determined through measurements made on marked fish, involving the recovery of fish whose length or weight was known at marking (e.g., fish marked with coded wire tags, as for Chinook salmon at the Sacramento–San Joaquin River estuary in California [Sommer et al. 2001]) or the consecutive recovery of PIT-tagged fish (e.g., steelhead at the Scott Creek estuary in California [S.A. Hayes et al. 2008]). There are also assumptions that the fish has remained in the estuary between consecutive detections and measurements, and that the tagging process itself has not affected growth.

Biochemical Methods

Indirect biochemical methods to assess estuarine growth, such RNA/DNA (e.g., coho salmon pre-smolts and smolts [Varnavsky et al. 1992]; Atlantic salmon smolts [MacLean et al. 2008]) and the insulin-like growth factor (e.g., Moriyama et al. 2000), show promise of working around some of the problems involved in assessing growth described in this chapter. Biochemical methods enable an estimate of the rate of metabolism, which can be correlated with growth. For example, studies of growth and associated hormone levels in feeding studies showed that slow-growing juvenile coho salmon were stunted and deficient in an insulin-like growth factor (Beamish and Mahnken 2001).

Range of Salmonid Growth Rates in Estuaries

Given an unlimited source of nutrition, growth in organisms proceeds almost exponentially, but depending on how long measurements are made for, growth rate continually decreases as size and age increase (Brett 1979). The juvenile life stages of salmonids are continually growing, although perhaps imperceptibly, as they complete their transition from the river to the ocean. Growth increments in the estuary are obviously related to residency time. For example, fish that spend only a day or two in the estuary would not achieve growth detectable by length or weight measurements. There are likely only a few instances where growth rate of salmonids in an estuary decreases or reaches an asymptote. An example would be the Atlantic salmon in the Koksoak River estuary in Québec, where some members of the population remain in the estuary and grow to adult size there (Robitaille et al. 1986).

A variety of metrics can be used to track growth rate. Growth rate data are often expressed as an increment in length (or weight) (e.g., $mm \cdot d^{-1}$ or $g \cdot d^{-1}$) or as the specific growth rate (SGR in percent, length or weight), where weight $SGR = ([\ln W_t - \ln W_0])/t \cdot 100)$, where $\ln W_t$ and W_0 are weight at time t and time 0, respectively. The instantaneous rate of increase in length or weight (I) (weight $I = \log_e W_t - \log_e W_0$) is a useful measure to track estuarine growth (Ricker 1975).

Representative incremental data (ranges) for salmonids in estuaries range from 0.07 $mm \cdot d^{-1}$ for Chinook salmon fry to 2.80 $mm \cdot d^{-1}$ for whitespotted char, but most of the incremental data fell into the 0.2–1.0 $mm \cdot d^{-1}$ range (see Appendix Table 1.5 at http://hdl.handle.net/2429/57062). The mean maximum estuarine growth rate of the iteroparous salmonids, which would be expected to be the larger post-smolts and sea-run migrants, was 1.45 $mm \cdot d^{-1}$, compared with 0.89 $mm \cdot d^{-1}$

for the semelparous species, mainly fry, parr, and pre-smolts. However, this difference was not statistically significant ($p > 0.05$) and the comparison is problematic because the larger iteroparous fish may have ranged into coastal habitat outside the estuary and data were not standardized by fish size. Data on specific growth rate were scarcer, but representative SGR (weight data) ranged from 0.28 percent • d^{-1} for cutthroat trout in Oregon to 1.49 percent • d^{-1} for Arctic char in northern Norway. Almost all of the authors cited gave growth data for spring and summer, or short periods within these seasons, but there may also be some growth in winter in parts of the estuary. Some juvenile Chinook salmon overwinter in the tidal freshwater part of the Columbia River estuary, Washington-Oregon, at which time their growth slows (Roegner et al. 2012). Growth in winter is likely affected by lower temperatures, and possibly reduced food supply.

Overview of Factors Affecting Growth Rates

Temperature is a key biotic factor affecting growth of salmonids in estuaries, but it probably interacts strongly with food supply (Brett 1979). It is therefore difficult to specify an optimum temperature for growth rate in the field. Laboratory and experimental data on salmonid growth and temperature can be instructive, especially for salmonids in the tidal freshwater zone. As an example, S. Larsson and Berglund (2005) found that Arctic char in fresh water that were fed natural food (the mysid *Neomysis integer*) grew best at 15°C. In experiments with Chinook salmon from the Big Qualicum River in British Columbia, when fish were fed at maximum daily ration the optimal temperature for growth was approximately 19°C, above which feeding and growth decreased, particularly above 22°C (Brett et al. 1982).

Salinity also needs to be considered for estuarine growth. There was a strong interaction between salinity and the size of coho salmon pre-smolts from the Cowichan River in British Columbia held at 10°C. Fish grew best at salinities between 5 and 10 psu (Otto 1971). Coho salmon pre-smolts at the Avacha River estuary in Russia showed a higher growth rate in the middle to outer estuarine zones (18–30 psu) compared with the inner estuarine zone (0.5–5 psu) (SGR 0.69 versus 1.42). The data are from the field, however, and it is possible that the variation in food supply between the estuarine zones was a confounding factor (Varnavsky et al. 1992).

There are a few regions and species where broad-scale comparisons of estuarine growth data and temperature from the field are available. Sea trout growing in estuaries and nearby coastal zones along the south coast of the North Sea (58°N; Ijsselmeer Lake, Rhine River estuary, Netherlands) grew at a rate of about 0.6 mm • d^{-1} during their growth periods in the estuary or coastal area, faster than those from the northern North Sea (64°N; Namsen River population in Norway) at 0.4 mm • d^{-1} (L'Abée-Lund 1994; De Leeuw et al. 2007), assuming that there was no growth in fresh water during the overwintering phase. Long-term temperature data are not available for the two estuaries, but the annual mean air temperatures at nearby cities (Amsterdam in the Netherlands and Namsos in Norway) are 10°C and 5°C, respectively, so the temperature factor clearly could explain the growth differences. Food supply was suggested as a possible difference between the populations (De Leeuw et al. 2007). Sea trout from the Rio Grande River estuary in Argentina showed growth rates comparable to those in the Rhine River estuary in the Netherlands, approximately 0.5 mm • d^{-1}; temperature was comparable to the Rhine River estuary. Better growth conditions in the south Atlantic coast may have led to larger adult sea trout relative to those in their native range in the northern North Sea (O'Neal and Stanford 2011). Species rearing in warmer estuaries do not necessarily grow faster than those in colder regions. Sockeye salmon fry at the Kamchatka River estuary in Russia grew at about the same rate as those at the Fraser River estuary in British Columbia (0.44 mm • d^{-1}), even though temperatures during growth were higher at the Fraser River estuary (maximum of 23°C [Birtwell et al. 1987], compared with a maximum of 18°C in the Kamchatka River estuary [Bugaev and Karpenko 1984]; temperature data from the Anapka River estuary, on the east coast of the Kamchatka Peninsula). There may also be variation in temperature effects for a particular species within a smaller geographic region. Levings and Bouillon (2008) showed that Chinook salmon fry at the glacier-fed Squamish River estuary in British Columbia grew faster relative to Chinook salmon fry at warmer coastal plain estuaries on Vancouver Island, British Columbia.

Conclusion

Growth rate for salmonids in estuaries can be species-, season-, and zone-specific. Because of the different patterns of residency, interpretation of differences and changes in growth rate relative to a specific habitat or subhabitat should be done cautiously, with attention to the methodology used. Growth can be a factor when considering shifts in life history stages of salmonids (e.g., fry/parr/pre-smolts to smolts). Growth rate was found to be a better indicator than size of the probability of sea trout smolting and migrating to the Sélune River estuary in France from a small third-order river (Acolas et al. 2012). In contrast, other authors have concluded that size may be the most important measure for predicting sea trout seaward migration (e.g., Jonsson and Jonsson 1993). The size when salmonids move from the intermediate-salinity estuary to the higher-salinity coastal zone is also species- and life history– specific. Estuary-reared Chinook salmon fry at the Nanaimo River estuary, British Columbia, moved to the coastal region at about 70 mm in length (Healey 1980a). Migration into and out of the estuary is closely associated with osmoregulatory preparedness, which will be discussed next.

9
Smolting and Osmoregulation

When salmonids enter the estuary from the river, they are faced with a major physiological challenge. Instead of retaining sodium and other ions in their body fluids, which is essential in the freshwater medium for metabolic processes, they go through a transition period and begin to pump excess salt out of the body. This process is known as the development of hypo-osmoregulatory capacity. Other modifications included collectively in the transition period, usually called smolting, are changes in morphology, pigmentation, migratory behaviour, and buoyancy. Most of the research on smolting has been conducted in either the laboratory to simulate the river-to-estuary or ocean transformation or in rivers before the fish reach the estuary. Very few papers deal with smolting in the estuary proper. Many papers cited in reviews of smolting deal with juvenile salmonids and were in support of culture operations (hatchery or aquaculture applications); fewer deal with kelts, returning salmonids, and field situations where novel populations might be found.

There is a spectrum of salmonid residency and size attained in the river or stream before smolting occurs. At one end of the spectrum is Atlantic salmon, which typically spend at least two years in freshwater residence before entering the estuary. At the other end of the spectrum are chum, pink, and ocean-type Chinook salmon, which spend only a few weeks or even days in fresh water after emergence as alevins. In addition to species-specific differences, ecosystem conditions, especially temperature, can affect the size and time at which salmonids complete smoltification. Osmoregulation is a problem for Atlantic salmon smolts at temperatures lower than 6–7°C (Sigholt and Finstad 1990). Both Atlantic salmon and sea trout smolts are larger at high latitudes than smolts originating from southern stocks (A.J. Jensen and Johnsen 1986), likely because the larger fish are better able to complete osmoregulatory adjustment in colder water relative to smaller fish. Arctic char may be an exception to this generalization if their smaller smolts can access an estuary and acclimate in lower-salinity water (e.g., in Nain Bay estuary, Newfoundland and Labrador [Dempson 1993]).

The role of estuaries as a critical location for completion of smoltification is frequently described in the extensive literature on smolting (e.g., McCormick et al. 2009; McCormick 2013). This chapter gives an overview of smolting and factors affecting the successful completion of smoltification in the salmonid estuary. It also discusses the process of desmoltification that must occur when salmonids re-enter the estuary on a spawning migration and need to retain salt while in fresh water.

Description of Smolting

The following general description of smoltification is based on Atlantic salmon, with information from other species included as appropriate. Smolting of Atlantic salmon is under genetic control, as shown by the results of translocation experiments in two sub-basins of the Tay River in Scotland (Stewart et al. 2006) and more recent specific ecophysiological-genetic data (Sundh et al. 2014). In Atlantic salmon, at a critical length of 12–17 cm, smolt transformation begins in spring and is related to photoperiod. There is a limited period of time during which the fish are at peak preparedness for seawater entry; this is known as the physiological smolt window (McCormick et al. 2009). Neuroendocrine-induced changes occur, and include increased numbers and size of chloride-excreting cells, increased thyroxine levels in the thyroid gland, and an increase in gill enzyme Na^+-K^+-ATPase levels (Boeuf 1993). These changes lead to a deposition of guanine and hypoxanthine pigments in scales to produce silvering, increased salinity tolerance, and an altering of visual pigments from porphyropsin to the rhodopsin that is characteristic of marine fish (G. Alexander et al. 1994). Salmonid fry and parr lose their vertical parr marks, spots, and yellow-brown colouration, and become more streamlined and silvery (McCormick et al. 1998). Loss of ability to swim against the current (decreased positive rheotaxis) and territorial behaviour in some populations can occur in the fall, accompanied by some downstream migration, but the most complete smolt transformation and development of schooling behaviour occur in spring. Behavioural changes also occur as territoriality is usually replaced by schooling. Condition factor also decreases (Boeuf 1993), which indicates that growth in weight relative to length is slowed, at least temporarily, while smolting is occurring.

Smolting is a process that begins in fresh water, continues in the river or estuary, and is completed in the estuary or the ocean. A salmonid physiologist stated that a smolt is a wild fish in active migration, just before reaching the estuary of the river (Boeuf 1993), and downriver migration has been found necessary to complete the physiological changes (Zaugg et al. 1985). In preparation for life in salt water, migrating coho salmon smolts decrease their density in more saline, more buoyant conditions by changing swim bladder volume and lipid content (Weitkamp 2008). Gradually increasing levels of gill Na^+-K^+–ATPase activity were observed in juvenile Chinook salmon, coho salmon, and steelhead undergoing parr-smolt transformation in artificial rearing facilities on the Columbia River, Washington-Oregon (Zaugg et al. 1985). Portions of the same populations released to migrate seaward, however, generally showed much greater increases in enzyme activity with time and distance from the release point. After migrating 714 km to the Columbia River estuary, Washington-Oregon, Chinook salmon had a mean gill Na^+-K^+–ATPase activity 2.5 times greater than fish retained at the hatchery, and 1.9 times greater than fish adapted to 28 psu seawater for 208 days.

Species and Estuarine Variation in Smolting

The ability to resist osmotic shock as juvenile salmonids encounter seawater varies widely between species (e.g., McCormick 1994) and is dependent on size and age of migration from fresh water. Some species show a distinct parr-smolt transformation in fresh water. Others show a more gradual transformation and require a transitional period in the mixed-salinity water that the estuary provides. The latter often benefit from longer estuarine residence and complete smoltification there, although there are few field data on the process from the estuary. Species that mainly exhibit a distinct parr-smolt transformation include all of the iteroparous species (see Appendix Table 1.6 at http://hdl.handle.net/2429/57062). Semelparous species that show a distinct transformation that is cued to photoperiod include amago, Chinook salmon yearlings, most coho salmon, masu, and lake-rearing sockeye salmon. Other species or life history types are insensitive to photoperiod and show a more gradual transformation. These are the species or life history types that benefit from gradual osmoregulatory adjustment in the estuary. Chum salmon, ocean-type Chinook salmon, and sea-type sockeye salmon are included in this category.

Smolting and the Salt Wedge

An astute early observer noted that "after congregating in great numbers at the extreme upper limit of the brackish water [at the Tay River estuary in Scotland], the [Atlantic] salmon smolts make a steady and comparatively rapid descent to the open sea" (Calderwood 1908, 20). The physiological changes that occurred at the upper edge of the salt wedge, where Calderwood was making his observations, have not yet been intensively investigated in the field. The location is important, because if a period of acclimation in intermediate-salinity water is required, this might determine residency in the various salinity regimes of the estuary. It may also relate to estuary type, as a weakly stratified estuary offers a large volume of mixed salt and fresh water compared with a highly stratified water body. It is not exactly clear when and where smoltification ends, but clearly salt pumping has to be operational when the fish is in the open ocean.

The literature is replete with experiments and observations comparing physiological status of the smolted fish relative to habitat conditions in fresh water and subsequent laboratory tests for survival and blood chemistry in water with relatively high salinity. However, there is a dearth of results from the highly spatially and temporally variable conditions in the saltwater-freshwater interface region of the estuary, which is where the transition to a marine environment actually occurs. Some data are available. Acclimation to isotonic estuary water (about 9 psu) for twelve hours was found to be sufficient for efficient adaptation of chum salmon fry to full seawater in an estuary in Japan (Iwata and Komatsu 1984). Abrupt salinity changes (fresh water to full seawater from the Barents Sea, salinity level not given) were tested in the laboratory with Atlantic salmon smolts from the Varzina River estuary in Russia (Chernitsky et al. 1993). Blood osmolarity was maintained within 10 minutes, through an "emergency" mechanism, possibly sequestering of sodium. At least 2.5 days were required for full seawater adaptation.

Distributional data and electronic tagging, combined with detailed oceanographic data, can be helpful in investigating behavioural responses to mesoscale salinity changes in the field. Chernitsky and colleagues (1995) found that sea trout moved up and down with tidal flows at the Varzina River estuary, maintaining their position in water masses not exceeding 12 psu, apparently to maintain a hyperosmotic condition of their body tissues. In a relatively small Québec estuary (4 km wide), salinity had a strong effect on Atlantic salmon smolts; exposure to more saline waters caused increased swimming speeds towards the ocean (Hedger et al. 2008).

Osmoregulatory Adjustment for Kelts, Adults, and Specialized Life History Types

Kelts

When kelts migrate back to the ocean, they may be weakened after having spent the winter in fresh water without feeding. Before entering the estuary and ocean, physiological processes for pumping out salt have to resume. Thirteen of the estuarine salmonids are iteroparous (see Appendix Table 1.6 at http://hdl.handle.net/2429/57062). Kelts or repeat sea-run migrants from some species could show numerous repeat entries into the estuary: twice for steelhead in California (S.A. Hayes et al. 2011), seven times at the Santa Cruz River estuary in Argentina (Pascual et al. 2001), and eight times for Sakhalin taimen in the Bogataya River estuary in Russia (Gritsenko et al. 1974).

Less ecophysiological data are available on repeat estuary adjustment relative to smolting per se, but clearly the kelts, and probably sea-run migrants, can adapt to seawater quickly. Sea trout in the Fowey River in England spawn in December (Bendall et al. 2005) and their Na^+-K^+-ATPase activity immediately after spawning was similar to that of sea trout parr in the river. While most of the sea trout kelts resided in the river for four to seven days post-spawning, a few larger fish moved to the estuary within a few days, indicating that salt pumping may have begun in the estuary when they arrived there. Repeat sea-run migrants of Arctic char lose their seawater tolerance during overwintering in fresh water and redevelop this ability prior to successive seawater migrations. As with smolts, this involves a shift from an energy-conserving overwintering state of the kelts and sea-run migrants to a condition where they become osmotically and metabolically prepared for migration to the estuary and the ocean (Aas-Hansen et al. 2005). In some sea trout, salinity tolerance during winter migration downriver to the estuary may be under genetic control (Larsen et al. 2008).

Adults

When adult or veteran salmonids return to the estuary and begin migrating upriver to spawn, they undergo "reverse smoltification." They need to retain salt and other ions instead of pumping them out (P. Persson et al. 1998). In Arctic char, sockeye salmon, and probably other salmonids, the upregulation of Na^+-K^+-ATPase in the gill occurs as with smolting, but a different isoform of the enzyme is involved (Bystriansky et al. 2007). Reverse smoltification may begin well before the fish reaches the estuary as, for example, low enzyme activity was found for sockeye salmon that already entered the brackish water environment at the mouth of the Fraser River in British Columbia (Shrimpton et al. 2005). Reverse smolting has been less studied than smoltification, but is becoming an important research area as the ecophysiological links between ocean, estuary, and river are explored as factors influencing survival and reproduction. New technology is being developed to help, such as hydroacoustic acceleration transmitters to help track energy costs during estuarine migration of returning adult sockeye salmon (S.M. Wilson et al. 2013).

Specialized Life Histories

There are numerous specialized life history forms with a range of osmotic capabilities in both the iteroparous and semelparous groups, mostly in the early life stages (see Appendix Table 1.6 at http://hdl.handle.net/2429/57062). A summary and detailed discussion of many of the specialized forms was provided by Pavlov and Savvaitova (2008). Most populations of Atlantic salmon display distinct parr-smolt transformation and can tolerate an apparent abrupt change in salinity at the estuary, but there are exceptions to this (e.g., parr at the Western Arm Brook estuary, Newfoundland and Labrador [Cunjak 1992]; smolts at the Varzuga River estuary, Russia [Kazakov 1994]). In most cases, the survivorship benefit of these adaptations in the specialized populations is not known and is difficult to determine. Nomad coho salmon fry are another example. These move into the estuary of their natal stream, rear for the spring and summer months, and then move back into fresh water or use the upper estuary for the winter, or perhaps a non-natal estuary or stream (K.V. Koski 2009). Other examples are the half-pounder steelhead in California that return to the river after only a few months in the ocean (Hodge et al. 2014); a complex of steelhead life history forms in estuaries on the Kamchatka Peninsula (Pavlov and Savvaitova 2008); and the Chinook salmon minijacks and jacks in the Columbia River estuary in Washington and Oregon (J. Johnson, Johnson, and Copeland 2012).

Conclusion

Much has been learned about salmonid smolting and ecophysiology, but clearly there is still scope to investigate how osmoregulation is actually achieved when salmonids enter the estuary as juveniles, adults, and kelts. Studies on the genetic basis of smolting are expanding rapidly.

Understanding species differences in osmoregulation is important in order to explain the evolution of anadromy, a fitness element that is best described as a threshold quantitative genetic trait. In sea trout and steelhead, this trait is controlled by multiple genes and environmental influences (Ferguson 2006; O'Neal and Stanford 2011). The genetic architecture of smoltification is now being explored (Nichols et al. 2008), and this rapidly expanding field will no doubt help further explain this fitness component in the estuary. The role of nutrition and food supply as factors affecting the osmoregulatory adjustment and growth of the various species and life history types also needs to be explored, and will be discussed next.

10
Habitat-Based Food Webs Supporting Salmonids in the Natural Estuary

In this chapter, after an introduction to food web concepts and methods used to study trophic links, I describe some of the important features of estuarine food webs for salmonids, in the context of their habitats. Food supply is a key habitat function for salmonids in estuaries, not only during the rapidly growing juvenile stages such as fry or subyearlings but also for sea-run migrants and kelts as they rejuvenate after overwintering in fresh water and for returning veterans as they begin their upriver migration. Klemetsen and colleague (2003) state that sea trout specifically migrate downriver to estuaries for feeding, a statement frequently found in the literature for iteroparous species, some of which can "choose" to stay in fresh water instead of becoming anadromous, possibly depending on food and growth in rivers and lakes (e.g., Arctic char [A.G. Finstad and Hein 2012]). This chapter provides background information and context for Chapter 14, which considers the implications of the loss of food-providing habitat for fitness components of survival, especially growth. In this chapter, I also describe the key characteristics of estuarine prey that appear to support the best growth of salmonids, and discuss possible gradients in food quality in the estuary.

Food Webs and Estuarine Habitats
A food web is an interacting network of feeding relationships that describes where each population or subpopulation gets its food (Kimmerer 2004). As with other ecosystems, food webs in estuaries are organized in trophic levels: (1) primary (basal) – plants (autotrophic production involving photosynthesis) and bacteria (heterotrophic production without photosynthesis); (2) secondary – mostly invertebrates; (3) tertiary – often juvenile fish and some predatory invertebrates; and (4) quaternary – apex predators such as mammals, larger fish, birds, and humans.

Salmonids are carnivores and feed at the secondary, tertiary, or quaternary trophic levels in the food web. Inorganic nutrients (nitrates, phosphates, and others) required for autotrophy (algae) and heterotrophy (bacteria) can be supplied to the estuary from the ocean as well as the river (C.A. Brown and Ozretich 2009). By residing in estuaries, juvenile and kelt-stage salmonids can take advantage of the increased secondary production rate of invertebrate food organisms available to them, relative to fresh water. Production of estuarine gammarid amphipods, a common food item for juvenile salmonids, can be nearly double that of similar taxa in fresh water. A gammarid amphipod in a stream in England produced about 13 g dry wt • m^{-2} • $year^{-1}$ (Welton 1979), whereas a similar amphipod in an estuary in British Columbia produced about 22 g dry wt • m^{-2} • $year^{-1}$ (Stanhope and Levings 1985). This generalization may not apply to the upper estuary or the freshwater tidal zone, where freshwater prey organisms for salmonids, such as insects, may be as abundant as in streams or lakes (e.g., dipteran insects in the Columbia River estuary, Washington-Oregon [Lott 2004]).

Studies of salmonid diets in estuaries using stomach analysis techniques to identify the prey species provide an important first estimate of energy flow direction and magnitude through the estuarine salmonid food web. There are many stomach content data available. Higgs and colleagues (1995) summarized estuarine feeding information for six species of Pacific salmonids (coho salmon, Chinook salmon, pink salmon, chum salmon, sockeye salmon, and masu), and data from estuaries are available for Atlantic salmon, sea trout, and char (Levings, Hvidsten, and Johnsen 1994; Klemetsen et al. 2003 and references therein; Rikardsen et al. 2004). The importance of prey species in the diet can be assessed using a variety of methods, including counts, weight (biomass) measurement, indices of relative importance, forage ratio (stomach content weight/fish body weight), index of relative importance, and others (see Hyslop 1980; C.A. Rice et al. 2005; Pasquaud et al. 2007). Temporal variation in food intake can be determined by weighing stomach contents at intervals over a twenty-four-hour period to determine peak feeding periods, which are often at dusk. Daily ration can be estimated by summing up the weight of food items consumed by individual fish over a twenty-four-hour period (e.g., chum salmon fry at the Anapka River estuary, Russia [Karpenko and Nikolaeva 1989]). Food consumption rate is a more complex issue that requires laboratory studies, such as gut clearance rate measurements (Jobling et al. 1995), use of estimated gastric evacuation rates to determine relative daily ration (Godin 1981), bioenergetics modelling (Gritsenko and Churikov 1977; P.C. Hanson et al.

1997), or special field measurements, such as the cesium clearance method (Rowan and Rasmussen 1996).

Estuarine food webs supporting salmonids are complex, but considerable knowledge is now available on the basic descriptions of the linkages. Juvenile salmonid tertiary production in some estuaries has been shown to be very significant and enough to account for large energy flows from basal sources such as vegetation and phytoplankton, to crustaceans, to chum salmon fry in the Nanaimo River estuary in British Columbia (Healey 1979). This species had an annual food requirement of 4,500 kg biomass at the estuary – in one year primarily from a single species of harpacticoid copepod (*Harpacticus uniremis*) (Healey 1982). However, the importance of food to salmonids in estuaries is not necessarily related to immediate increased growth and production resulting from increased food supply, and some of these subtle benefits are only now being discovered. Juvenile Chinook salmon in the Columbia River estuary in Washington-Oregon benefit from short-term residency in the tidal channels, where they feed on insects and crustaceans (Lott 2004). Physiological studies showed that even though growth was not evident during residency, the triglyceride level, a measure of energy storage, was maintained. However, this was not found in fish using riverine channels, where food may have been less available. The maintenance of triglycerides was thought to be a benefit to the species when they reach the open ocean (K.C. Hanson, Ostrand, and Glenn 2012) and may require more energy for burst swimming to avoid predators or swim against ocean currents. In general, knowledge of estuarine food webs supporting salmon has progressed from the descriptive phase to research emphasizing dynamics and pathways of energy flow, as described below.

One of the first conceptual diagrams of an estuarine food web that involved the basal, secondary, and tertiary production levels supporting juvenile salmonids was developed at the Squamish River estuary in British Columbia (Department of Environment 1972; Figure 17). This diagram was based on feeding data for chum salmon fry, coho salmon smolts, and Chinook salmon fry and smolts in the brackish marshes of the estuary, as well as the riverine channels and the contiguous fjord (Howe Sound). The web is much more complex than shown, and there would be hundreds of links if food species were described at the species level. Subsequent estuarine salmonid studies in British Columbia employed more sophisticated techniques. Sibert and colleagues (1977) studied chum salmon fry feeding habits in the field and laboratory at the Nanaimo River estuary. They used radioactive tracers to determine whether the food web supporting chum salmon fry was different from

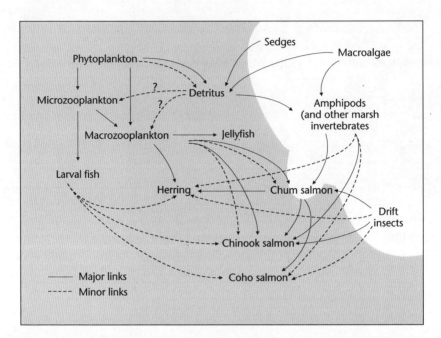

Figure 17 One of the first portraits of an estuarine food web supporting juvenile salmonids.
Notes: Solid lines represent major links; dotted lines represent minor links. Middle estuarine subhabitats providing detritus (primary level – brackish marsh and macroalgae) and secondary production (amphipods and drift insects) are not shaded grey. Riverine channel subhabitats and outer estuarine (pelagic) zone providing primary-level detritus and algae from phytoplankton, secondary production (micro- and macrozooplankton), and tertiary production (Pacific herring) are shaded grey.
Source: Data for the Squamish River estuary, British Columbia, from Department of Environment 1972.

the classic autotrophic, phytoplankton-based one found in coastal and ocean waters. Based on carbon-14 labelling of detritus from vascular plants and subsequent uptake into *H. uniremis,* the important food source for the chum salmon fry documented by Healey (1979), Sibert and colleagues (1977) showed that that the salmon fry in the estuary were part of a detrital food web based on heterotrophic processes. Detritus is a mixture of bacteria and broken-down vascular plant tissue, and it is seen as a key source of the organic material fed into the food web. Detritus production therefore relies on both heterotrophic and autotrophic processes to generate organic matter and energy flow. This concept was initially developed in a brackish marsh in the US state of Georgia (e.g., Odum 1988), but was transferred to marsh ecosystem studies elsewhere, including salmonid estuaries.

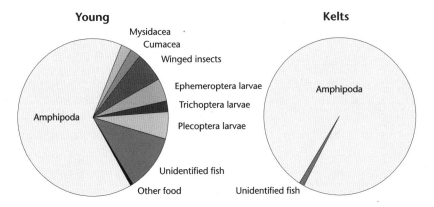

Figure 18 Food of whitespotted char downstream migrating young and kelts at the Bogataya River estuary, Russia.
Note: Areas of the sectors are proportional to percentage of food items by weight.
Source: Data from Gritsenko and Churikov 1977.

Relying on the classic descriptions of feeding habits of invertebrate prey (i.e., determination of detritus in their gut contents), detrital-based food webs have also been found for estuarine salmonids around the world: Chinook salmon fry and smolts at the Fraser River estuary, British Columbia (Northcote et al. 1979); chum salmon fry in the Skagit River estuary, Washington (Congleton et al. 1982); Arctic char in a Beaufort Sea lagoon, Alaska (P.C. Craig et al. 1984); whitespotted char downstream-migrating young (likely first sea-run migrants) and kelts; Figure 18) at the Bogataya River estuary, Russia (Gritsenko and Churikov 1977); sea trout and Atlantic salmon in the Gironde River estuary, France (Lobry et al. 2008) and numerous other estuaries around the world, including those in the Southern Hemisphere. Sea trout and juvenile Chinook salmon in New Zealand estuaries are linked to detrital food webs via retropinnid smelts, secondary-level consumers whose primary diet consists of amphipods and polychaetes (Boubee and Ward 1997). Mysids (*Tenagomysis macropsis*) were also recorded as a primary food item for juvenile Chinook salmon (James and Unwin 1996) in Akaroa Harbour (an embayment partially influenced by fresh water from the Rakaia River as well as smaller rivers within the bay), New Zealand. Power (1969) found that Atlantic salmon kelts in the Koksoak River estuary in Québec fed on amphipods, polychaetes, sand lance, and cottids, and hence were part of the detrital food web. The detrital food web is clearly a key pathway in many estuaries around the world that support salmonids.

Technological Advancements in Dietary Investigations

Newer technology has made possible detailed investigation of estuarine food webs, and the results have changed the detrital paradigm to some degree. In recent years, the use of natural isotopic signatures has vastly increased our knowledge of estuarine food webs involving salmonids. The conservative or stable transfer of naturally occurring carbon (^{13}C), nitrogen (^{15}N), and sulphur (^{34}S) isotopes through the food web can be very useful in tracing food webs in systems where there are food sources with large differences from a standard (described as ∂^{13}C, ∂^{15}N, and ∂^{34}S; see Michener and Schell 1994). For example, marine food webs with a basal source of algae would have a different (higher) ∂^{13}C value relative to a food source from detritus from brackish marsh plants. These signatures are transferred up the trophic levels and can be detected in invertebrates and juvenile salmonids, thus revealing their food sources. Beginning in the 1980s, studies using stable isotope analyses have been conducted to trace the importance of alternatives to detritus as energy sources in salmonid food webs in estuaries of the northeast Pacific (Puget Sound, Washington [Simenstad and Wissmar 1985]; Fraser River estuary [Levings 1994b]; San Francisco Bay, California [Kimmerer 2004]; Columbia River estuary, Washington-Oregon [Maier and Simenstad 2009]).

Stable isotope analysis increased the scope of our understanding and highlighted the importance of the highly variable sources of energy flows in estuarine salmonid food webs. In addition to detritus, autotrophically produced organic material (e.g., benthic algae) was identified as an important source. Significant variation in stable isotope signatures was found between estuaries and within species. A summary and interpretation of the available data led Wissmar and Simenstad (1998, 226) to conclude that "it is becoming apparent that organic matter production and the food web processes supporting juvenile salmon differ among estuaries, often irrespective of seemingly deterministic food web pathways." Large estuary-embayment systems, such as the Sacramento–San Joaquin River estuary and San Francisco Bay in California, showed two predominant food web pathways: a zooplankton-phytoplankton autotrophic pathway found in riverine channel subhabitat led to juvenile Chinook salmon, whereas a detritus-based web from riparian vegetation and macroalgae led to resident fish (Grimaldo et al. 2009). In other estuaries where turbidity reduces primary production by phytoplankton (e.g., Fraser River estuary [Kistritz et al. 1983]), the detritus pathway is clearly more important. As a further example of the variation, in short streams in Alaska the organic material and nutrients from decomposing pink salmon and chum salmon adult carcasses are important in stimulating autotrophic primary production in the estuary (Gende et al. 2004). All habitats in salmonid estuaries function to provide organic matter through bacteria and plants, in different ways and degrees.

Variation in Salmonid Food Webs in the Estuary

In this section, I give an overview of the variation in food web linkages to salmonids in the estuarine habitats. It is a challenge to identify usage of specific subhabitats for feeding, and stomach content analysis is often used for this interpretation. That is, if a prey species produced in a subhabitat dominates in the salmonid stomach, then preferred use of that subhabitat by the fish is recognized. This may be correct, but because the fish has likely roamed over various habitats or subhabitats, the observation reveals only a snapshot of the diet (Sibert and Obreski 1977). As well, the disappearance of food from the intestinal tract is temperature-dependent, and food species may be digested more rapidly in warmer temperatures. Thus, further information may be required. Most of the data on food pathways described below were determined using classic stomach analysis methods. However, the use of stable isotope analyses, together with more powerful mixing models using two or three isotopes to analyze the data, is expanding our knowledge of salmonid estuarine food webs.

Beaches

The basal sources of food webs on beaches are usually a mixture of detrital and algal primary production. In some instances, microalgae may grow in the interstices of sandflats, providing an autotrophic source for the invertebrate prey of salmonids (e.g., the gammarid amphipod *Eogammarus confervicolus* at the Fraser River estuary [Pomeroy and Levings 1980]). In other instances, detritus may collect under cobble or boulder, providing a food source for invertebrates (Levings and McDaniel 1976). There are numerous varieties of food sources from cobble. Microbenthic algae attached to cobble in the streambed of the Porcupine Creek estuary in Alaska was the food source for *E. confervicolus* in this estuary (Murphy 1984). At the gravel-based Utkholok River estuary in Russia, a gammarid amphipod (*Anisogammarus kygi*) migrated upstream and fed on salmon carcasses – an example of an estuarine invertebrate feeding at the tertiary level. In a subsequent zonal shift, the amphipods moved downstream and were a major food source for juvenile chum and coho salmon, Dolly Varden, steelhead, and whitespotted char in the estuary (Thompson 2007). Tube-dwelling amphipods such as *Corophium* spp., which live in sandflats and mudflats, were a food source for juvenile Chinook salmon at numerous northeast Pacific estuaries (Sixes River estuary, Oregon [Reimers 1973]; Fraser River estuary [Levy et al. 1982]). Concentrations of salmonid food near the bottom of the estuary can change dramatically at the mud-water interface, as shown by Sibert (1981), who discovered dense concentrations of harpacticoid copepods potentially available to chum salmon fry in mudflats at the Nanaimo River estuary. Biofilms on mudflats are important for energy flow to food webs supporting juvenile salmonids. Invertebrates such as harpacticoid copepods feed on the bacteria and algae in the biofilm complex (Sahan et al. 2007). Food linkages on beaches can also be indirect, with linkages at the quaternary level. For example, forage fish for salmonids spawn on sand and gravel beaches (e.g., surf smelt, *Hypomesus pretiosus* [C.A. Rice 2006]).

Channels
Basal food from pelagic phytoplankton and other algal primary production in riverine channels is dependent on the surface area of the system, its perimeter, and the stability and residency of the water masses. Low estuarine flushing rates may make possible a relatively permanent resident phytoplankton community, as was found in part of the Sacramento–San Joaquin River estuary (Grimaldo et al. 2009) or the Neva River estuary in Russia (Telesh, Golubkov, and Alimov 2008). In far northern or southern estuaries, under-ice algae in channels can be a basal food source (Hiwatari et al. 2011).

Juvenile salmonids, especially fry, migrating in riverine channels are strongly oriented to the surface and feed in the shallow water column or surface habitats above the halocline (Birtwell and Kruzynski 1989). Larger life history stages occupy deeper habitats and feed on zooplankton with an autotrophic phytoplankton link (e.g., Chinook salmon smolts feeding at the Squamish River estuary, Figure 17; Chinook salmon juveniles feeding on zooplankton in the salt wedge [J.S. Macdonald, Birtwell, and Kruzynski 1987]) or nekton, such as Atlantic herring (*Clupea harengus*), in the water column (e.g., Atlantic salmon post-smolts feeding in Norwegian fjords [Rikardsen et al. 2004]). The estuary turbidity maximum at the upstream face of the salt wedge is also a concentrating mechanism for potential near-bottom food in channels. An example is the concentration of copepods in the Columbia River estuary, which increases by several orders of magnitude in the estuary turbidity maximum (C.A. Morgan, Cordell, and Simenstad 1997). At the Orkla River estuary in Norway, Atlantic salmon post-smolts ate a detrital-feeding gammarid amphipod produced in the shallower water of a riverine channel (Levings, Hvidsten, and Johnsen 1994). There are examples of sea-run

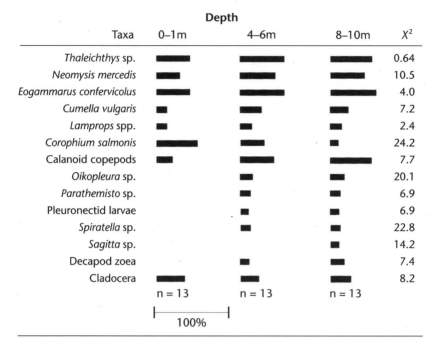

Figure 19 Vertical distribution of potential salmonid prey in the stratified Fraser River estuary, British Columbia.
Notes: 0–1 m is the freshwater layer, 4–6 m is close to the halocline, and 8–10 m is in the salt wedge. Length of the bars is proportional to percent occurrence in plankton drift nets at depths indicated. Column on far right shows χ^2 values to test for homogeneity among the three depths. $p \leq 0.05$ for all taxa except *Thaleichthys* sp.
Source: Reproduced from Levings 1980a.

migrants or returning veteran salmonids feeding on juvenile salmonids (quaternary-level feeding), especially on salmonid fry migrating downstream in riverine channels. At the Bol'shaya River estuary in Russia, sea-run migrants or returning veteran whitespotted char fed on pink salmon fry (Tokranov and Maxsimenkov 1995). Riverine channels also bring invertebrates produced upstream to juvenile salmonids in the estuary (e.g., drift insects [D.D. Williams 1980]). Marine or brackish water animals (e.g., larvaceans, *Oikopleura* spp.) move upstream in the channel's salt wedge on flood tides, and freshwater zooplankton (e.g., cladocerans) are flushed into the estuary from upstream in surface layers (Figure 19). Tidal creeks in the Fraser River estuary were found to be key feeding areas for juvenile salmonids. Coho salmon smolts and sockeye salmon fry fed on insect larvae and semi-terrestrial invertebrates such as springtails (Collembola) (Levings, Boyle, and Whitehouse 1995). Salmonids in side channels with brackish marsh fed on dipteran insects produced in the marsh (e.g., Fraser River estuary [Levy et al. 1979]).

Vegetation
The basal food or primary production source in this habitat is the vegetation itself, when decomposing leaf material adds to the detrital complex, although in some instances microbenthic algae (epiphytes) growing on the leaves is an additional carbon source (Heck and Valentine 2006). Eelgrass can be one of most productive basal-level plants in the estuary. Eelgrass beds provide structure for a diverse suite of salmonid prey species, including crustaceans (e.g., harpacticoid copepods) and forage fish (e.g., sticklebacks). Species feeding on invertebrates and fish in eelgrass range from whitespotted char in the Akkeshi Lagoon (Bekanbeushi River estuary), Japan (Watanabe et al. 1996), to chum salmon fry at the Fraser River estuary (Webb 1991). Brackish marshes have been intensively studied as sources of detritus as well as habitat for invertebrates fed upon by juvenile salmonids. Many of the data are from estuaries in the northeast Pacific. Macrodetritus from brackish and freshwater marsh was considered the most important indirect driver for productivity of juvenile salmonids in the Columbia River estuary (Sherwood et al. 1990). The invertebrates produced in brackish marshes or linked to the detrital food and eaten by juvenile salmonids include a complex of dipteran insects (e.g., Columbia River estuary [Lott 2004]) and mysids (e.g., Fraser River estuary [Levy et al. 1979]). At the Squamish River estuary, juvenile salmonids fed on *E. confervicolus*, which was both abundant and productive under a skirt of sedge rhizomes along the edge of the brackish marsh (Stanhope and Levings 1985). In the coastal zone, a basal source may be riparian vegetation growing above the high-tide level (Romanuk and Levings 2005). Riparian vegetation in the freshwater tidal zone and the river also contributes to the detrital web of the estuary (Naiman and Sibert 1978). Terrestrial carbon from peat was shown to be important in the detrital food webs supporting Arctic char at estuary lagoons in the Beaufort Sea (Carmack and Macdonald 2002; Dunton et al. 2006). Based on stable isotope analyses, adult pink salmon and Arctic char diets in these lagoons overlap when terrestrial plant signatures at the basal level of the food web are considered (Hoekstra et al. 2003). Detritus from kelp enters into the detrital food of intertidal invertebrates living on rocky shores (Schaal et al. 2009). Rockweed harbours gammarid amphipods (Levings et al. 2004). It also enters the detrital food web at the basal level as wrack as the macroalga decomposes (Romanuk and Levings 2010).

Artificial Habitats and Habitat from Invasive Ecosystem Engineers

Artificial beach rock subhabitat is developed by placement of riprap along riverine channels (Morley et al. 2012), and in some instances contributes nutrition at the basal level. Microbenthic algae can grow on riprap, but the surface area of the material is low relative to the natural sand and mud substrate it often replaces (Pomeroy and Levings 1980). In some instances, artificial habitat has been colonized by vegetation (e.g., the Yolo Bypass channel, a seasonal floodplain transformed into a 24,000 ha leveed basin that conveys excess flow for flood protection and during low flows into the tidal freshwater zone of the Sacramento River estuary, California [Sommer et al. 2001]; sand islands with brackish marshes [Levings 2004]). Some industrialized estuaries, with their artificial walls of wood or dredged material, support riparian vegetation (e.g., Thames River estuary in England [Francis and Hoggart 2009]). In turn, the riparian vegetation supports insects, potentially available as juvenile salmonid food (Hoggart et al. 2012).

In summary, the food web support system for juvenile salmonids is complex, and clearly salmonids are relying on a variety of food sources, depending on the habitat and subhabitat they are in while residing in or migrating through the estuary. In terms of basal and secondary production, the estuary is not a "food island" unto itself. Salmonids in the estuary receive basal and secondary trophic subsidies from multiple sources, including the river flow from upstream in the watershed, the adjacent riparian or terrestrial vegetation, the ocean, and even adult salmon carcasses bringing nutrients in from the ocean. The estuarine ecosystem is thus a classic example of a landscape receiving multiple trophic subsidies (Polis et al. 1997). In addition to food availability, however, additional prey characteristics are important factors that determine an estuarine salmonid's feeding choices.

Prey Characteristics Important for Growth

Thousands of invertebrates and fishes are produced in the habitats described in this book and all are potentially available to grow salmonids, with due consideration to factors affecting their uptake: abundance, size, behaviour, colour, nutrition, and others (Table 3; Gill 2003). A review of all these factors is beyond the scope of the book but nutritional aspects are worthy of inclusion because of their possible linkages to estuarine zonation.

For this discussion, I draw on the vast literature on growth of cultured salmonids in a variety of feeding regimes (temperature, salinity, density,

Table 3

Selected examples of biological factors affecting food acquisition by salmonids in estuarine habitats

Biological factor	Examples	References
Prey size	Chum salmon show marked shift in prey size at 60 mm.	Okada and Taniguchi 1971
	Atlantic salmon show increased feeding response to oblong-shaped food that is 2.2–2.6% of body length.	Jobling 1989
	Maximum size of prey consumable is about 50% of the predator body size.	Pacific Northwest Hatchery Scientific Review Group 2009
Previous experience of predator	Hatchery-reared juvenile coho salmon switched to natural food after less than 24 hours in the laboratory.	Paszkowski and Olla 1985
	Hatchery-reared coho salmon post-smolts using channel habitats in an Oregon estuary ate more larval fish relative to wild coho salmon.	Myers 1978
Temporal abundance of prey species	Diurnal migration of zooplankton affects timing of feeding for coho salmon.	Pearcy et al. 1984
	Seasonal peak in zooplankton abundance matches Chinook salmon migration into the Strait of Georgia.	Healey 1980b
Prey visibility, colour, and contrast	Large black eyes of hyperiid amphipods enhance their visibility to coho salmon.	Peterson, Brodeur, and Pearcy 1982
Biomechanical and nutritional aspects	Requirements for protein, amino acid, and fatty acid composition may be specific for the post-smolt stage.	Literature summarized in Higgs et al. 1995
Perceived risk by predator on post-smolt	Chinook salmon avoided surface feeding when predator threat was perceived.	R.S. Gregory 1993
Risk avoidance by schooling with other fish affects habitat where food is obtained	Chum salmon fry switched to surface food when schooling with three-spined sticklebacks, even though a predator was present.	Tompkins and Levings 1991

Source: Updated from Levings 1994a.

and others) and food quality programs (see Higgs et al. 1995; Pennell and Barton 1996). Certain critical nutritional factors explored in this literature may be involved in salmonid estuarine food webs.

Among the various factors affecting the importance of specific food taxa growth, nutritional status, using lipid content as a metric, should rank fairly high in terms of importance to salmonids (Higgs et al. 1995). Food containing higher lipid levels should in general result in better growth. However, use of lipid content as an index does overlook the possible prey species-specific differences in certain highly unsaturated fatty acids (HUFAs) (Higgs et al. 1995). Within this general lipid class, HUFA type can determine the nutritional value for marine fish (e.g., within the harpacticoid copepod taxon [Coull 1999]).

The nutritional value of prey species may show a general increase on a gradient from the freshwater tidal zone to the coastal zone. Energy-rich prey at the seaward edge of the lower estuary or in the adjacent coastal ocean may generate more growth relative to upper-estuary prey (e.g., northern anchovies, *Engraulis mordax*, for juvenile Chinook salmon at San Francisco Bay and the seaward edge of the Sacramento–San Joaquin River estuary [MacFarlane 2010]). Prey brought into the estuary from the ocean, such as the calanoid copepod *Calanus finmarchicus*, an important food item for Arctic char in the estuary of a short stream draining a lake in Norway (Storvatn) (Rikardsen et al. 2000), typically have very high lipid levels, up to 50 percent dry weight when carrying eggs. Clupeid fishes, such as capelin (*Mallotus villosus*), are also brought into the estuary and are consumed by estuarine salmonids, such as whitespotted char in the Kamchatka Peninsula, Russia (Tokranov and Maxsimenkov 1995). These fishes are high in lipids, with an average content of about 32 percent. Benthic crustaceans (harpacticoids, amphipods) and aquatic insects (dipterans, especially chironomids) from beaches and vegetation in the middle sectors of the estuary may have lower lipid levels (e.g., the harpacticoid *Nitocra spinipes*, with 18 percent dry weight [Weiss et al. 1996]). The tube-dwelling amphipod *Corophium volutator*, representative of this widely consumed amphipod genus, showed a relatively low lipid content at about 6 percent body weight (annual mean) in an estuarine population in the Baltic Sea. As with other crustaceans, the lipid content was higher during the reproductive period, when female amphipods were carrying eggs (Dobrzycka and Szaniawska 1995). Lipid content for *E. confervicolus*, representative of free-swimming estuarine amphipod prey, averaged 10 percent (Higgs et al. 1995). Lipid content for the larval forms of chironomid insects may be higher (17–23 percent was reported in Hamburger et al. 1996). Terrestrial insects dropping from riparian vegetation, such as spiders, have relatively low lipid content (5 percent) (Walters et al. 2010). Such insects may be typical of prey organisms in the freshwater (upper estuary) zones of the salmonid estuary as well as in marine riparian vegetation in the outer estuary (Romanuk and Levings 2003).

Although scarcely investigated, some data show that growth of salmonids may reflect a decreasing up-estuary trend, possibly reflecting a nutritional gradient. Juvenile Chinook salmon may grow more slowly in the tidal freshwater reaches of the Columbia River estuary relative to reaches closer to the ocean (Roegner et al. 2012). Size is a complicating factor for a general theory relating to a nutritional value trend, as by the time salmonids have moved into the coastal zone, they are large enough to consume energy-rich prey such as herring (e.g., Atlantic salmon in the Baltic Sea [Salminen et al. 2001]).

Conclusion

The direct role of feeding and food supply for salmonids in estuaries as possible factors affecting survival has been the focus of many important salmonid studies around the world, in support of the idea of bottom-up control of a species' dynamics. This basic ecological concept implies that a lower trophic level has an effect on a trophic level above it. However, the related top-down control system, where a higher trophic level has an effect on a trophic level below it, is also possible (J.A. Miller et al. 2013). Predation and competition are important agents of top-down, or sometimes "within trophic-level," control. I discuss these concepts next in the context of biotic interactions between salmonids, as well as with other species in salmonid estuaries.

11
Biotic Interactions in the Natural Estuary

Predation and competition, two of the most recognized biotic interactive processes in ecology, have been established as factors influencing the survival of salmonids in a variety of ecosystems. It is important to give an overview or baseline perspective of these biotic processes in the undisturbed salmonid estuary before discussing the effects of human intervention, with the caveat that many of the data are in fact from partially disturbed systems. On the west coast of North America, where there have been many salmonid estuarine research projects, almost all studies in the past thirty years have been conducted in estuaries between Vancouver Island, British Columbia (latitude 50°N) and Los Angeles, California (latitude 34°N). This region coincides with most of the estuarine and watershed development in this part of North America. Even sixteen years ago, most of these estuaries had been disrupted to some extent (Emmett et al. 2000), and development has proceeded apace since then. Estuaries in the higher latitudes are likely the least disrupted and represent the natural situation.

In lake and stream fishes, the presence of predators often results in increased use of vegetated or other physically complex habitats by prey fish (e.g., B.C. Harvey 1991). If this cover-seeking behaviour is shown by juvenile salmonids when they are prey in estuaries, it would indicate that predator avoidance, a mechanism for reducing the number of individuals removed from the population, is a selective force for the species. Predator avoidance may be especially important for juveniles of the semelparous Pacific salmonids that produce fry and parr that undergo osmoregulatory adjustment in the estuary. Predation on returning veterans or adults in the estuary may have an even greater effect on the population, as the reproductive-ready individuals have already survived predation threats throughout the majority of their lifespan. Similarly, if intraspecific or interspecific competition is affecting survival, this process deserves attention. Studying biotic interactions of salmonids in the estuary experimentally can be a challenge but laboratory and observational data can provide some insight.

Predation
Predation is a fundamental process in ecosystems and evolution (Begon et al. 2005) and is generally assumed to select against substandard individuals – the young, the senescent, the sick, or individuals in poor physical condition (Genovart et al. 2010). I use the definition of predation from Abrams (2000): an interaction in which individuals of one species kill and are capable of consuming a significant fraction of the biomass of individuals of another species. Predation shapes food webs and can have a cascading effect on ecosystem properties and functions (Hawlena and Schmitz 2010). For example, in a three-step food web, a predator at trophic level one may have an indirect effect on trophic level three, owing to its consumption of organisms at trophic level two (Shurin et al. 2002).

As a generalization, the topic of predation on salmonids (outside of harvesting, which also removes individuals) has been one of the most actively researched areas of applied salmonid ecology, especially in fresh water (D.H. Mills 1989). As early as 1904, some commentators in England were discussing the control of predators, specifically birds, to reduce their consumption of young salmon and trout in rivers and estuaries (see Marston 1904); controversy over the practice is still strong (e.g., Cowx 2003). Scientific literature dealing with bird consumption of Pacific and Atlantic juvenile salmonids in estuaries extends at least from the 1930s (e.g., Munro and Clemens 1937; White 1936). Marine mammal predation on adult returning salmon also has a long history of study (see Middlemas et al. 2006). Here I discuss approaches to assessing predation, comment on the importance of schooling as behaviour relating to predation, and give an overview of the available data on predation in the salmonid estuary.

Approaches to Assessing Predation
Stomach content analysis is possibly the most common direct method of documenting predation on salmonids in estuaries, but the impact of the predation on the total population of the prey fish is difficult to estimate with these data alone. Even if the prey occurs in only a few stomachs of the predator, if a large predator population is present the impact could be significant, particularly if the prey population is small

(Beamish, Thomson, and McFarlane 1992). Therefore, small proportions of salmonid prey in predator stomachs should not necessarily be dismissed as unimportant for salmonid survival. For this reason, population estimates of the predator are required in order to obtain a better idea of the impact of predation. Newer tagging technologies such as PIT tags enable the direct estimation of the relative vulnerability of different species of juvenile salmonids to predation (e.g., Caspian terns, *Sterna caspia*, and double-crested cormorants, *Phalacrocorax auritus*, preying on juvenile Chinook salmon, coho salmon, and steelhead in the Columbia River estuary, Washington-Oregon [Ryan et al. 2003]).

Indirect estimates of predation rates have been obtained using methods such as examination of scarring. Scars from predator attacks on salmonids in estuaries have often been used to indicate a lethal attack, and the proportion of scarred individuals is used to estimate mortality rate. Scars have been recorded on salmonids from bird strikes on, for example, Atlantic salmon smolts in the Margaree River estuary in Nova Scotia (White 1936), and from Japanese lamprey (*Lampetra japonica*) preying on pink salmon and chum salmon fry in the Amur River estuary in Russia (Novomodnyi and Belyaev 2002). However, scars may not necessarily indicate that the attacked fish will die, as shown for seal predation on adult Chinook salmon in the lower Columbia River (Naughton et al. 2011). Some returning adult chum, sockeye, and pink salmon in the Kamchatka River estuary in Russia had at least five seal scars on their bodies, indicating that the fish had been attacked multiple times (Shevlyakov and Parensky 2010). If the attacked fish survived, then the predator may in fact be classified as a parasite. The trophic classification of some predators that scar salmonids is equivocal: the Pacific lamprey (*Lampetra tridentate*) has been called a parasite (e.g., Beamish and Levings 1991), whereas the Japanese lamprey was described as a major predator, consuming 93–96 percent of the chum salmon fry moving through the Amur River estuary (Roslyi and Novomodnyi 1996).

Predation rate by birds and mammals can be assessed by observing the number of predators and their consumption rates, usually from direct observation or experimental data. Using direct observation, C.C. Wood (1987) estimated that mergansers (*Merges merganser*) consumed less than 6 percent of the migrating wild chum salmon fry at two streams and estuaries on Vancouver Island. Consumption can also be estimated by determining diet composition and measuring prey hard parts (e.g., bones, scales, and otoliths) recovered from scats (feces) and by reconstructing prey biomass (e.g., harbour seal predation on Atlantic salmon smolts in a Scottish estuary, Cromarty Firth [Middlemas et al. 2006]).

Field or laboratory experiments are frequently used in freshwater salmonid predation studies to determine the importance of risk factors such as size and species behaviour. The variation in water levels, currents, and salinity changes in the estuary mitigates against field experiments such as blocking off an area to study a habitat with and without a predator. However, some investigators have successfully used floating pens to study predation (e.g., coho salmon smolt predation on chum salmon fry at the Yakoun River estuary, British Columbia [Hargreaves and LeBrasseur 1986]). It is also a challenge to conduct laboratory estuarine predation experiments, although mesocosms that can simulate the salinity stratification (e.g., the vertical tank described in Birtwell and Kruzynski 1989) can help. Small tank experiments with artificial cover and water at 10 psu to simulate the estuary were conducted by Abrahams and Healey (1993). Chinook salmon smolts were willing to take greater predation risk in order to feed relative to smaller coho salmon smolts, pink salmon fry, and chum salmon fry, but differences in risk taking were not explainable by size. This was unexpected, as usually larger prey have a more successful escape response if they are within the size range that can be consumed (Keeley and Grant 2001), usually about 50 percent of the predator body size (Pacific Northwest Hatchery Scientific Review Group 2009). As with all laboratory experiments, there are caveats, such as possible unnatural responses because of tank effects, and some investigators have turned to modelling estuarine predation as another study strategy.

Bioenergetics modelling essentially builds on the physiological premise that growth can be equated to food consumption, but growth rate may either increase or decrease depending on the nature of the food/metabolism/temperature relation (Brett and Groves 1979). The metabolic rate determines how energy is channelled within the body. The ability to catch food and the energetic costs involved consist of a series of behaviours, including locating, identifying, chasing, capturing, handling, and consuming. If energy taken in to support these functions is in surplus, growth can occur. An early bioenergetics model applied to chum salmon migrating in estuaries showed that the metabolic costs of maintenance and migration lie in a delicate balance with food intake and growth (Wissmar and Simenstad 1988).

A more recent and widely used bioenergetics model (P.C. Hanson et al. 1997) quantifies consumption demand while accounting for temporal changes in body size, temperature, and food. The model estimates how much consumption is required to satisfy the observed growth of consumers over a specified time interval, given changes in diet and temperature. As an equation, total energy consumption (C) over a particular time frame equals the sum of growth (G, positive or negative), metabolic costs (M), and waste losses (W):

$$C = G + M + W$$

Although a very versatile methodology, the bioenergetics approach has some limitations. In particular, the physiological variables needed have been determined for only a few of the salmonid species that use estuaries, and calibrations are often with related taxa.

An example of the application of the bioenergetics model of P.C. Hanson and colleagues (1997) to the estuarine salmonid predation issue was provided by Duffy and Beauchamp (2008), who estimated cutthroat trout consumption of juvenile salmonids (Chinook salmon smolts, coho salmon smolts, chum salmon fry, and pink salmon fry) in Puget Sound, Washington. The primary model inputs were thermal experience (temperature experienced by the cutthroat trout), diet, prey and predator energy densities, and growth. For the cutthroat trout model, Duffy and Beauchamp (2008) used the model's default physiological parameters for coho salmon, except that they modified two of the temperature-dependent parameters' effects (critical thermal optimum and critical thermal maximum). Results showed that consumption rates of salmonids (grams of prey per month) varied with location and predator size. Cutthroat trout predation in northern Puget Sound, with its numerous salmon-producing streams and hatcheries, was higher relative to southern Puget Sound, where few juvenile salmonids are available.

Predation in Estuarine Habitats

There are different region- and species-specific hypotheses concerning how habitats may or may not function to lower predation rates on juvenile salmonids in the estuary. The hypothesis that the estuary is a predator refuge for salmonid juveniles has been raised by estuarine salmonid ecologists in the northeast Pacific (Simenstad and Cordell 2000). This hypothesis is not necessarily supported by data on juvenile salmonid species in northwest Pacific estuaries (Koval and Gorin 2013) or the north Atlantic (Thorstad et al. 2012). Estuarine predation therefore needs to be viewed as a species-specific (Thorpe 1994) and context-specific issue (Mather 1998; D.M. Ward and Hvidsten 2011). It is difficult to generalize from the few data found in the literature, but in what follows I give an overview of the possible relative role of predation in the salmonid estuary.

Beaches
Tagged sea trout in the Aurland River estuary in Norway used rocky beaches less than 1 m deep and within 4 m of the shoreline. The authors of the study concluded that this was to avoid Atlantic cod (*Gadus morhua*) predation, which they observed in deeper water (Lyse, Stefansson, and Ferno 1998). Beaches are characterized by shallow water where larger piscivorous fish are usually not found. There were differences in predation rates between northeast and northwest Pacific estuarine beaches. Potential predators, such as staghorn sculpins (*Leptocottus armatus*) and northern pikeminnow, were commonly caught on mudflat beaches in the upper Fraser River estuary with pink salmon fry and chum salmon fry, but fry were never found in the predator stomachs (Northcote et al. 1979). Staghorn sculpins did prey on chum salmon and chum salmon fry in tide pools on gravel beaches at the Big Qualicum River estuary in British Columbia, but only a few salmon were found in sculpin stomachs (Mace 1983). On beaches at Karaginsky Bay (the outer estuary of the Maminkvayam and Uka Rivers and numerous other rivers) in Russia, Dolly Varden, cottids, Arctic char, and Asiatic smelt (*Osmerus mordax dentex*) were major predators on chum salmon fry and pink salmon fry, and also ate sockeye salmon smolts. Cottids preyed on young coho salmon. At the estuary of the nearby Khaylyulya River, predators removed between 5 and 29 percent of the pink salmon fry and less than 30 percent of the chum salmon fry (Karpenko 1990). Thus beaches in the outer estuary may not be a refuge in these and other estuaries in the northwest Pacific (Koval and Gorin 2013). Reasons for the differences in predation between the northeast and northwest Pacific beaches are not known, but in some cases could relate to turbidity as a predator refuge. Turbidity levels are higher in the Fraser River estuary than in the estuaries at Karaginsky Bay (R.S. Gregory and Levings 1998; Wernand et al. 2013).

Chum salmon and pink salmon spawn in shallow gravel/cobble beaches, where they can be prey to bears, such as black bears (*Ursus americanus*), at an estuary in British Columbia (Reimchen 1998). The black bears at the estuary preyed on chum salmon at night and apparently used tactile behaviour to find the salmon.

Channels
For Atlantic salmon smolts and post-smolts, riverine channels have been described as places where predation by Atlantic cod and other marine

fish species is high (mortality rate of 0.6–36 percent • km^{-1}) in both northeast and northwest Atlantic estuaries (literature summarized in Thorstad et al. 2012). Predation on salmonid fry in northwest Pacific riverine channels can also be substantial in some instances. An example of the contextual habitat effects for estuarine predators in riverine channels was presented by Dobrynina and colleagues (1989) in their study of predation by steelhead smolts, starry flounder (*Platichthys stellatus*), Asiatic smelt, and whitespotted char on pink salmon fry at the Utka River estuary in Russia. The riverine channel at this estuary is 40–50 m tidal, and channel flows are constricted. The authors found that maximum predation rates in the riverine channel of this estuary occurred when incoming tides (up to 2.7 m amplitude) concentrated the fry, possibly on the upstream side of the salt wedge. The incoming flood tides occurred at night, coinciding with the nocturnal immigration of fry into the estuary and leading to large schools of vulnerable prey for predators during early morning daylight hours. Predation rates may have been lower on ebbing tides, when river and tidal currents would sweep the majority of the pink salmon fry out of the estuary. Flows are a component of riverine channel subhabitat and thus are important in regulating predation rates.

Very few data are available on predation in riverine channel subhabitats in northeast Pacific estuaries. Information suggests the predation rate is lower, but the data are dated, equivocal, and scarce. Staghorn sculpins were found to prey on chum salmon fry in Hood Canal, Washington. In the same study, cutthroat trout were predators on chum salmon fry, but the fry ranked fourth in importance after several invertebrate taxa. Stomach contents of other potential pelagic or demersal predatory fish (representatives of Squalidae, Clupeidae, Gadidae, and Pleuronectidae) were examined, and none fed on juvenile salmonids (Bax et al. 1980). On the other hand, cutthroat trout in Puget Sound, Washington (which connects to Hood Canal), fed on chum salmon fry and other juvenile salmonids (Duffy and Beauchamp 2008). Steelhead smolts (about 20 cm long) were abundant in riverine channels and co-existed with Chinook salmon fry (about 10 cm long) at the Columbia River estuary, but McCabe and colleagues (1983) found no evidence that the steelhead ate the Chinook salmon.

Reasons for the differences in these various scenarios are not clear. However, when comparing the predation rates on salmonid fry in the riverine channels of the Utka River estuary with those in the Columbia River estuary, channel width may have been a factor. Lower predation by steelhead smolts may have been related to the broader dispersion of flows as the riverine channel of the Columbia River estuary is much wider (up to 1 km [McCabe et al. 1983]) than that of the Utka River estuary, enabling more space for escape by the subyearling Chinook, thus reducing availability. Turbidity may also have been a factor. In addition, many of the steelhead smolts at the Columbia River estuary were hatchery-reared fish and may not have switched to natural food at this point in their seaward migration.

Alternative prey from the river or ocean immigrating into riverine channels may be important for buffering predation on salmonids. This was a role ascribed to alewives (*Alosa pseudoharengus*), blueback herring (*Alosa aestivalis*), American shad, and rainbow smelt (*Osmerus mordax*) as alternative prey for predators such as double-crested cormorant (*Phalacrocorax auritus*), river otter (*Lontra canadensis*), and osprey (*Pandion haliaetus*) on Atlantic salmon smolts migrating through riverine channels in Maine (Saunders et al. 2006). In another example, at the Rivière à la Truite, just upstream of the Moisie River estuary in Québec, W.L. Montgomery and colleagues (1983) documented highly synchronized migrations of nonsalmonids (lamprey, cyprinids, etc.) with Atlantic salmon parr and smolts and sea-run migrant brook trout into the estuary from the watershed. Data on lesser sand eels as alternative prey for marine fish such as Atlantic cod, saithe (*Pollachius virens*), whiting (*Merlangius merlangus*), haddock (*Melanogrammus aeglefinus*), and sea trout at the Tana River estuary in Norway was striking: no Atlantic salmon smolts (or otoliths, indicative of predation if the smolts had been digested) were observed in over 500 fish predator stomachs (Svenning et al. 2005). Other forage fish species from the ocean (e.g., Pacific capelin, *Mallotus villosus socialis*, at the Bol'shaya River estuary in Russia) may be alternative prey for both salmonid and nonsalmonid fish predators (Tokranov and Maxsimenkov 1995). Provision of alternative prey in riverine channels from upstream and downstream is another form of trophic subsidy.

The functional response of different species of fish predators in riverine channels to estuarine salmonid prey abundance is another contextual aspect of predation. The study of functional responses encompasses data on the effects of predators and how their habitat utilization, behaviour, and consumption patterns impact prey populations at various population sizes. Predation by steelhead smolts on pink salmon fry at the Utka River estuary showed a linear relationship with the prey abundance (Dobrynina et al. 1989; Figure 20). The rate of predation did not seem to level off at high prey densities (at least up to the level of about 250 fry · m^{-3}) as might be expected in a Holling type 2 response, where

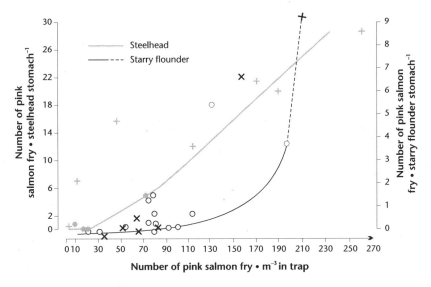

Figure 20 Predation by steelhead smolts and starry flounder in relation to pink salmon fry abundance at the riverine channel of the Utka River estuary, Russia.
Note: Data were obtained on various dates between May and July in 1978, 1979, and 1981. Symbols: O, x = starry flounder; ●, + = steelhead; – – – = extrapolated by present author.
Source: Data from Dobrynina et al. 1989.

consumption rate remains constant regardless of prey density (Begon et al. 2005). On the other hand, starry flounder showed a partial type 3 response with an accelerating phase at low densities of pink salmon fry (about 190 fry · m^{-3}), and consumed fewer pink salmon fry. This was likely because of the very different detection and consumption phase of the bottom-feeding flounder relative to the steelhead smolts, which tend to prey in mid-water or the surface. This example shows how the spatial relationships of the prey and predator are important contextual features of predation in riverine channels, and likely other habitats. Context at the macrohabitat scale may also be important. There may be a gradient of predation rate on juvenile salmonids along the river-estuary-ocean continuum, but this has not been investigated in detail. The Neva River estuary in Russia is an example, with the following potential predator array: pike (*Esox lucius*) in freshwater tidal zone, European flounder (*Platichthys flesus*) in the middle estuary, and Atlantic cod in the outer estuary (Telesh, Golubkov, and Alimov 2008). The physiological stress response of prey to predation risk has also been proposed as a mechanism that can help explain context dependency in ecosystem functioning (Hawlena and Schmitz 2010), including predation. This concept may have application to salmonid smolts moving to the ocean in riverine channels. They are stressed during this transition period because of osmoregulatory challenges. Handeland and colleagues (1996) found in laboratory experiments that the greatest predation on Atlantic salmon smolts by Atlantic cod occurred during the first days of exposure to seawater, when osmotic perturbations were greatest.

Schooling as a refuge is a point to consider when discussing predation in channels. Being a member of a group is thought to reduce predation risk in many animals and may allow fish to devote more time to feeding and less to vigilance (Pitcher and Parrish 1993). As salmonids school in most estuarine habitats, this behaviour may not be a habitat- or species-specific advantage. Laboratory experiments with chum salmon fry in coastal zone conditions (salinity 30–32 psu) showed that chum salmon fry had the ability to alternate between schooling and solitary agonistic behaviour. This may enable juvenile chum salmon to respond to local patterns of food distribution and predation risk (Ryer and Olla 1991). Adult salmonids in estuarine channels are vulnerable to mammals such as harbour seals (*Phoca vitulina*) (Middlemas et al. 2006), California sea lions (*Zalophus californianus*), and beluga whales (Seaman et al. 1982). At the Klamath River estuary in California, Chinook salmon upstream migration from a lagoon channel into the river was related to the presence and predation threat of sea lions (Strange 2013). Deeper riverine channels where the adult stages migrate may afford some refuge relative to beaches. In the shallow water habitats, the salmonids may be more vulnerable to mammalian predators.

Vegetation
Vegetation is hypothesized as a refuge habitat from predation for several species of juvenile salmonids in northeast Pacific estuaries (especially Chinook salmon fry and chum salmon fry), but researchers note the weakness of the data in support of the hypothesis (e.g., Simenstad and Cordell 2000). Case studies of predation on estuarine salmonids in vegetated habitat are scarce, but an example was provided by C.C. Wood (1987), who showed that merganser predation on chum salmon fry and Chinook salmon smolts in the lower Rosewall Creek estuary in British Columbia, characterized by brackish marsh, was lower than in the narrow freshwater reaches of the stream. Similar low predation rates by mergansers on Atlantic salmon smolts, sea trout, and Arctic char were noted in the Hals River estuary in Norway, although comparisons with

other habitats were not made in the study. The shallow water [0–5 m (Kålås et al. 1993)], of the Hals River estuary was probably characterized by brackish marsh and eelgrass.

Fish predation on juvenile salmonids in estuarine vegetation has not been studied in detail in the field, but some laboratory data are available. In freshwater mesocosm experiments, adult cutthroat trout predation on Chinook salmon fry, chum salmon fry, sockeye salmon fry, and cutthroat trout fry was less in artificially vegetated (plastic rope stems to simulate brackish marsh plants) tank habitats than in tanks with no vegetation (R.S. Gregory and Levings 1996). Other aquarium experiments, also with artificial vegetation, were conducted to simulate the freshwater tidal zones of the Fraser River estuary. In aquaria with and without a predator (Chinook salmon smolts, 15–20 cm), chum salmon fry and pink salmon fry less than 10 cm long were given a choice of two habitats: open water with high prey density, and vegetation with low prey density. Pink salmon fry were found more frequently in the vegetation when predators were present than otherwise. When hungry, however, they occupied the open area to a greater extent than when satiated. Chum salmon fry did not show as much difference in habitat choice as pink salmon fry (Magnhagen 1988).

Artificial habitats and habitat from invasive ecosystem engineers
Predation in artificial habitat and habitat from invasive ecosystem engineers is discussed in Chapter 14 in the context of effects of loss of natural habitat.

Predation within the Salmonid Trophic Guild in Estuaries

Predation in estuaries by larger salmonids on smaller fry or smolts of other species of salmonids is an aspect of predation within the salmonid trophic guild (Polis et al. 1997). Intraguild predation may be a selective force within the salmonid estuarine community but it is a complex problem that has not been investigated in detail. Steelhead smolt predation on pink salmon fry at the Utka River estuary (Dobrynina et al. 1989; Figure 20) is an example of the phenomenon. There are other examples from both the northwest Pacific (e.g., whitespotted char feeding on pink salmon fry and Dolly Varden sea-run migrants at the Bol'shaya River estuary in Russia [Tokranov and Maxsimenkov 1995]) and the northeast Pacific (e.g., coho salmon smolts feeding on pink salmon fry at the Bella Coola River estuary in British Columbia [Parker 1971]; and coho salmon smolts feeding on chum salmon fry in restored vegetation habitat at the

Table 4

Summary of aspects or factors to consider when selecting methods of assessing predation on salmonids in estuaries, with comments on their application

Aspect or factor	Influence on assessment of predation rates in the estuary or on other aspects
Approaches to measuring predation	Each method (scarring, stomach contents, bioenergetics, laboratory experiments, and consumption rates combined with predator population size) has varying accuracy.
Intraguild predation, especially older life stages of iteroparous species feeding on younger stages of semelparous species	Competition is an interacting biotic factor and may be alleviated by size differences of food eaten by prey and predator; alternative prey is a likely influence also.
Refuge	Evidence from the laboratory and limited field data show that vegetation use and dispersion into shallow water by juveniles can provide a refuge, as can (interacting) turbidity; flow may stimulate migration to lower predation zone for some species; generally context is important.
Risk when feeding	Estuarine laboratory experiments showed that Chinook salmon smolts were willing to take greater risk in order to feed than coho, pink, and chum salmon.
Schooling	Transformation from agonistic behaviour to schooling and shoaling can provide refuge.
Osmotic stress	Increases in levels of stress hormones can result in greater vulnerability.

Grays River estuary, Washington [Roegner et al. 2010]). Salmonid intraguild predation appears to be infrequent in Atlantic estuaries. Sea trout fed on Atlantic salmon smolts in the estuary of a short stream draining Lough Furnace, Ireland (Thurow 1966).

In summary, predation is clearly an important biotic interaction between salmonids and other members of the salmonid estuarine community. However, our knowledge of the phenomenon can be influenced by methodological approaches and contextual factors (Table 4). Behaviour and use of specific habitats at various life stages can condition predation rate and are probably adaptations to decrease the removal of individuals from the population.

Predation is an ecosystem and community process with implications for estuarine survival that relates to interactions between salmonids, between salmonids and nonsalmonid fishes, and between salmonids and birds and mammals. Competition, another important biotic interaction, adds another layer of complexity. Predation and competition can interact, affecting prey communities such as juvenile salmonids in an estuary in a variety of ways that may interact. Predation may also mediate changes in prey behaviour that in turn mediate the intensity of competition among prey species (Kotler and Holt 1989). An overview of competition in the salmonid estuarine community is required to further elaborate on biotic interactions in the estuary.

Competition

Competition among fish was defined in a seminal paper by Larkin (1956) as the demand by two or more individuals of the same or different species for a resource that is actually or potentially limiting in their ecological niches. Usually it is a challenge to identify whether competition is actually limiting a key process such as growth. Typically, it is described as competition for the same food (exploitative competition) or habitat (interference competition) (Grant and Imre 2005). Competition can be examined by field experiments in freshwater habitats, and has been in numerous papers dealing with a wide variety of salmonids in streams and lakes (e.g., Heggenes, Bagliniere, and Cunjak 1999; H.B. Rich et al. 2009), but the topic has been much less studied for salmonids in estuaries (Fresh 1997; Emmett 1997). For some species, it has been suggested that the evolution of estuarine rearing phases and completion of the smolting process in the estuary (e.g., Chinook salmon [Reimers 1973]) are an adaptation to avoid competition and to avoid exceeding the carrying capacity in freshwater habitats (including the freshwater tidal zone). However, the lability of adaptation to a specific carrying capacity level can be seen in the invasive Chinook salmon populations in the Rakaia River in New Zealand, where there was interannual variation in the proportion apparently reared in the river and that then migrated to the hapua-type estuary as fry or after about two to three months in the river as subyearlings. Over a six-year period, the ocean type that may have resided in the estuary accounted for between 54.4 percent and 78.8 percent of the population (Quinn and Unwin 1993). The Chinook salmon juveniles in the Rakaia River hapua-type estuary fed on dipteran larvae and an amphipod, while sea trout fed on these organisms as well as fish.

It is likely that the productivity of these food sources varies from year to year, and data on whether the dipterans and amphipods were produced within the estuary or were imported from upstream are not available, complicating the issue. Nevertheless, the lagoon appeared to support significant numbers of Chinook salmon and sea trout and there was no evidence of competition. Twenty-one thousand tagged hatchery Chinook salmon were released directly to the Rakaia River lagoon, and some resided there for 140 days. Sea trout annual growth rates were 0.3 mm $\cdot$ d^{-1}, comparable to sea trout found in other estuaries and coastal zones in their native range (Eldon and Greager 1983).

Exploitative competition is more common in the estuary than interference competition. This is a reasonable assumption given that territorial behaviour and defence of space for salmonids appears to be limited or transient in the estuary. Generally, anadromous salmonids reduce agonistic behaviour when they begin to smoltify. For example, nipping decreased in masu parr when they began to smoltify (Hutchison and Iwata 1997). Most juvenile salmonids school together when they are in the estuary.

Approaches to Assessing Competition

Methods for investigating inter- and intraspecific competition in fishes often rely on feeding data to approximate conditions in the ecological niche (Rachlin et al.1989; Marshall and Elliott 1997). Interactions between fish populations will typically be a complex mixture of feeding interactions where small stages of a piscivorous species (e.g., Chinook salmon fry in the estuary at the tertiary trophic level) may experience competition with planktivorous and benthivorous nonsalmonid species at the same trophic level (e.g., sticklebacks). In the same fish community, piscivorous species at the quaternary level (e.g., cutthroat trout) prey on both the Chinook salmon fry and the sticklebacks, and the competitive outcome between the latter two species might affect the growth of the cutthroat trout (see L. Persson and De Roos 2006). A possible approach to competition could involve investigation of stress hormones such as cortisol, which increased in Atlantic salmon juveniles in fresh water when they interacted with invasive species (brown trout and rainbow trout), since stressed fish feed less and therefore are at a competitive disadvantage in the estuarine ecosystem (van Zwol et al. 2012). However, diet overlap is perhaps the most common method of inferring exploitative competition within the salmonid community in the estuary, as well as between salmonids and nonsalmonids.

Exploitative Competition in Estuarine Habitats

The literature on exploitative competition between salmonids or between salmonids and nonsalmonids in the salmonid estuary is very sparse. I could find only a few examples that discussed exploitative competition on a habitat basis.

Beaches
Potential interspecific competition for food between pink salmon fry, chum salmon fry, and coho salmon parr and smolts was studied at gravel and cobble subhabitat in the Porcupine Creek estuary in Alaska. Diet overlap was greater between pink salmon fry and chum salmon fry relative to coho salmon fry. Size-related differences in prey selection were reported, with the smaller pink salmon fry and chum salmon fry eating smaller invertebrate prey. Coho salmon parr grew larger than the other two salmonid species, so overlap in prey size decreased over the spring period. Exploitative competition between the three juvenile salmonids appeared to be minimized by prey size differences (Murphy, Thedinga, and Koski 1988).

Channels
The distribution of Dolly Varden sea-run migrants in the inner parts of riverine channels in Beaufort Sea estuaries, Alaska, may be a mechanism to avoid competition with the more abundant whitefish species, whose coastal distribution seems limited to the outer riverine warm and less saline estuarine areas. Dolly Varden sea-run migrants were found in very cold offshore water feeding on *Apherusa glacialis*, a gammarid amphipod living under ice floes (Fechhelm et al. 1997). Thus, habitat segregation and utilization of food sources from different habitats may be adaptations to avoid exploitative competition between Dolly Varden and whitefish.

Diet overlap between Atlantic salmon, sea trout, and Arctic char (length 33–56 cm) in the riverine channels at the Å River estuary in Norway was studied by Grønvik and Klemetsen (1987). Diet similarity, measured with correlation coefficients, ranged from 0.69 to 0.74, because all three species fed heavily on Atlantic herring. Differences in similarity values were related to Arctic char feeding on invertebrates (especially crab larvae and benthic amphipods) and sea trout feeding on sand eels (*Ammodytes* spp.). Feeding specialization suggested by the feeding behaviour of Atlantic salmon, sea trout, and Arctic char may be a mechanism for minimizing competition for food. The diet of juvenile chum salmon overlapped strongly with the food of juvenile fat greenling (*Hexagrammos otakii*) in

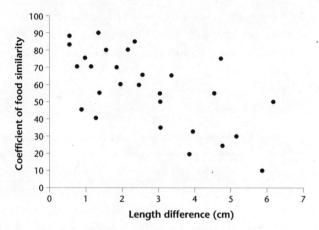

Figure 21 Relationship between size differences of subyearling and yearling Chinook salmon and a coefficient of food similarity at the Bol'shaya River estuary, Russia.
Source: Modified from Leman and Chebanova 2005, with permission.

the inshore waters off northern Hokkaido, Japan, which partially included the outer riverine channels of the Syokanbetsu River estuary. The investigators concluded that there was potential competition between the two species (H. Kawamura et al. 1998), but no evidence was given for food limitation for juvenile chum salmon or juvenile fat greenling.

Intraspecific competition between Chinook salmon smolts and subyearlings was reduced by size-specific prey species differences in a riverine channel in the freshwater tidal zone of the Bol'shaya River in Kamchatka. If the juvenile Chinook salmon differed in length by at least 3 cm, the coefficients of food similarity were reduced (Figure 21). The subyearlings fed primarily on cumaceans (*Lamprops korroensis*) and aerial insects, whereas smolts also ate aerial insects but diversified their diet by consuming larger invertebrates such as oligochaetes, cumaceans carrying eggs, mayflies, and caddisflies (Leman and Chebanova 2005).

Vegetation
In brackish marshes of the Columbia River estuary, exploitative competition was unlikely to occur between subyearling Chinook salmon and three-spined stickleback because the availability of benthic prey was not limited by the foraging behaviour of either fish species. Competition between the two species was reduced by habitat partitioning: Chinook

salmon displayed a greater tendency to feed near the surface and in mid-water, whereas three-spined stickleback fed mostly on benthic prey (Spilseth and Simenstad 2011). In the brackish marshes of the Fraser River estuary, diet overlap was greatest between Chinook salmon and chum salmon (90 percent) and lower between pink salmon and the other two species (chum salmon, 32 percent; Chinook salmon, 30 percent). Spatial, temporal, and diet differences among the cohabiting species suggested segregation of resource use during their estuarine residency (Northcote et al. 2007).

A study with connections to the wrack line (and possibly wood debris) at the Kalininka River estuary in Kamchatka, Russia, examined potential competition between hatchery-reared juvenile chum salmon and six species of fish: Japanese smelt (*Hypomesus japonicus*), saffron cod (*Eleginus gracilis*), dusky flounder (*Liopsetta obscura*), Pacific redfin (*Tribolodon brandtii*), whitespotted char, and three-spined stickleback. The chum salmon were less than 5 cm in length, whereas most of the other species were adult or older juvenile fish greater than 10 cm in length. Fifty-two species of amphipods were found in the stomachs of the seven species of fish; chum salmon ate twenty-five species. The chum salmon were feeding primarily on two small species of talitrid amphipods, a group of amphipods living at the wrack line. The other fish species were feeding on two larger gammarid amphipods, which live lower in the intertidal zone in cobble subhabitat. However, the size separation was not clear, as some of the larger fish also ate juveniles of the larger gammarid amphipods. The author concluded that there was no "sharp" food competition for amphipods (Budnikova 1994), another example of possible habitat and resource segregation.

Artificial habitats and habitat from invasive ecosystem engineers
Competition in artificial habitat and habitat from invasive ecosystem engineers is discussed in Chapter 14, in the context of the effects of loss of natural habitat.

Interference Competition
There are some examples of possible interference competition in the estuary. Pink and chum may use the same gravel habitat at spawning beaches in some estuaries (Murphy 1985). Territoriality may be maintained by the juveniles of some species in the upper estuary or freshwater tidal zone (e.g., Chinook salmon fry at the Sixes River, Oregon [Reimers 1968]). These are likely to be individuals that have not completed smolting.

In summary, food and habitat should be looked at together in any examination of competition between salmonids and other estuarine species, but this has not been done very often. The variety of ecological and habitat niches in the estuary implies that habitat segregation may occur because of adaptations that enable exploitation of different food sources by different species. Interactions upstream and downstream of the estuary may also be influential. Competition in fresh water may have consequences for species in the estuary. McCarthy and Waldron (2000) found that some brown trout in Loch Eck, Scotland, which is connected to the ocean by the Eachaig River, showed $\partial^{13}C$ and $\partial^{15}N$ stable isotope signatures indicative of marine feeding despite their appearance (colouration) as residents of fresh water. The authors concluded that these fish were misidentified sea trout. They also concluded that it was possible they were brown trout that had adopted an estuarine feeding strategy to avoid interspecific competition for food with Atlantic salmon, powan (*Coregonus clupeoides*), and Arctic char within Loch Eck. This opens the door to the interpretation that there is less competition for the species in the estuary but possibly more in the ocean, which the fully anadromous sea trout also uses. On the other hand, competition for food in fresh water was implicated in the development of anadromy and estuarine use by introduced brown trout in the Rio Grande River in Argentina (O'Neal and Stanford 2011). After about twenty years of population growth when the species was only in the freshwater form (brown trout), a portion of the population began to use the estuary and coastal region as sea trout, with dramatic results in terms of growth and size (O'Neal and Stanford 2011; Figure 13).

Conclusion

It is likely that biotic interactions between salmonids and other fish in natural estuaries are key fitness components affecting salmonid survival. Predation has received considerable attention as a selective force that has shaped juvenile salmonid use of the estuary as a refuge, but the species and context for this inference need to be qualified. Predation on Atlantic salmon smolts is thought to be high in riverine channels, and river flow that moves them quickly out of the channels may provide a type of shelter. There are regional differences in apparent predation pressure on juvenile Pacific salmonids. Estuaries on the northeast Pacific are thought of, or at least suggested, as predator refuges from fish predation on juvenile salmonids. In northwest Pacific estuaries, fish predation was found to be a major factor reducing survival. Competition is poorly understood for salmonids in estuaries, and fewer data are available.

Exploration of the competitive relationships between species using ecological guilds (M. Elliott et al. 2002; Lobry et al. 2003) and niche structure ideas (Lappalainen and Soininen 2006) are possible methods for expanding investigations. Experimental manipulation of densities and observation of performance-related response variables (e.g., growth success as a consequence of competition or lack of it) in relatively sedentary species such as estuarine cottids can help identify the relative degree of saturation of major habitat types (e.g., Polivka 2005). Similar experiments have not been attempted with the more mobile estuarine salmonids.

The adaptive significance of food and growth, osmoregulatory success, and biotic interactions is becoming increasingly clear. However, it is likely that all three factors have been affected by a wide range of human activities in salmonid estuaries, ranging from acute habitat loss to pollution to community change from invasive species. The next chapter will give an overview of these changes in salmonid estuaries. The discussion is required to provide context for Chapters 13 and 14, which will review our current understanding of the relationship between estuarine habitat loss and fish community change with salmonid survival.

12
How Have Habitat and Water Properties Changed for Salmonids in Estuaries?

Although a worldwide overview of the number of salmonid estuaries totally decimated or partially disturbed is not available, clearly the degradation has been significant on numerous coastlines. Using eelgrass loss as an index, major disruption has occurred within the original salmonid range around the world (Waycott et al. 2009). Looking at the problem on a smaller geographic scale, 88 percent of British estuaries (136 out of 155) have experienced habitat loss to agricultural reclamation. In 17 major estuaries in England, most of them on salmonid rivers, 88,800 ha of habitat have been lost since Roman times (Healy and Hickey 2002). Along the coast of Europe, 1 km of shoreline per year was developed between 1960 and 1995 (Airoldi and Beck 2007). In northeast Asia, degradation has occurred over centuries (Figure 2). In North America, the same pattern has been observed since Europeans began to live on the various temperate coastlines. While the focus of many estuarine degradation studies has been on the estuary itself, salmonids are affected by habitat disruption in their watersheds and changes in the adjacent ocean, so they are susceptible to upstream and downstream forces as well. The rate of environmental change in estuaries is now greater than during the Pleistocene, when salmonids may have been expanding into estuaries. For example, soil eroded by various agricultural, construction, and mining practices is transported into rivers and then to estuaries (Wilkinson and McElroy 2007); these human-induced geological changes are even greater than some single natural catastrophic changes in estuaries from earthquakes in Japan (Tanaka et al. 2012) and Alaska (Thorsteinson et al. 1971).

A number of books, papers, journals, websites, and regional groups deal with estuary disruption around the world and provide extensive information on how the industrial and urban developments have changed estuarine habitat and water properties (McLusky and Elliott 2004; Kennish et al. 2008; Eionet 2014). This chapter provides an overview of some of the generic threats and activities that affect estuarine salmonid habitat and water properties. Generally these are known as stressors, which can be defined as "an environmental or biotic factor that exceeds natural levels of variation" (Breitburg et al. 1999).

Habitat Loss

Habitat loss in estuaries was identified by Odum (1970) as "insidious alteration." The following are some examples of the phenomena specific to salmonid estuaries.

Beaches

Beaches have possibly suffered the most loss or change in many salmonid estuaries, because, lacking visible vegetation, they appeared unproductive and featureless in early ecological surveys. Beaches are frequently filled in with dredged material to develop shipping facilities and housing and industrial developments. Beach habitat is lost when riprap is placed on sand- and mudflats (Pomeroy and Levings 1980). Modifications of channels can affect nearby beaches because channels connect with them and supply sediment. The deflection of sediment by construction of a river training wall to maintain a riverine shipping channel at the Fraser River estuary in British Columbia reduced the capacity of the system to maintain sandflats (Levings 1980b). Dredge spoil disposal and silt spills from gravel mining operations, road construction, and related land-use activities can change gravel or cobble to mud or sand (P.J. Wood and Armitage 1997). Log storage on sand- and mudflats can result in coarsening, and sunken wood debris changes the nature of the beach material (e.g., Nanaimo River estuary, British Columbia, where log storage formerly affected almost half of the sand- and mudflats [Bell and Kallman 1976]). An additional factor for sand and gravel beaches is the change in wave or current energy level that occurs when a sea wall or coastal protection structure is built (Figure 22). The shoreline is steepened as the fine sediment erodes, and the grain size changes to coarser material. When structures reduce movement along drift cells (specific reaches of a beach where sediment is moved along a shoreline) from nearby sources such as gravelly bluffs, the resulting decrease in coarse sediment supply is also a factor (Shipman 2008). Large pulsed reservoir water spill events from rivers can also create and destroy sand beaches in the estuary as river forces redistribute sediment (Dalrymple and Rhodes 1995).

Figure 22 Sea wall at the St. Lawrence River estuary (Île d'Orleans) near Québec City. Woody debris from fragmented riparian vegetation has collected at the base of the seawall. *Note:* Colour photo in Appendix 5 at http://hdl.handle.net/2429/57062.
Source: Author photo.

Channels

Channel habitat is changed in an acute sense when channels are filled in and covered for dock construction and creation of land for commodity storage (e.g., a coal loading facility in the Fraser River estuary (tidal creeks) [Levings 1985]). Dredging and channel training walls to deepen riverine channels for shipping have modified salmonid estuaries around the world, ranging from the major ports on the Scheldt River estuary in Belgium (van den Bergh et al. 2005) to smaller ports elsewhere in northern Europe (e.g., Orkla River estuary, Norway; shown in Appendix 5 at http://hdl.handle.net/2429/57062). Depth is one of the key geomorphological characteristics of channels because it controls several oceanographic features, such as the salt wedge and surface temperature. Shallow water can warm faster than deeper water bodies. Dredging to deepen channels can result in increasing the capacity of the lower end of the estuary to receive sediment. Infilling may occur, and the cross-sectional area of the riverine channels may be reduced (e.g., Kushiro River estuary, Japan [Nakamura et al. 1997]).

Diking for flood control on the margins of estuary riverine channels is common in disrupted salmonid estuaries. Hydraulic connectivity to the main estuary is lost. Interactions among diking, sedimentation, and depth are important for side channels and tidal creeks. In the Skagit River estuary in Washington, diking increased sedimentation outside the dike, resulting in the direct loss of tidal channels that provided low-tide living space (Hood 2004b). Water flow into channels from the river and ocean is a key factor for riverine channels. Dams and other water diversions (e.g., irrigation) affect water flow in channels since the amount of fresh water entering an estuary, together with tidal flows from seaward, determines the degree of flooding into key salmonid subhabitats, such as side channels (Kukulka and Jay 2003). Horizontal changes in salinity distributions and current patterns in the estuary can result from dike construction or harbour development (e.g., Levings 1980b; Hvidsten et al. 2000). In watersheds with reservoirs, the seasonal patterns are disrupted. Spills are required for flood control and dam maintenance on rivers with reservoirs, and flows into the estuaries of these rivers vary dramatically from the natural seasonal flow pattern (Johnsen et al. 2011; Figure 7). The freshet is the process that carves channels through terrestrial habitat (Figure 9). New channels can be built or existing channels filled in by pulsed (sometimes daily) flood events, or seasonally changed. Erosion and accretion processes are also halted or modified when freshet and sedimentation patterns are changed by dams (Kukulka and Jay 2003). Sea level rise from the rebound of the land owing to glacier melt (Beechie, Collins, and Pess 2001) and climate change also changes the configuration of riverine channels as well as flow patterns from the river. In a Puget Sound, Washington, salmonid estuary, sea levels may have risen 5 m in the past 5,000 years (Beale 1991), possibly creating more channels, but natural sediment aggradation may have filled in the created habitat. Climate warming is also likely to cause a shift in discharge patterns through channels in northeast Pacific estuaries, with more rainfall-generated flow occurring in winter months (Cuo et al. 2009).

Water pumped from riverine channels for cooling industry can also cause problems for migrating salmonids (e.g., at a nuclear power station on the Baltic Sea [Bryhn et al. 2013]). Even after a deflective screen was placed at a power station intake on the Severn River estuary in England, migrating Atlantic salmon smolts were killed by the pumps (Claridge and Potter 1994).

Interference with processes that bring ocean water into riverine channels also needs to be recognized. Some types of barrages are dams that totally block flow in an estuary, such as that on the Petitcodiac River in New Brunswick (Aubé, Locke, and Klassen 2005). Other types of tidal barrages are clamshell-like rotating weirs used to partially or totally block water from moving in and out of an estuary during storm surges to protect cities, such as London, on the Thames River estuary (S. Gilbert and Horner 1986). When deployed, the structures drastically change flow patterns in riverine channels (e.g., Thames River estuary [Lavery and Donovan 2005]). Tidal barrages have also been proposed for tidal energy projects (e.g., Severn River estuary, England-Wales [Wolf et al. 2009]). Floating bridge pontoons are another example of a structure may obstruct water flow in salmonid estuaries (e.g., Hood Canal, Washington [M. Moore, Berejikian, and Tezak 2013]).

Vegetation

Vegetation can be affected by the same activities that modify beaches, especially brackish marsh and eelgrass, which depend on a gently sloping platform for growth. Acute loss can occur from filling for industrial use and urbanization. Macroalgae require a silt-free substrate to attach to rocks or other hard strata. Vegetation succession, the natural process where marshes gradually move to the front of the estuary as natural sedimentation occurs, is also stalled by constraining structures such as seawalls (Figure 22). In the heavily developed Schelde River estuary in the Netherlands, there is no space for new vegetative community development. Vegetation in the later successional stages is missing, and the marsh is a thin strip backed by seawalls (Van den Bergh et al. 2005). Dredged channels often have steep armoured shorelines that impede growth of riparian vegetation. Typically, wharf construction for large ships or dredging for marinas results in deeper water and loss of vegetated habitat that was present initially (D.L. Ward et al. 1994). Sea level rise is a collateral effect of climate change that could affect estuarine habitat by increasing erosion, especially in diked estuaries where structures may prevent adjustment of tidal elevation–related subhabitat such as marshes (Kirwan and Murray 2008).

Artificial Habitats and Habitat from Invasive Ecosystem Engineers

Artificial habitats and habitat from invasive ecosystem engineers are clearly implicated in the loss of natural habitats, but can also be changed after the development or construction of the artificial habitats. For example, some constructed riverine channels are wider and deeper at high flows but can become narrower when flows decrease when water is extracted for irrigation (e.g., the Yolo Bypass on the Sacramento River estuary, California [Benigno and Sommer 2008]). The direct habitat loss brought about by invasive ecosystem engineers also needs to be considered. The invasive water hyacinth (*Eichhornia crassipes*) is an example. Water hyacinth in the Sacramento–San Joaquin River estuary in California slows water flow in riverine channels by clogging them, and ultimately replaces open water in the channels (Toft et al. 2003).

In summary, human-induced changes to the landforms in salmonid estuaries result in geomorphological modifications in the structure, function, and areal extent of estuarine habitats. The most catastrophic human-induced change, total loss of habitat, occurs internal to the estuary when infilling is conducted for industrial development. External issues include dams on the river or tidal barriers near the river mouth. Collateral damage in channel configuration and capacity to support vegetation can also occur. In addition to physical habitat change, significant modifications to the estuarine water properties can result from human activity, as described next.

Water Quality

Water quality is said to change when conservative or nonconservative water properties diverge from natural values. The literature on estuarine pollution is very extensive, and only an overview of some of the major water quality issues can be given here.

Temperature

Water temperatures can be raised in estuaries where heated effluents are discharged (e.g., cooling water from power plants and pulp mills [Marchand et al. 2002]). Models and empirical data showed that temperatures as high as 36°C could occur at the discharge point of the Connecticut Yankee nuclear power station on the Connecticut River estuary (J. Richardson and Dixon 2004). In some regions, cold river water is released into estuaries from upriver dams in winter, in contrast to natural flows on unregulated rivers in spring and summer; this increases the chance of ice in the estuary (Johnsen et al. 2011). Climate change has the most pervasive long-term effect on salmonid estuarine temperatures, not only in situ but from warming of the adjacent ocean (Cheung et al. 2013) and incoming river water (Meybeck 2003). Climate models indicate that change in upwelling will not be sufficient to bring up more cool upwelled water in the northeast Pacific (Mote and Salathé 2010).

Salinity
Extremely low river discharge over the long term can result in hypersalinity in estuaries. The estuaries around the Aral Sea (e.g., the Syr Darya River estuary in Kazakhstan and the Amu Darya River estuary in Uzbekistan) provide a classic example (Pavlovskaya 1995).

Dissolved Oxygen
The dissolved oxygen content of estuarine waters can be reduced by the addition of oxygen-demanding organic material such as pulp mill effluent, wood debris, other industrial effluents, sewage, and excessive biological material, such as macroalgal blooms. These pollutants are among the most common in salmonid estuaries around the world. Tide gates are used in some salmonid estuaries to regulate water flow into flooded areas behind dikes. Periodic low-flow or stagnant waters caused by tide gates are major problems on numerous estuaries (e.g., Carneros Creek estuary, Elkhorn Slough, California [Gee et al. 2010]; Columbia River estuary, Washington-Oregon [Roegner et al. 2010]). Dams immediately upstream of the estuary can affect dissolved oxygen levels. An example is the estuary dam on the Nagara River in Japan, which has promoted the development of a low dissolved oxygen water mass in the estuary (Funakoshi and Kasuya 2009). Lower dissolved oxygen levels were observed in channels of the Columbia River estuary in years when offshore upwelling was weak (Roegner, Needoba, and Baptista 2011a), a phenomenon possibly related to climate change.

Turbidity
Turbidity in salmonid estuaries can be increased by land-use practices in the watershed, such as excessive timber harvest (Levings and Northcote 2004), by road/railroad construction within the estuary itself, or by spills from decommissioned dams (Shaffer et al. 2008).

pH
In low-salinity estuaries, acidic runoff can change pH values. Several square kilometres of shallow estuarine bays in Finland were made permanently acidic by the discharge of low-pH drainage water resulting from changes

in forest cover and draining for agriculture (Kyronjoki River estuary, Finland [Hildén and Rapport 1993]). The pH in salmonid estuaries can also be lowered by acid rain (Watt 1987), discharge of acid mine drainage (Barry et al. 2000), and pulp mill effluent (Servos et al. 1996). Ocean acidification is an increasingly important problem for estuaries and results from enhanced anthropogenic carbon dioxide in seawater and reduced pH, causing dissolution of calcium carbonate (Fabry et al. 2008).

Other Important Water Properties
Dissolved nutrients such as nitrogen and phosphate are necessary for biological production at the basal level (Lalli and Parsons 1997), but excessive nutrients from agricultural run-off and other sources can elevate levels. In the Baltic Sea, total nitrogen and total phosphorus introduced from rivers between 1996 and 2000 were about 773,000 t and 39,000 t, respectively, likely far in excess of the natural condition (Mörth et al. 2007). Data from European rivers indicate a ten- to twentyfold increase in phosphate and a five- to tenfold increase in nitrate in heavily populated areas since the beginning of the twentieth century (U. Larsson, Elmgren, and Wulff 1983). Excessive plant growth can then occur; decomposition of plant material causes a higher than natural oxygen demand, resulting in low dissolved oxygen in salmonid estuaries (Stringfellow et al. 2009). This condition is known as eutrophication. Upwelling offshore and increased penetration of the salt wedge can influence nutrient supply by bringing nutrient-rich water into the estuary (Roegner, Seaton, and Baptista 2011b).

Increased light at night and decreased light during the day are also characteristic of many disrupted estuaries. Nighttime illumination on bridges and docks can make light level equivalent to daytime (J.W. Curran and Keeney 2006). Conversely, docks and wharves can reduce daytime light levels by shading (Able et al. 2013). Noise levels in estuaries due to ships such as large tankers are about 170 dB and are higher from seismic surveys (for comparison, air guns are 210–250 dB [Gausland 2000]). These noise levels are higher than those expected in a natural estuary.

Contaminants: The Chemical Soup
Since the Industrial Revolution, salmonid estuaries have received an ever-increasing mixture of chemicals from factory discharges, sewage, agriculture, mariculture, energy-producing industries, and other sources. Only a few of these contaminants can be touched on here. In Europe, oil refineries discharge an average of 9 t • yr^{-1} of oil during routine production (Wake 2005; data from 2000), but major spills from pipeline breaks and

ship accidents can introduce more oil into salmonid estuaries. For example, the tanker *Exxon Valdez* spilled 35,000 t in Prince William Sound, Alaska, in 1989, and Carls and colleagues (2004) showed that bunker oil from the tanker remained in intertidal pink salmon spawning beds in Alaska for a decade. Metals from historical mining activities (lead, cadmium, zinc) are retained in the sediments of salmonid estuaries on a long-term basis (e.g., Humber River estuary, England [Macklin et al. 1997]) and are now included in the "legacy contaminants" category. Organic chemicals are also part of this category. Pesticides used in farming move into rivers and eventually the estuary; many of these chemicals degrade very slowly and build up in sediments (e.g., twenty-three active ingredients from pesticides in a tributary of the Danube River, Germany [Schulte-Oehlmann et al. 2011]). Radioactive substances can spread into estuarine sediments from very distant sources; for example, plutonium from the Chernobyl nuclear plant was found in sediment cores of the Pettaquamscutt River estuary in Rhode Island (Lima et al. 2005). A host of contaminants are included in sewage, notably contaminants related to human health. Pharmaceuticals are one of the categories whose levels have been rising in recent decades (Spromberg and Johnson 2008), because of human population growth in the vicinity of many salmonid estuaries.

Conclusion

It is a well-known fact that many salmonid estuaries around the world have been affected by human-induced physical changes, and water quality has been lessened by the chemical soup discharged into them. Much of the physical modification has resulted from the development of shipping ports. Thus the history of salmonid estuarine disruption is directly related to international commerce and human consumption or delivery of goods, both on the coast and inland. In general, the degradation of salmonid estuarine habitats follows a latitudinal gradient (possibly in both the Northern and Southern Hemispheres) that mirrors a trend in human population density.

In Chapter 14, the stressors described in this chapter will be related to the three key fitness components associated with salmonid survival in estuaries: osmoregulation, growth, and biotic interactions, with specific discussion and examples. First of all, however, it is necessary to describe the meaning of fitness components for survival in the context of salmonids in estuaries, and to elaborate on some of the experimental and methodological approaches that have been used to assess survival in the estuary.

13
Salmonid Survival in Estuaries

The fitness of an organism can be defined as reproductive power, the rate of conversion of energy into offspring (J.H. Brown, Marquet, and Taper 1993). The survival of an individual salmonid in the estuary has multiple fitness components, which are the factors affecting survival (e.g., growth, osmoregulatory adjustment, biotic interactions). The reproductive power of the surviving individual then needs to be interpreted in terms of variables such as fecundity and spawning behaviour, which influence the transmission of genetic material to the next generation – in other words, survival. Fitness therefore has an obvious genetic context. This chapter provides a general overview of fitness assessment and genetic aspects that are especially relevant to salmonids in estuaries, and comments on the methodology used to measure survival in the estuary, a metric needed to ultimately assess the fitness of estuarine salmonids.

Fitness Components of Survival in the Estuary
A measure of relative fitness from one generation to another, focused on the impact of salmonid habitat change but taking into account estuarine conditions, is required. There are examples from other arenas. For instance, to assess the survival of hatchery-reared versus wild salmonids, lifetime fitness can be measured, from a spawning adult to its surviving adult progeny (i.e., whole life cycle) (Araki et al. 2008). To test whether the estuary had reduced fitness components for salmonids relative to a disturbed estuary or another macrohabitat such as an embayment, a comparable experiment would require measuring the survival within an experimental area without an estuary and within a reference estuary. Alternatively, if the intent is to test for the effects of changed habitat within an estuary, a reference estuary and a degraded estuary would be needed. Large-scale experiments are a challenge with any ecosystem,

but especially in estuaries, which are dynamic and open to the upstream river and the downstream ocean. As alternatives to whole life cycle experiments, salmonid ecologists working in estuaries have focused on fitness elements that are key estuarine components of the salmonid life cycle.

Migrations are adaptations to ensure the fine-tuning of growth, survival, and reproduction in relation to environmental conditions experienced by parents and offspring (Dodson 1997). For salmonid migrations into the estuary, there are some empirical data showing how fitness may actually be achieved by the migratory experience. Adaptive migration into the estuary may be correlated with growth in fresh water, as shown by Riddell and Leggett (1981) for the autumnal migration of Atlantic salmon "large parr" into the Miramichi River estuary in New Brunswick. The fall movement of the "large parr" may facilitate a more precise timing of their movement through the estuary and result in less predation-related mortality relative to the main body of the relatively smaller smolts migrating earlier (Riddell and Leggett 1981). However, there can be interannual variation in selection for smolt size in Atlantic salmon (Hendry et al. 2003). Thus growth, size, morphology, and migration patterns are key fitness components to consider, but some variation can be expected.

Of all phenotypic traits examined, variation in body size in Atlantic salmon (or in correlated characters such as growth rates, age of seaward migration, or age at sexual maturity) generally shows the highest heritability as well as a strong effect on fitness (Garcia de Leaniz et al. 2007). For sockeye salmon, the smaller ocean/river ecotype that migrates to estuaries for rearing is better adapted than the larger lake or nonanadromous forms (kokanee) or ecotypes for persisting in unproductive glaciated streams, because the ocean/river ecotype relies less on oligotrophic freshwater productivity for growth and nutrition (C.C. Wood et al. 2008). Some of the migratory adaptations interact with morphology, which in turns affects behaviour. Age-0 river-ecotype sockeye salmon were deeper-bodied and had deeper but shorter tail structures and larger eyes, whereas age-0 lake-type fish were shallower-bodied with a shallow yet longer tail structure and smaller eyes, although the pattern was not totally consistent in the Alaskan systems studied (Pavey et al. 2010). The morphology of the riverine sockeye salmon probably enables more efficient feeding in heterogeneous river environments. The river/ocean form of sockeye fry found feeding in the vegetation of tidal creeks in the upper Fraser River estuary in British Columbia (Levings, Boyle, and

Whitehouse 1995) may be another example. Other adaptations relate to osmoregulatory preparedness for saline conditions (Chapter 9).

Estuarine fitness components of adult salmonids returning to spawn may offer a more specific measure as salmonids are in a state of reproductive readiness and loss of gametes and reproductive individuals may more easily be translated into loss of reproductive power. This is an area of research where specific genetic mechanisms are being explored. Genetic variation in run timing has now been correlated with specific functional genes in Chinook salmon returning to the Columbia River in Washington-Oregon, some of them associated with immunological functioning and muscle differentiation (Hess and Narum 2011). How these factors interact with estuarine habitat is unknown, but they could be correlated with specific temporal changes in water quality (e.g., disease suppression) and currents (e.g., energy utilization for swimming) as the fish enter the river mouth from the ocean. O'Malley and colleagues (2010) found that a specific gene appeared to play a key role in mediating seasonal adaptation in several, but not all, Pacific salmonid species on the west coast of North America. While the gene (*OtsClock1b*) probably influences the timing of reproduction in chum, Chinook, and possibly pink salmon – and hence their appearance at the estuary for migration upriver to spawn – there are apparently alternative ecological factors and genetic mechanisms regulating the timing of these life history events in coho salmon. Presumably arrival from the ocean is timed to match optimal conditions for each species or population migrating as a school through its natal estuary. Early research provided an insight to the problem as transfer experiments with adult Atlantic salmon in England indicated a strong heritable component for the run timing trait (Solomon 1973) and timing was tributary-specific (Stewart et al. 2002).

Understanding which evolutionary or ecological factors influence genetic structure, and quantifying their spatial scale (e.g., between watersheds) of influence (landscape genetics), is becoming an important field of salmonid conservation biology. Homing behaviour of salmonids is thought to be responsible for the genetic differentiation generally observed among river systems. In Atlantic salmon in eastern Canada, a regional genetic structure was observed. This raised the hypothesis of a lower success of interregional immigrants relative to local immigrants. However, temperature was also found to be important so that interactions between gene flow (distance between populations) and additional factors are probably important (Dionne et al. 2008). Similarly, the broad dispersal of introduced sea trout in the Southern Hemisphere may also have been partially facilitated by an ability to disperse between estuaries, using specific genetic adaptations to landscape features such as habitat types and freshwater plumes (e.g., multiple estuaries on the Kerguelen Islands, District of the French Southern and Antarctic Lands [Launey et al. 2010]). Steelhead, on the other hand, have been recorded in only one Southern Hemisphere estuarine system, the Santa Cruz River estuary in Argentina (Riva Rossi, Lessa, and Pascual 2004), although the species' freshwater counterpart (rainbow trout) has been extensively stocked in the region (Arismendi et al. 2011). The genetics of the introduced populations are of course a likely factor in the invasion scenarios. The role of the estuary is not clear in these scenarios, but since salmonids may imprint on fresh water from their natal stream or groups of streams, it could be important. The low-salinity water masses originating from the estuarine outflow, an aspect of the landscape, may influence homing and facilitate dispersal or invasion.

On a long-term evolutionary time scale, the emergence of estuaries in glaciated areas such as Alaska created portals for colonization of rivers along the coastline. For sockeye salmon, Ramstad and colleagues (2004) found that temporal isolation based on spawning time and founder effects (decreased genetic variation owing to colonization by small populations) associated with ongoing glacial retreat and colonization of new spawning habitats contributed significantly to genetic population structure and fitness, while geographic distance and spawning habitat differences did not have significant influence.

While understanding fitness is a key issue in salmonid estuarine ecology, it is also important to measure estuarine survival itself, which is the ultimate fitness metric. A number of methods have been used in studies to estimate juvenile salmonid survival in estuaries.

Methods to Assess Survival in the Estuary

If conditions within the estuary differ from those that salmonids have adapted to, an individual's survival, and hence its fitness, is likely to be affected during its time in the estuary when it encounters anthropogenic disruption or changes in biotic interactions. The null hypothesis is that the changes have had no effect. There are two major groupings of studies that relate to the hypothesis. The goal of the first group, which I call the estimation group, has been to actually measure survival. The two basic approaches of the estimation group have been as follows: (1) measure the survival rate of subyearlings or smolt to the adult stage when the salmonids are excluded from an estuary, and (2) directly measure survival after subyearlings or smolts have migrated from the lower reaches of the river into the estuary and have attained the post-smolt stage. The goal of the second grouping of studies has been to use statistical or modelling approaches to infer, but not actually measure, survival from habitat observations or assumed habitat preferences within the estuary.

Estimating Survival from Experimentation or Direct Observation

The estimation group includes a number of experiments that test juvenile-to-adult survival over a coarse gradient (e.g., whether or not an estuary is present) (Figure 23). This approach can incorporate all the estuarine factors together. Several of the experiments were conducted with smolts from Chinook salmon, thought to be a species reliant on estuaries (Magnusson and Hilborn 2003). Perhaps the first experiment of this type was conducted by Kjelson and colleagues (1982). Working in California on the Sacramento River and its estuary, they found that Chinook salmon smolts released in the outer estuary survived eighty times better than those released in the river, but the authors assumed that ocean conditions were comparable between the different years of the experiment. Somewhat similar results were found in Chinook salmon smolt transfer experiments at the Campbell River estuary in British Columbia, but this experiment was repeated for three years. The data showed that while the estuary did confer survival benefits in two out of three years (1983 and 1984), a factor outside the estuary, possibly in the river or the coastal ocean, appeared to have overwhelmed estuarine effects in the third year (1985) (Levings et al. 1989; Levings and Macdonald, n.d.; Table 5). Release site, year, and an interacting factor were all statistically significant ($p < .05$). Ocean conditions can interact strongly with estuarine factors and affect survival of juvenile Chinook salmon that disperse widely into the ocean (Lindley et al. 2007, 2009).There was some evidence of better ocean conditions for the fish released in 1985, but there were also differences in river flows between the three release years. There were other confounding effects. The Campbell River estuary was also undergoing a major restoration project at the time, so habitat conditions such as food supply were in a transitional state (Levings and Macdonald 1991). Notwithstanding these interactions, when data were summarized over the three years, fish deprived of the estuary showed a 50 percent decrease in survival relative to those that had estuary access. Another experiment to test smolt-to-adult survival with juvenile Chinook salmon at the lower Sacramento River, California, showed the importance of arranging the release groups relative to the habitat effect being tested (Sommer, Harrell, and Nobriga 2005). The Yolo Bypass runs parallel to

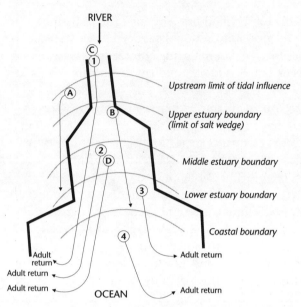

Figure 23 Idealized design of reach survival tracking and experimental transfer approaches to estimate survival of Chinook salmon smolts.

Notes: A = survival of Chinook salmon smolts within the entire estuary; B = survival within the saline part of the estuary; 1–4 = smolt-to-adult survival when transferred to (1) river, (2) middle estuary, (3) outer estuary, and (4) coastal zone; C and D = smolt-to-adult survival when transferred to river and middle estuary or to an artificial habitat and a reference river reach.

Table 5

Number of adults recovered from a Chinook salmon smolt transfer experiment at the Campbell River estuary, British Columbia

Year	River (1)	Middle estuary (2)	Outer estuary (3)	Coastal (4)	Totals
1983	223	283	72	50	628
1984	127	190	65	50	432
1985	226	186	179	252	843
Totals	576	659	316	352	1,903

Note: Numbers following release areas refer to the site codes in Figure 23.
Source: Levings and Macdonald n.d.

the main river. Uniquely coded wire-tagged Chinook salmon smolts were released into the bypass and the river to test survival differences. Results were mixed over the three years (ratios of bypass fish survival to river fish survival were 2.1, 0.9, and 0.6 in three consecutive years), possibly owing to variation in river flows. These case histories illustrate some of the opportunities and challenges involved in experimental design when attempting to study the juvenile-to-adult survival of Chinook salmon in the estuary.

Transfer experiments, or release experiments with a design similar to them, have been conducted with other Pacific salmonids and with Atlantic salmon. B.R. Ward and Slaney (1990) released steelhead smolts into four stations on the Keogh River watershed in British Columbia: the upper river, the lower river, the estuary, and a nearby ocean site. There were no significant differences between the ocean, estuary, and lower river sites, and the fish released into the upper river survived at the lowest rate (B.R. Ward and Slaney 1990). Coho salmon transfer experiments around the Columbia River estuary were conducted in 1983–87 (Solazzi et al. 1991). Their experimental design is not comparable to the other experiments, but is still instructive. The reference or control location was in the river at 235 km from the mouth, which is within tidal influence and considered the upper limit of the estuary (R.N. Williams 2006), but about ten times further from the ocean than most of the previously mentioned experiments. Coho salmon released in the lower estuary, within the zone influenced by brackish water, survived best. Solazzi and colleagues (1991) also found increased ocean survival for all release groups in the 1985 releases. They further suggested that better ocean conditions during the La Niña year (cooler ocean temperatures during spring) improved survival. Results from a transfer experiment with Atlantic salmon suggested better survival of fish excluded from the river. Atlantic salmon smolts were released into the Surna River in Norway, into the estuary, and at an offshore site about 100 km from the estuary (Heggberget et al. 1991). Survival rates for the three release groups were 2.0, 2.9, and 3.7 percent, respectively, suggesting that the estuary in fact conferred some benefit relative to the river, but releases were not replicated, so detailed statistical analyses were not possible. Transfer experiments have the potential to shed light on survival patterns relative to the river, estuary, and ocean, but are expensive and complex and require thousands of marked fish. Hatchery-reared fish have to be used, meaning that the results may not be applicable to wild salmonids. To avoid these problems, most investigators have focused on survival estimates of younger life stages.

Direct estimation of juvenile survival rates from entry into the estuary to departure from the estuary, or during residency within an estuary (e.g., smolt entry–to–post-smolt departure or within-estuary survival) is a quasi-experimental approach to measuring survival (Figure 23). Attention to species differences and the habitat context is required because as some fish are arriving into the estuary, others are leaving it, depending on their size and osmoregulatory status. In most cases, emigration is to the ocean, but in some instances juvenile stages move back upstream into the river on rising tides or for overwintering. Therefore, following the numbers of a discrete population of fish is necessary and the task is getting easier as tagging technology improves. Examples of major efforts to determine within-estuary survival by following samples of marked fish include the studies of Chinook salmon smolts in the Sacramento River and Columbia River estuaries. Newman and Rice (2002) released adipose fin–clipped and coded wire–tagged Chinook salmon smolts in the Sacramento River at seven locations in the inner estuary and one upriver site, and recovered them at the outer estuary in a trawling program. Recovery rate was only 0.0007 percent, reflecting the challenge of sampling in the riverine channel. Hydroacoustic tags were used to estimate survival of Chinook salmon within the Columbia River estuary. Survival probability ranged from 0.584 to 0.824 for yearling Chinook salmon smolts and from 0.185 to 1.005 for subyearling Chinook salmon (Harnish et al. 2012). It is generally easier to measure survival or mortality for species with large smolts that can be tagged externally or fitted with internal devices (hydroacoustic or PIT tags) and that show directed emigration through the estuary. Atlantic salmon smolts meet these criteria and a number of studies have been conducted with this species. Atlantic salmon smolt mortality in the estuary was 0.6–36.0 percent • km^{-1} (median 6.0 percent • km^{-1}) in a summary of data from a variety of tagging studies (Thorstad et al. 2012).

Carefully designed estimation and direct observation approaches to measure survival are becoming more common in the literature, and the availability of miniature tags will enable survival estimates on smaller salmonid life stages such as fry in the future. Until further technological developments are completed and further experimental studies are performed, the descriptive or inferential approach, as described in the next section, will likely continue to be the main method for investigating factors affecting salmonid survival in estuaries.

Statistical Inference and Modelling
Researchers have often used statistical inference to reach conclusions about relative fitness when they have found differences in abundance (see, for example, the habitat differential discussed as an overview in Chapter 4, based on assumptions that habitat quality varies between habitat types). In essence, researchers conclude that high abundance in a particular habitat means that salmonids are adapted to it. Conversely,

low abundance suggests that the habitat is substandard, reducing salmonid survival. The general conceptual scheme follows that of the theory of the "habitat template" set out by Southwood (1977): "The favorableness of the habitat will depend on the level of resources ... the number of natural enemies ... and the density of the organism" (Southwood 1977, 344). Below, I give a brief overview of some of the statistical methods and models used to make these inferences with salmonids in estuaries. Usually several habitat or water property metrics are included as variables and the statistical methods are used to infer the most important variable. To paraphrase from numerous papers, often the stated aim is to relate these catches of salmonids to various biological and physical environmental variables.

Spatial and temporal differences in abundance and distribution are usually based on parametric correlative statistical analyses, which do not prove cause and effect but provide strong inference when they are significant. In the simplest case, the null hypothesis of no difference in abundance between two habitat types is tested with an ANOVA. More sophisticated multivariate approaches are frequently used to identify patterns of habitat use, as they enable links to a variety of habitat or water property measurements. Numerous texts and web-based resources are available for general guidance on ecological statistical analyses. Examples include those of R.H. Green (1979) and C.A. Rice and colleagues (2005), and the Plymouth Routines in Multivariate Ecological Research (K.R. Clarke and Gorley 2006). J.S. Macdonald, Birtwell, and Kruzynski (1987) provide an example of multivariate discriminant analysis with juvenile Chinook and coho salmon and water property variables at the Campbell River estuary in British Columbia. They developed a data matrix of physical variables (temperature, dissolved oxygen, water depth, light, water velocity, and salinity) from various depths in the water column, with coincident estimates of fish counts (using scuba) and potential prey abundance at each depth. The discriminant analysis was used to identify the most important variables correlated with juvenile Chinook and coho salmon abundance. A study in upper San Francisco Bay, California (including the Sacramento–San Joaquin River estuary), related fyke net catches of estuarine fish communities, including juvenile Chinook salmon, to salinity, temperature, and channel morphology as well as biological features such as residency behaviour, whether the fish species was invasive or not, and whether the species was demersal or pelagic (Gewant and Bollens 2012). Multivariate analysis was used, with each fyke net catch comprising a row in the data matrix. Thus, numerous variables can be related to salmonid catches in estuaries to habitat variable and water property metrics with these multivariate statistical techniques.

Alternative methods outside the realm of conventional statistics include nonparametric and nonlinear approaches. Distributional data on species can generate presence or absence data, which can be useful in broad-scale studies. An example of this approach was a study comparing the predictive power and explanatory insight provided by traditional, linear approaches (e.g., logistic regression analysis and linear discriminant analysis) and alternative, nonlinear approaches (e.g., classification trees and artificial neural networks) for modelling presence/absence of fish species in Canadian north-temperate lakes (Olden and Jackson 2002). To my knowledge, nonlinear approaches have not been used to investigate salmonid habitat relationships in estuaries, but the method could be important for analyses of coastwide distribution of salmonids over a range of estuaries.

Models forecasting abundance, survival, and species presence/absence of salmonids from physical and biological data are potentially significant tools for investigating habitat and survival questions concerning salmonids in estuaries and other habitats (E.E. Knudsen and Michael 2009). Although models have been used extensively in fresh water (e.g., Fukushima et al. 2011) and marine studies (e.g., Keefer et al. 2008), they have been used relatively infrequently to explore salmonid ecology and survival in estuaries.

For example, the environmental predictors of return rates of ocean-type Chinook salmon in the Skagit River basin, Washington, were investigated by Greene and colleagues (2005) using a multiple regression model. Factors considered were number of spawners in a particular year, number of adults that produced them, fecundity, freshwater survival, middle-estuary survival, outer-estuary survival, and ocean survival, together with a density-dependent adjustment. As with all models, a number of assumptions were required and estimated survival rates had to be used, but the results were instructive to estuarine ecologists in that the best predictors of return rate included the magnitude of floods experienced during incubation, a principal components factor describing environmental conditions during outer-estuary residency, a similar factor describing conditions experienced during the third ocean year, and an estimate of egg production. Another regression model, developed by Magnusson and Hilborn (2003), tested the relationship between hatchery fall Chinook and coho salmon survival rates and three habitat metrics (estuarine area, fraction of estuary in natural condition, and presence of oyster aquaculture) in twenty estuaries in Washington, Oregon, and

California. A significant positive relationship was found for fall Chinook salmon but not for coho salmon, probably reflecting the higher degree of estuarine dependency of Chinook salmon. Spatially explicit modelling was used by Fukushima and colleagues (2011) to investigate the historical distribution of Sakhalin taimen in Japan. Fourteen variables were included in the model, including the presence of estuarine lagoons, and the method was found to be a reasonably good predictor of distribution of this endangered species. Interestingly, none of the models to help forecast abundance and survival actually used mesoscale habitat data; instead, they employed large-scale data, such as proportion of the whole estuary disrupted. This probably reflects the difficulty in assigning survival values to specific estuarine habitats used by Sakhalin taimen and the uncertainty about residency.

Another quantitative effort to statistically link salmonid abundance in estuaries to habitat is the Habitat Suitability Index (HSI) method, originally developed and widely used in stream habitats. A study testing the method with salmonids in an estuary was conducted with hatchery-reared Atlantic salmon smolts in the Kennebec River estuary in Maine (S.K. Brown et al. 2000). Suitability Indices (SIs) ranging from 0 to 1.0 were assigned by expert panels to various species' use of areas characterized by particular salinities, temperatures, substrates, and depths. A value of 1.0 equates to high usage in the field and high survival in the laboratory; 0 equates to no occurrence in the field and high mortality in the laboratory. The equation

$$HSI = (SI_{salinity} \cdot SI_{temperature} \cdot SI_{substrate} \cdot SI_{depth})^{1/4}$$

was used to calculate a set of HSI values that could be mapped using a geographic information system (GIS). Although the system worked well in the Kennebec River estuary for sedentary invertebrate species with relatively simple life histories, such as the softshell clam (*Mya arenaria*), S.K. Brown and colleagues (2000) concluded that Atlantic salmon in estuaries have very complex patterns of habitat use that apparently could not be accounted for using the four variables they considered. In addition, data on Atlantic salmon smolt usage in the estuary were scarce.

An additional potential method of linking salmonids to habitats involves the use of probability density functions or kernel estimator methodology, following J.A. Rice (1993), who used the technique for Atlantic salmon in a stream. Kernel estimators, a type of nonparametric probability density estimation technique, can use available data to estimate the probability density function of abundance, given specified habitat conditions. This method showed promise for identifying the preferred salinity for juvenile Chinook salmon at the Campbell River estuary in British Columbia (Levings and Bouillon 1997). Results showed that while there is a 20 percent probability of beach seine catch of 15 per unit effort at 0.5 psu, the probability of this catch per unit effort at 5 psu is 10 percent. The pattern could reflect high abundance in the intermediate-salinity zones of the estuary. This method can be applied to other habitat characteristics.

Conclusion

Because of the complexities of the estuarine ecosystem and the salmonid life cycle, it is likely that inference, supported by the increasingly sophisticated statistical methods and models available to the ecologist, will be important for elucidating the connections between salmonid habitat and survival in the estuary. Where time and resources do not permit estimation of survival, comprehensive data collection, and detailed statistical analyses, estuarine habitat managers and conservationists will rely on traditional abundance and distribution sampling, again supported by inference (see Boyd 1989). As part of their evaluation process, there will be continuing reliance on past findings on biological and biophysical estuarine factors. The following chapters will review some of these findings and relate them to key fitness components of salmonid survival in the estuary.

14
Effects of Habitat and Community Change on Fitness Components for Survival in the Disrupted Estuary

This chapter discusses how human-induced changes in salmonid estuaries may have affected the key fitness components of survival of salmonids. The question requires a temporal perspective because of the huge changes that have occurred to salmonids in estuaries in the past few centuries. A consideration of the effects of changes in habitat, water properties, and biotic factors affecting the fitness components of survival can therefore be divided into two eras: (1) the preindustrial era, before the Industrial Rvolution (about 1750), when natural forces were mainly shaping salmonid distribution and abundance, and (2) the postindustrial era, when pollution from factory operations and major port developments in estuaries began to affect habitats and water quality. There is a wealth of oral and sometimes written history from the preindustrial era on salmonid estuarine conditions from First Nations and medieval cultures, but its interpretation is beyond the scope of this book. The following information from the postindustrial era should provide context for later chapters dealing with natural factors determining how estuaries are exploited by salmonids as well as discussions of management strategies and future conservation actions.

There are probably cascading effects arising from physical disruption and water property changes in estuarine ecosystems that affect biotic interaction processes such as predation and competition. Biotic factors such as invasive species also need to be considered, as physically disrupted habitats are more susceptible to invasions (e.g., rivers transformed into reservoirs [Marchetti et al. 2001]). In other situations not connected with disruptions, the estuarine invaders are anadromous salmonids themselves, introduced intentionally into rivers or streams. Competition between wild and domesticated salmonids is another concern in estuaries in the postindustrial era. The use of estuaries by hatchery-reared salmonids around the world (e.g., Pacific salmonids [Rand et al. 2012]) and possibly by escaped farmed salmonids (Morris et al. 2008) may be an important biotic factor. There is also a debate as to whether biotic interactions exert a dominant role in governing species distributions at macroecological scales, such as those important for climate change (Araújo and Luoto 2007). Thus, it is important to provide perspective on biotic interactions, as they relate to salmonid community change, at several scales.

To provide a framework for a comprehensive review, I discuss the anthropogenic effects using the key fitness components of survival (osmoregulatory adjustment, growth and food, and biotic interactions) described in Chapters 8 to 11.

Compromises to Successful Completion of Osmoregulatory Adjustment in the Estuary

It has been known for some time that salmonids are under stress during smoltification (Fagerlund, McBride, and Williams 1995), so it is germane to describe how additional stress related to habitat change affects osmoregulatory adjustment. The ultimate location of final osmoregulatory adjustment is probably at the seaward end of the estuary, where low-salinity water mixes with ocean water and salmonids enter or exit the ocean. For this reason, the discussions here relate primarily to channel habitat, and focus on changes in water quality.

Salinity Change as a Stressor

Hormonal changes caused by salinity change effects on osmoregulatory adjustment can relate to stress and can affect survival (e.g., McCormick et al. 2009; Figure 24). As a stress-related hormone, cortisol facilitates reallocation of energy away from growth towards activities directed at survival (Wendelaar Bonga 2011) (e.g., predator avoidance), while also promoting seawater acclimation through special salt-secreting cells in the gills and other biochemical changes (McCormick 2013). Increased cortisol levels can therefore indicate stress. As an example, an experimental release to test the importance of estuarine residency to Chinook salmon smolt survival required a thirty-minute helicopter trip from a hatchery. A subsample of one release group was held in cages at a middle-estuary site (salinity 10 psu) and another at a coastal site (salinity 33 psu). All fish had successfully met a seawater challenge test. There was a transitory increase in stress resulting from transport that interacted with location where the fish were held, even though transport time was equal between the two release sites. Blood tests showed that the

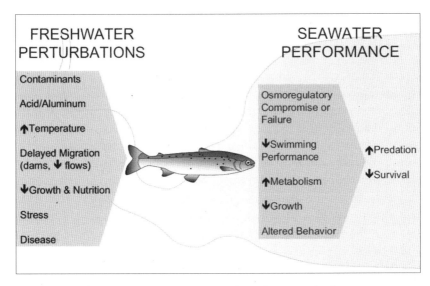

Figure 24 Exposure to poor environmental conditions in fresh water may affect seawater performance of smolts, with effects on survival and predation in the estuary modulated through osmoregulatory compromise or failure. Decreased swimming performance, increased metabolism, decreased growth, and altered behaviour can result.
Source: Reproduced from McCormick et al. 2009, with permission.

coastal-held smolts had high cortisol levels and their kidney cells were characterized by larger nuclear diameters than smolts held at the estuarine site. Mortality rates were highest at the coastal site, perhaps because these fish did not have an opportunity to complete smoltification in the estuary (J.S. Macdonald et al. 1988). It is difficult, however, to extrapolate these types of cage experiment results to the actual estuary, and so it is necessary to consider migrations in the field.

The role of osmotic shock as a salmonid stressor in the modified ecosystems of modern-day estuaries may be underestimated. Increased temperature and river discharge are thought to be two key drivers associated with smolts leaving fresh water and entering the estuary (e.g., Atlantic salmon smolts [Jonsson and Ruud-Hansen 1985]). The migration may be very finely tuned with river conditions (especially flow), and if there has been alteration by dams or other impairments, the recruitment of smolts to the estuary could be impaired (e.g., Whalen, Parrish, and McCormick 1999). On the other hand, if fish that have not completed the smoltification process leave the estuary prematurely (e.g., as fry or parr or pre-smolts from a life history type adapted to rearing in an estuary) because of flushing from reservoir storage water spills, they may be exposed to direct salinity toxicity, exacerbating the stress response. Temperature conditions in the river may also contribute to the problem (see review in A. Richter and Kolmes 2005). Chinook salmon smolts from the Sacramento–San Joaquin River estuary in California reared at 21–24°C (above expected temperatures in this river) experienced impaired smoltification indices (Marine and Cech 2004). These fish were clearly not ready for estuarine conditions and their osmoregulatory ability would likely be compromised if they were flushed to the estuary by water spilled from a dam. As expected, there are species differences in ability to cope with habitat changes. Takami (1998) found that river temperatures higher than 16°C reduced the osmoregulatory ability of whitespotted char smolts, but temperature lower than 4°C did not. Hypersalinity is another water property factor that may compromise osmoregulation in the estuary (Largier, Hollibaugh, and Smith 1997).

As more data become available on the details of osmoregulation ecophysiology in estuaries, it may be possible to rate the various species and their juvenile life history types according to the ability of the fry, parr, and smolts to resist osmotic shock – a fitness factor affecting survival. A "seawater hazard" rating could be developed within and between taxa, following McCormick and Naiman (1984) and McCormick (1994). The former authors arrayed sea-run brook trout according to their length-related osmoregulation, concluding that 18 cm was the minimum size at which this species could withstand "full-strength" seawater. However, their findings have been supplanted by results from a specialized form found by Curry and colleagues (2010) (see Appendix Table 1.6 at http://hdl.handle.net/2429/57062), which showed that sea-run brook trout shorter than 18 cm can survive and grow in estuarine water (26 psu). An osmotic shock rating according to size or growth rate might help identify populations that are more susceptible to the effects of salinity poisoning.

Water Quality as a Stressor on Smolting Salmonids
The effects of deleterious water quality and contaminants as an additional stress on smolting juvenile salmonids in the estuary were recognized in the early eras of research on smolting. This is now a field unto itself, and this section provides an overview with some key references.

Change in major water properties can have direct effects on smolting salmonids. Dissolved oxygen levels can be low in eutrophic estuaries, with deleterious effects on smolting salmonids (Lotze et al. 2006). Also, high water temperatures can accelerate the physiological processes involved in osmoregulatory adjustment, to the point that the fish may die. The upper lethal temperature for juvenile salmonids in fresh water ranges from approximately 24°C for chum salmon fry to 26°C for Atlantic salmon smolts (Rounsefell 1958). Most smolting salmonids suffer sublethal effects if they are subjected to temperatures above about 20°C, but it is difficult to generalize for estuarine habitats due to the strong interaction with salinity and oxygen as well as local adaptation. Salmonids in some estuaries are adapted to warmer temperatures, such as Chinook salmon smolts at the Sacramento River estuary in California, where field data showed the upper lethal temperature to be about 23°C (Baker, Ligon, and Speed 1995). Sea trout in the lower Danube River in Romania may experience temperature up to 28°C if they are present in summer (Sandu and Oprea 2013). Physiological responses to heat stress at the cellular level are now known to be under genetic control (Prunet et al. 2008), but acclimation effects indicate partial phenotypic responses.

As our knowledge of the effects of water quality on salmonid osmoregulation grows, the data are reinforcing the idea that contaminant levels need to be very carefully controlled in the salmonid estuary. Reviews by Wedemeyer, Saunders, and Clarke (1980) and McCormick and colleagues (2009) cover many acute physiological problems with contaminants and smolting, and discuss the numerous smolting-related experiments showing the interaction between exposure to contaminants in fresh water and the estuary. An early experiment showed that coho salmon smolts exposed to copper in fresh water began to show severe mortality when they were moved into salt water (30 psu) (Lorz and McPherson 1976). Sand and wood particulate matter that could be stirred into the water column from dredging or log storage in estuaries has been shown to directly affect gill functions and respiration in juvenile sockeye salmon (Werner and Robinson 1978). Damage to chloride-excreting cells on the gills could occur. A number of field experiments in estuaries have documented the direct mortality of contaminant mixtures on fry or smolts in estuaries (e.g., chum salmon fry and pulp mill effluent [Birtwell and Harbo 1980]; Chinook salmon smolts and acid mine drainage [Barry et al. 2000]). The possibility that mortality was related to osmoregulatory failure was not investigated in the latter studies, but was suspected. Many water quality monitoring programs are established to test salmonid smolts in freshwater/pollutant mixtures and controls. Assessments such as death of fish in a seawater challenge test or changes in blood ion content indicate that the salmonids' osmoregulatory mechanisms are impaired (Blackburn and Clarke 1987), and are often used as a measure of effects (e.g., of pH and aluminum in streams on Atlantic salmon [Kroglund et al. 2007]). The tests are a practical experiment to simulate the estuary.

More recent and sophisticated methods demonstrate the sublethal effect of the modern chemical soup on smolting salmonids. Toxic stressors can act directly by damaging the osmoregulatory processes. In particular, toxic metal ions in the water can inhibit active ion transport and endocrine mechanisms that remove sodium and chloride (Wendelaar Bonga 2011). Contaminants that have been studied with smolts of several salmonid species include environmental estrogens (McCormick et al. 2005), oil, polyaromatic hydrocarbons, pesticides, polychlorinated biphenyls (PCBs), fungicides, metals, and flame retardants (McCormick et al. 2009). Some research has focused on salmonid immune systems and related sublethal tests for the effects of organic contaminants (Palm et al. 2003). Radioactive materials such as cesium from Chernobyl have been recorded in salmonids in lakes (Forseth et al. 1991), and if contaminated fish migrated to estuaries, smoltification could be affected. Ionizing radiation could also affect osmoregulation. Rainbow trout gill cells from untreated fish developed a specific cell type characteristic of irradiated fish and responded as though they had been exposed to radiation ("bystander effect") (Mothersill et al. 2006).

The effect of estuarine residency on contaminant exposure is an important consideration for effects on osmoregulatory adjustment. A. Moore and colleagues (2003) found that seven days of laboratory exposure of Atlantic salmon smolts to a mixture of an estrogenic chemical and a pesticide caused deleterious effects on osmoregulatory performance. The effects were not seen in fish held in cages in a contaminated estuary for five days – which the authors considered to be a realistic estuarine residency time for this species – but they showed that barrages or dams that cause migration delays would be problematic. Ohji, Arai, and Miyazaki (2007) and Ohji, Harino, and Arai (2011) provided field case histories on organotin accumulation and compared the anadromous type of masu and whitespotted char to the nonanadromous type. The former type accumulated more of the contaminant. Given that estuaries are often a sink for pollutants, the studies demonstrated that some life history types of salmonids may be more susceptible than others to contaminant stress in estuaries compared with rivers.

Osmoregulation difficulties in the contaminated estuary are also possible with older life stages. Kelts or sea-run migrants returning to the ocean are also exposed to contaminants in the modern estuary and, probably already under stress, may be in poor condition after overwintering in the river without feeding. The effects of contaminants on kelts or returning sea-run migrants in the estuary, or "pre-smolting" kelts or sea-run migrants in the river, appear to be undocumented. Kelts moving through urbanized estuaries, such as the Sacramento–San Joaquin River estuary, are probably exposed to a variety of contaminants (e.g., steelhead kelts [Null, Niemelä, and Hamelberg 2013]).

Effects on "Reverse Smolting"

While navigating through riverine channels on their upstream migration, it is possible that adult salmonids and returning veterans are stressed during "reverse smolting," or a resumption of salt retention. This osmoregulatory adjustment can begin in the ocean. If estuary conditions are degraded, additional stress may result if the fish encounter physical and chemical barriers. Increased flow and current speeds may impair upstream migration of adults (Levy and Cadenhead 1995). Returning Atlantic salmon were held up in the Ribble River estuary, England, with low dissolved oxygen conditions (Priede et al. 1988). Temperature in the estuary can also have an indirect effect on osmotic stress for returning adults of some species. An interesting example was seen in the older age classes of sea-run brook trout in the upper Saguenay River estuary in Québec, where because of freezing the salinity can increase up to 22 psu. As the temperature decreased, sea-run brook trout lost their osmoregulatory ability and moved into the river as a "sheltering" behaviour (Lenormand, Dodson, and Menard 2004). Similar findings were reported for sea trout in Europe, such as in the Ribe River estuary in Denmark (D.S. Thomsen et al. 2007). Adult salmonids or returning veterans in the estuary are at a crucial stage in their life history as they are beginning their spawning migration, so any compromise to osmoregulatory adjustment as a result of contaminants or habitat changes may reduce their reproductive success.

In summary, three interrelated factors likely affect the osmoregulatory fitness components of survival in many of the world's salmonid estuaries today: (1) timing of migration and capability to remove salt upon reaching the estuary; (2) degree of osmotic shock and behavioural responses in various habitats within the estuary, possibly reflecting the degree of salt water mixing with fresh water; and (3) contaminants as an additional stress element.

Given the changes in salinity regimes in salmonid estuaries from dredging, water diversion projects and dams, estuary-specific or comparative information on smolt behaviour relative to the salt wedge is important for future considerations. Some changes can be acute. Temporary streams that do not flow every year are predicted in regions where rainfall will decrease due to climate change (Moyle 2014), possibly resulting in temporary estuary salt wedges. Estuary morphology can also play a role in salmonid susceptibility to contaminant exposure, and this may be related to the salt wedge. Body burdens of PCBs were higher in deeper industrialized estuaries than in shallow estuaries along the northwest coast of the United States (L.L. Johnson et al. 2007a); the deeper estuaries would be expected to be more saline. Experimental approaches for estimating survival from pollution effects can be instructive. In large-scale exposure experiments, sublethal effects of oil on pink salmon fry released into the Campbell River estuary in British Columbia did not influence fry-to-adult survival (Birtwell et al. 1999). However, pink salmon fry do not reside in riverine channels for more than a few days (Heard 1991), and different results might be found for species such as juvenile Chinook salmon, which reside longer. Most pollutant effects on survival of estuarine salmonids are complex and require modelling and analytical methods such as Bayesian networks for analysis (Lecklin, Ryömä, and Kuikka 2011).

Survival in disrupted estuaries relative to food intake and biotic interactions also need to be considered and are discussed next.

Implications of Estuarine Habitat Loss and Community Change for Salmonid Feeding and Growth

The implications of habitat loss for salmonid food webs and possible reduction in growth and survival in estuaries have been mainly investigated through laboratory studies, inferential or statistical approaches, modelling, and carrying capacity estimates.

Laboratory Studies

Laboratory studies have examined whether prey species differences affect salmonid growth in controlled experiments, where temperature, salinity, and dissolved oxygen mimicked estuarine conditions. One example is the difference in growth in chum salmon fry when fed three different crustacean prey (Volk et al. 1984). Growth was positively correlated with food ration, and food conversion efficiency was much higher for fish fed the harpacticoid copepod *Tigriopus californicus* than for those fed the calanoid copepod *Pseudocalanus minutus* or the gammarid amphipod *Paramoera mohri*. Considering the subhabitats where the prey species are available (eelgrass, riverine channel, and cobble or gravel, respectively), it might be possible to conclude that eelgrass is the priority habitat for conservation or restoration, but extrapolation from the laboratory tank to the field is not always feasible. Larger mesocosms do offer advantages over smaller laboratory tank feeding experiments, but scale effects can be significant even in larger containers. For example, in the Controlled Ecosystem Enclosure Experiment program, it was estimated that containers of over 1,000 t of seawater would be needed to simulate juvenile salmonid feeding and growth (100 fish up to ninety days) in a simulated estuarine ecosystem (Parsons 1982).

Inference of Fish Food Loss from Field Studies of Habitat and Community Changes

As a general rule, increased food rations do equate to better growth within the physiological limits of the fish (Brett 1979), so loss of food species productivity is a concern in the salmonid estuary. Feeding habits suggest possible prey selection, but the hypothesis that salmonids are generalized or opportunistic feeders (within limits; see Table 3) cannot be rejected (e.g., Atlantic salmon post-smolts [Andreassen et al. 2001]). Models to help predict availability may be useful in this regard (Esteban and Marchetti 2004). It has therefore been a challenge to demonstrate for salmonids in the estuarine ecosystem the paradigm that certain food types produced in specific habitats are critical for growth.

Habitat loss

Currently available technology limits our ability to measure how much food a fish has gathered and the resulting accrued growth from contrasting subhabitats, such as sandflats compared with an adjacent brackish marsh. However, the food argument has often been used in inferential or weight-of-evidence arguments regarding salmonid survival in the estuary.

Beaches: Atlantic salmon parr in the Western Arm Brook, Newfoundland, were found feeding on crustaceans living in mud and sand flats. Loss of mud and sand flats and, by implication, loss of associated food organisms were cited as a possible reason why estuary use of Atlantic salmon parr is a relatively rare finding in the region (Cunjak 1992).

Channels: I could not find specific studies dealing with the direct effects of channel habitat loss on feeding. Filling in channels results in decreased availability of food species' secondary production to salmonids. Flow changes from spills could result in loss of drift insects, as they are moved into the estuarine system by river currents (D.D. Williams 1980) and could be flushed to deeper water in major flows. Availability of salt wedge food items such as calanoid copepods is contingent on channels having a connection to the ocean, as found by Busby and Barnhart (1995) in a study of juvenile Chinook salmon at the Mattole River estuary in California.

Vegetation: Loss of vegetated habitat from a variety of industrial developments has very often been linked to food reduction for estuarine salmonids. There is usually a positive relationship between biomass or density of vegetation and salmonid prey that live on or in the vegetation (e.g., the gammarid amphipod *E. confervicolus* and biomass of Lyngbye's sedge, *Carex lyngbyei*, at the Squamish River estuary, British Columbia [Levings 1976]). As but one example, loss of *E. confervicolus* productivity from log storage was linked to salmon survival at this estuary (Stanhope and Levings 1985). Salmon food at the tertiary level may also be lost to development (e.g., sticklebacks associated with vegetation at the Western Arm Brook, Newfoundland [Cunjak 1992]).

Artificial habitats and habitat from invasive ecosystem engineers: Some artificial habitats that provide food species for estuarine salmonids may in turn be destroyed in other industrial developments, but there are no data on this topic in the literature.

Water quality: Pollutants and decreased water quality can reduce the availability of food species for estuarine salmonids by decreasing productivity of prey species. In clear-water estuaries affected by turbidity from industry, the ability of salmonids to obtain prey can be reduced because prey species are less visible or behaviour may be affected. For example, in laboratory experiments, reaction distance of juvenile coho salmon was decreased in turbid water (30 and 60 NTU [L. Berg and Northcote 1985]).

Community change
The effects of community change on food acquisition by estuarine salmonids are tightly connected with competitive effects.

Hatchery-reared fish: After their release from the facility, hatchery-reared salmonids require time (hours to days) to switch to feeding on natural prey (C. Brown and Laland 2001). During this period, they are not likely to affect feeding by wild salmonids in the estuary. After they switch, however, it is possible that the feeding of wild salmonids is affected by their displacement by the large schools of hatchery fish moving through the system.

Invasive species: Interactions between invasive species and estuarine salmonids may result in losses of natural energy flow patterns via interactions at several trophic levels, but here we are concerned about the direct link to salmonids (secondary trophic level). The change in species composition of calanoid copepods, which are potential food for juvenile salmonids in riverine channels at the Columbia River estuary in Washington-Oregon (Dexter et al. 2015), is representative of a food web issue that might be important. It is not known whether the invasive species are as available or as nutritious as the native copepod species. Other invasive taxa are involved in different estuaries. An invasive species of gammarid amphipod, *Pontogammarus robustoides*, is potentially available in large quantities as salmonid food in the Neva River estuary in Russia (Panov et al. 2008).

Modelling

Bioenergetics models may be particularly useful for estimating water quality effects on food requirements, metabolic costs, and growth effects on salmonids in estuaries. Major deviations from normal temperatures – for example, from heated power station effluents – have implications for food demands, and growth and effects can be modelled. Other environmental variables can be explored with bioenergetics models. The metabolic cost of swimming or holding in the Chehalis River estuary in Washington was estimated using a bioenergetics model for coho salmon smolts using the ambient temperature. Estimated mean total metabolic cost incurred by migrating smolts was 0.027 calories $\cdot$ s^{-1} $\cdot$ smolt^{-1} (Moser, Olson, and Quinn 1991). Estimated daily caloric expenditure by smolts in 1988 was twice that of smolts tracked in 1989, due to greater flow-related water velocities in 1988. While there is evidence of salinity interactions of temperature-mediated effects on growth and metabolism of salmonids (e.g., W.C. Clarke, Shelbourn, and Brett 1981), oxygen consumption rates in coho salmon smolts acclimated for six weeks to freshwater, isotonic saline (10 psu), and seawater (28 psu) were not significantly different (J.D. Morgan and Iwama 1998).

Bioenergetics models that account for habitat-mediated food availability as well as temperature have been applied in a few cases. Using P.C. Hanson and colleagues' model (1997), Sommer and colleagues (2001) investigated the feeding success of juvenile Chinook salmon foraging in the Yolo Bypass compared with the natural lower Sacramento River. Observed growth rates and temperature data suggested that feeding success (maximum theoretical food consumption) was greater in the Yolo Bypass than in the river, despite the greater metabolic costs of rearing in the warmer Yolo Bypass region. P.C. Hanson and colleagues' bioenergetics model (1997) was also used to investigate whether there were differences in the growth potential for juvenile Chinook salmon between restored brackish marshes and natural brackish marshes within the Salmon River estuary in Oregon. The potential food availability differed between the marshes, and there was evidence that restored and natural marshes were converging in terms of food supply, although growth potential differences were found between the two marsh types (Gray 2005).

Food web modelling has been integrated into large-scale and complex ecosystem models such as Ecopath, which can simulate changes in productivity at several trophic levels (Christensen and Walters 2004). A complex ecosystem model using Ecopath was proposed for Puget Sound, Washington, by C.J. Harvey, Williams, and Levin (2012) and could possibly be used to explore top-down effects of predation on salmonid growth. Thus, a variety of ecosystem models can be useful in developing testable hypotheses concerning the importance of food on other growth-related metrics, such as carrying capacity.

Carrying Capacity Estimates

Food-related reasons for differences in salmonid growth rates between or in various estuarine habitats can be examined in the context of two closely related concepts: carrying capacity and density-dependent growth. Carrying capacity is an ecological concept that has been in the literature for over a century. The term was originally used in population dynamics studies, where it is well known as a factor that limits population growth in the logistic equation (Kashiwai 1995). However, the term has evolved to functional use by ecological modellers to estimate the food demand by salmonids in coastal feeding areas (e.g., Cooney 1993; Saito et al. 2009) and the ocean (e.g., Batchelder and Kashiwai 2007). I use the following working definition of carrying capacity: "the maximum abundance or biomass of species of concern that can achieve adequate somatic growth needed to support population growth given the accessible quantity and quality of food available through time" (Naiman et al. 2012). This definition enables the consideration of temporal variation in carrying capacity; the metric definitely should not be considered a constant (see Sayre 2008). Density-dependent growth is a well-established phenomenon in many animals, ranging from Protozoa to mammals. In salmonids, density-dependent growth can be related to territorial behaviour as well as carrying capacity limits mediated through food (e.g., brown trout in streams [Jenkins et al. 1999 and many other papers in the freshwater salmonid literature]). Density-dependent growth of salmonids in estuaries is more likely related to food, however, since the estuarine salmonids do not generally show territoriality.

In the following discussion, I give an overview of some examples where salmonid estuarine carrying capacity may have been limited by food supply, as reflected by density or density-dependent growth. For logistic and study design reasons, the more analytical capacity studies have usually been approached on a "whole-estuary" basis, without consideration of specific habitats. I discuss these case histories first.

Estimating carrying capacity on an estuary-wide basis
A few detailed studies have shown that the capacity of an estuary to grow salmonids may be limited by the abundance of a specific food organism. For example, the harpacticoid copepod *Harpacticus uniremis* is a key food item for chum salmon fry at the Nanaimo River estuary in British Columbia, and is found on beaches (mud-sand) and vegetation (eelgrass). A very high proportion of the annual production of *H. uniremis* was taken by chum salmon fry (Healey 1979), and the seasonal pattern of abundance of such fry in the estuary was the same as the seasonal pattern of abundance of this apparently preferred food (Sibert 1979). The chum salmon fry population required up to about 105 kg $\cdot$ d^{-1} of harpacticoid biomass for growth, resulting in annual fry production of 0.2–0.4 g $\cdot$ m^{-2}. Growth rates of the chum salmon fry were similar in two consecutive study years (about 6 percent body weight $\cdot$ d^{-1}) and the fry consumed most of the estimated production of the harpacticoid. Significantly, at peak population densities of the chum salmon fry, there was a drop in average stomach contents of all food species (*H. uniremis*, gammarid amphipods, shrimp larvae, insects, and other invertebrates) in both of the study years, even though food species other than *H. uniremis* were more abundant in one of the study years. This suggested that alternative food resources were not available to the chum salmon fry, and that *H. uniremis* abundance could limit their productivity.

A study with chum salmon fry in a different estuary also showed some evidence for carrying capacity limitations, at least compared with the Nanaimo River estuary. At Netarts Bay, Oregon (estuary of Whiskey Creek and a number of other small creeks; riverine channels and sandflats or mudflats), Pearcy and colleagues (1989) calculated that annual chum salmon fry production was 0.01–0.03 g $\cdot$ m^{-2}, lower than production at the Nanaimo River estuary, even though Netarts Bay is about 3.5 km^2 larger. Growth rates of chum salmon fry at the Netarts Bay estuary were about the same among years, even though there were several-fold differences in numbers of chum salmon fry released from the hatchery and estimated total biomass of juvenile chum salmon in the estuary. The reduced growth rate of chum salmon fry at Netarts Bay (1.6–2.3 percent body weight • d^{-1}) relative to chum salmon fry at the Nanaimo River estuary may have been due to the short residence times of large hatchery fish released late in the spring, as well as environmental factors (e.g., temperature) rather than direct competition for food (Pearcy et al. 1989).

The energetic cost of obtaining different food items in the two estuaries was raised as a hypothesis by Pearcy and colleagues (1989), who suggested that gammarid amphipods, the main chum salmon fry food at Netarts Bay, required more energy to obtain than the harpactiod copepod *H. uniremis* at the Nanaimo River estuary, resulting in reduced growth and production. However, chum salmon fry and other juvenile salmonids consume gammarid amphipods in numerous estuaries. It is not known whether consumption of gammarid amphipods restricts growth or whether their nutritional factors (e.g., specific fatty acids) are limiting. As well, a variety of sampling methods are used to sample harpacticoid copepods and gammarid amphipods to assess their availability to salmonids in estuaries. Sampling protocols range from coring and quadrat sampling in the beach at low tide (e.g., Levings 1976; Spilseth and Simenstad 2011) to plankton pumping at high tide (Sibert 1981). Therefore, an unresolved issue for studies of food relationships in relation to carrying capacity remains: which methods measure actual food availability for salmonids in estuaries?

The size of the estuary, inclusive of the various habitats within it that can set its food potential, may in turn control the density or the number of juvenile salmonids it can support for growth. Estuarine salmonid density may therefore be a surrogate for carrying capacity, but density data are a function of sampling methods if the data are not calibrated for effectiveness. Estimates of juvenile salmonid densities in estuaries vary widely, but in some instances the data appear to show a relatively narrow range when compiled for certain species and areas. Juvenile Chinook salmon densities assessed with uncalibrated beach seining showed a range of 0.1–1.0 fish • m^{-2} in eleven British Columbia estuaries (Levings 1984). Mean abundance of sea trout was estimated at 0.0007 per square metre for thirty-one estuaries in Europe (Portugal to Scotland), but this was likely underestimated with the bottom beam trawls used to sample (Nicolas et al. 2010). The species was found at only two estuaries.

A few studies have investigated juvenile salmonid carrying capacity relative to estuary size. Data reported by Herring and Nicholas (1983 draft, cited in Wissmar and Simenstad 1998) showed a negative trend between juvenile Chinook salmon size and areas of eleven Oregon estuaries (25 mm over an area ranging from 10 to 3,500 ha). A similar trend was shown for sea trout in five Irish estuaries, where length approximately doubled when estuary length increased by a factor of 10 (Fahy 1981).

Estimating carrying capacity in specific habitats
There have been a few studies where carrying capacity has been examined in estuarine habitats using a more general inferential approach.

Beaches: Working on the sandflats at the Sixes River estuary in Oregon, Reimers (1973) analyzed growth of juvenile Chinook salmon using tagged fish, length data, and scale patterns. His seminal paper suggested that reduced food supply resulted in growth limitation of juvenile Chinook salmon in this small estuary. Reimers (1973) proposed that the abundance of the amphipod *Corophium salmonis,* which lives on sand and mud flats, may have limited the capacity of the estuary to support estuary-rearing Chinook salmon. Subsequent research in the Sixes River estuary, using otolith microstructure analysis and data on amphipod biomass, supported the hypothesis, although increased temperatures in summer could not be ruled out as a factor (Neilson, Geen, and Bottom 1985). The growth rate of tagged juvenile steelhead was assessed at the Scott Creek lagoon-sandflat estuary in California (S.A. Hayes et al. 2008). Growth rate varied inversely with steelhead density when the lagoon was enclosed by a seasonal sandbar.

Channels: There have been few studies of estuarine carrying capacity focusing specifically on salmonid channel habitat. Maximum carrying capacity for juvenile Chinook salmon in side channels at the Skagit River estuary in Washington was estimated at about 1.4 fish • m^{-2}, based on the observation that catch per unit effort reached an asymptote as fry moved into the side channels from riverine channels (Beamer et al. 2005). Coho salmon fry in channels at the Carnation Creek estuary in British Columbia were about 40 mm when they migrated to the estuary in April, and had grown to approximately 80 mm by September. Growth rates in 1979 and 1980 were similar even though the fry population in the estuary in 1980 was double that in 1979, suggesting that resources were not limiting in the second year (Tschaplinski 1982).

Vegetation: The literature contains no data on salmonid carrying capacity estimates in estuarine vegetation.

Artificial habitats and habitat from invasive ecosystem engineers: The literature contains no data on salmonid carrying capacity estimates in artificial habitats and habitat from invasive ecological engineers.

Effects of water quality on carrying capacity in salmonid estuaries: Water quality that affects prey species for salmonids in estuaries could affect growth and carrying capacity by reducing the productivity in the food web. For example, acid mine drainage was toxic to gammarid amphipods and chironomid larvae at the polluted Britannia Creek estuary in British Columbia, and fewer of these salmonid prey species were found in the affected area than in a reference site (Levings et al. 2004). Laboratory or mesocosm culture studies can be illustrative. An example was the reduction in carrying capacity of the harpacticoid copepod *Tisbe battagliai* after exposure to pentachlorophenol (Sibly, Williams, and Jones 2000), a toxic chemical used as a wood preservative in salmonid estuaries. More macroscale water quality changes are also possible. For example, ocean acidification is affecting major estuaries such as Puget Sound, Washington, and is a concern for calcareous organisms such as bivalves (Feely et al. 2010), whose larvae are sometimes eaten by salmonids in estuaries (e.g., McCabe et al. 1983).

Effects of invasive species on carrying capacity in salmonid estuaries
Changes in food web structure due to invasive species at several trophic levels will likely change carrying capacity. The overbite clam, *Potamocorbula amurensis*, an invasive ecosystem-altering invertebrate at the secondary food web level, decimated the zooplankton community in the Sacramento–San Joaquin estuary, shifting the food web from pelagic production towards a benthic base and forcing nonsalmonid fish to find alternative prey (Feyrer et al. 2003). The reduced food supply for salmonids implied a change in carrying capacity. Similar concerns have been raised in the Neva River estuary in Russia, where the zebra mussel (*Dreissena polymorpha*), another filter feeder well known for its ecosystem effects, is an invasive species (Panov et al. 2008).

In summary, approaches to the study of food as a limiting factor range from laboratory studies to modelling for estuarine salmonids, and each method has merits and limitations (Table 6). Comparisons of growth and feeding can be done by inter-estuary comparisons, which can be an important strategy for studying limiting factors for estuarine salmonid survival. Although long recommended as an experimental design strategy (e.g., Merrell and Koski 1978), comparison between estuaries within a region is an underutilized approach. In the future, technological developments or innovative approaches may enable experimentation to test

Table 6

Approaches for assessing the effects of habitat-related food deficits on estuarine salmonid growth

Approach	Advantages/limitations
Laboratory studies	Factors can be controlled, but space for interaction could be severely affecting results.
Inferences of fish food loss from field studies of specific habitat types	May have a solid foundation in adaptive evolutionary theory, but general factors affecting food species selection need further exploration, although abundance and prey size are widely recognized.
Bioenergetics modelling	Close to ecophysiological reality but requires major data sets for calibration.
Carrying capacity	Integrates population and bioenergetics data but requires major data sets for development of energy flow; surrogates can be used.

food limitation in salmonid estuaries. Bioenergetics modelling of the growth potential of various estuarine habitats using the model of P.C. Hanson and colleagues (1997) or alternatives such as those used to model net energy intake (e.g., Urabe et al. 2010) show promise, but will require new data sets to calibrate them as well as information for undisrupted estuaries to verify model output.

Most of the detailed ecosystem studies that investigated food as a limiting factor were conducted at least twenty years ago in northeast Pacific Ocean estuaries. Estuarine carrying capacity for salmonids has been investigated in limited detail elsewhere. Comparable studies have not been conducted in several regions, nor has the research been updated or extended for Pacific salmonids. Effects at the basal trophic level could involve major shifts in carrying capacity of entire estuaries. For example, climate-driven increases in terrestrial primary production are expected to increase primary production in lakes and ultimately reduce the prevalence of anadromy in Arctic char populations, which will reduce this species' use of the estuary (A.G. Finstad and Hein 2012). As with survival data, inference from rather dated and localized studies therefore remains the main method for determining the importance of food for survival. Our lack of knowledge of growth and trophic relationships has implications for studies of biotic interactions such as predation and competition, which will be discussed next.

Effects of Estuarine Habitat Loss and Community Change on Biotic Interactions

Predation and Habitat Loss

A number of studies have directly or indirectly investigated how habitat loss or disruption has affected predation in salmonid estuaries.

Beaches

There are no specific studies that give data on predation rates in modified mudflats, sandflats, or other beach subhabitats. If beaches are lost by filling to create land for industrial activities, juvenile salmonids adapted to rear in the shallow water might be forced to use deeper water outside the estuary, where large potential fish predators are present, such as saithe preying on Atlantic salmon post-smolts (Jepsen, Holthe, and Økland 2006) and North Pacific hake, *Merluccius productus,* on sockeye salmon smolts (C.C. Wood et al. 1993).

Channels

Changes in the hydrodynamics of channels, such as reduced flow, may decrease the water volume available. Adult and juvenile salmonids can be stranded and more susceptible to bird or mammalian predators if water flow drops suddenly due to reservoir operations upstream (Nagrodski et al. 2012). Reduced flow can also decrease transit time through the estuary, possibly making juvenile salmonids adapted to moving quickly through the estuary more susceptible to predators. At the Orkla River estuary in Norway, the mortality rate of Atlantic salmon smolts was 20 percent • km^{-1}, with Atlantic cod and saithe implicated as major predators (Hvidsten and Lund 1988). If river flow is reduced, salmonids may reside longer in estuaries, increasing the predation threat (Kocik et al. 2009). When the Atlantic salmon smolt migration and mortality rate data of Thorstad and colleagues (2012, Table 2) are combined with similar data from the Penobscot River estuary in Maine (Renkawitz, Sheehan, and Goulette 2012), there was a significant negative correlation between distance in the estuary over which mortality was recorded and mortality ($n = 11$, $r = 0.55$, $p < 0.05$; range 0.6–36 percent • km^{-1}). Therefore, it is highly likely that slowing the migration speed increased mortality, although the interaction with increased predation due to other physical changes, such as channel deepening, also needs to be considered. Flow barriers may also be deleterious. Pontoons on floating bridges can result in increased predation and reduced survival, as shown with migrating steelhead smolts at Hood Canal, Washington

(M. Moore, Berejikian, and Tezak 2013). On the other hand, for different estuaries, increased flow that moves smolts to the coast more quickly by increasing flushing rate may put them at risk in the ocean. At the Nehalem River and Alsea River estuaries in Oregon, steelhead smolt survival was inversely correlated with river discharge (Romer et al. 2013). A further example might be the Columbia River estuary, Washington-Oregon, where Chinook salmon smolts showed greater survival probability in the lower river and estuary (0.69–1.00) than in the coastal ocean (0.04–0.29) (Rechisky et al. 2012). The coastal habitat off the mouth of the Columbia River is known as an area of high potential predation by a variety of marine fish, including North Pacific hake and mackerel (*Trachurus symmetricus*) (Emmett, Krutzikowsky, and Bentley 2006).

Change in salinity gradients, temperatures, and deepening and narrowing of the riverine channel may also affect the susceptibility of estuarine salmonids to predation. Smolts and kelts moving through the disrupted riverine channel are already stressed from osmoregulatory adjustment, so additional or extended salinity-related stress, such as changes in the vertical distribution (e.g., increased salt wedge penetration resulting from reduced flows [Jassby et al. 1995]) is a potential concern. Horizontal changes in surface channel salinity distribution also have the potential to increase predation. An example of the latter might be the potential spread of pike further into the Baltic Sea from the Bothnian Bay, where the piscivorous pike can tolerate salinities of up to 6–7 psu (Raat 1988). In the same region, colder temperatures delayed the migration of Atlantic salmon smolts through the Simojoki River estuary in Finland, which may have exposed them to more predation by pike relative to warm years (Jutila, Jokikokko, and Ikonen 2009). There are likely more potential marine fish predators such as Atlantic cod found in deeper water, outside the estuary (Hvidsten and Lund 1988; Hedger et al. 2011). Deepening and narrowing of riverine channels, which enables predators to more fully exploit the riverine channels via the salt wedge and reduces prey dispersal, are possible serious concerns. Some of the modified estuaries where heavy Atlantic cod predation on Atlantic salmon smolts has been observed are examples (Orkla River estuary; shown in Appendix 5 at http://hdl.handle.net/2429/57062). An analogy from two contrasting systems in British Columbia is instructive. Spiny dogfish (*Squalus acanthias*) are predators on hatchery-reared salmonids in deep water (>10 m) off the riverine channels of the Big Qualicum River estuary (Beamish, Thomson, and McFarlane 1992). At the shallow (<5 m) riverine channels at the Squamish River estuary, spiny dogfish were caught only in low abundance (Levy and Levings 1978).

Thus concentration of predators through the deepening of river channels, combined with reduced dispersal of the salmonid prey due to habitat loss in channels, can be deleterious. However, concentration of prey can also yield negative trade-offs for predators (e.g., predator swamping (Ims 1990) or reduced success of predator attack (Neill and Cullen 1974).

An example of the interacting effects of harbour maintenance dredging on riverine channels and predation can be found in the Columbia River estuary, where sand islands were created from dredged material and were subsequently colonized by large numbers of predatory nesting Caspian terns (*Sterna caspius*) and other predatory birds (Good et al. 2007). A somewhat similar situation was observed at the Skjern River estuary in Denmark, where cormorants preyed on sea trout smolts and Atlantic salmon smolts at an artificial lake developed in a restoration project (Koed, Baktoft, and Bak 2006). Bioenergetics models to predict consumption of salmonids by predatory birds have been used (Roby et al. 2003). The models estimated that Caspian tern consumed between 9 and 16 million juvenile salmonids in the Columbia River estuary in one year. Shipping channels that direct flow and salmonids through specific channels close to the sand islands were suggested as a cause for lower survival of Chinook and coho salmon smolts within the Columbia River estuary (Harnish et al. 2012). Another example of the likely effect of harbour development is seal predation on Chinook salmon smolts at the Puntledge River estuary in British Columbia (Yurk and Trites 2000). The upper part of the estuary is very constrained by diking as well as urban/industrial development.

Changes in modified riverine channels can also affect predation on adult salmonids. Sand islands can create haul-out sites for mammals, increasing the predators' habitat. The directed upstream migration rate and pattern can be affected by river training structures, possibly increasing the potential for additional predation relative to the natural estuary. Sea trout at the Rhine River estuary in the Netherlands have fewer entry routes now than in the past, and high velocities at the channels leading in and out of the diking system complicate their upstream migrations (bij de Vaate et al. 2003). At the Columbia River estuary, cutthroat trout (likely returning veterans) migrating along shorelines moved out into deeper channel water where structures intersected river flow (Zydlewski et al. 2008), possibly increasing their predation risk from larger fish.

Vegetation
I could find no specific examples of altered predation on salmonids where vegetation has been modified. Direct loss of possible refuge habitat

and displacement of juvenile salmonids into deeper water would be a primary concern. The height and density of vegetation may be a factor influencing cover-seeking behaviour of estuarine salmonids. Vegetation height can range from a few centimetres (rockweed) to a metre (brackish marsh) (Table 2) and stem density in brackish marshes can be up to a few hundred stems per square metre (R.S. Gregory and Levings 1996). If sediments from dredging operations are spilled into brackish marsh (e.g., Squamish River estuary [Lim and Levings 1973]), stem height would be reduced, but there are no data on the effects of the possible decrease in refuge function.

Artificial habitats and habitat from invasive ecosystem engineers
Overwater structures, such as wharves, have the potential to attract fish predators that might use them as refuge or cover habitat. This has been a concern in Puget Sound, Washington. However, scuba and echo sounder surveys did not record many instances in which large water-column fish that are potential salmonid predators, such as quillback rockfish (*Sebastes maliger*) and copper rockfish (*S. caurinus*), or demersal fish such as lingcod (*Ophiodon elongatus*) or staghorn sculpin (*Leptocottus armatus*), were associated with ferry terminal structures (G.D. Williams et al. 2003). As well, juvenile salmonids may be less available under docks, as fewer were found under structures in the estuary (Munsch et al. 2014).

Water quality
Reduced water quality can affect the vulnerability of juvenile salmonids by changing their swimming performance and other anti-predator behaviour. Depletion of dissolved oxygen has long been recognized as an acute problem. A classic case is a study in the 1930s of the highly polluted and oxygen-deficient Tees River estuary in England: "In the Tees estuary, death of migratory fish is common. [Atlantic] Salmon are often seen floundering near the surface or floating dead, while salmon and sea-trout smolts die in great numbers during their passage to the ocean in spring" (W. Alexander, Southgate, and Bassindale 1936, 720). Obviously, the salmonids are vulnerable to greatly increased predation from birds and mammals in this distressed state. The effects of acute water quality degradation on salmonid mortality from predation can sometimes be detected visually, as with dissolved oxygen in the Tees River estuary, but the sublethal effects of minute concentrations of contaminants in the chemical soup require more sophisticated detection and analytical methods. For example, in a study of subyearling Chinook salmon in the

Columbia River estuary, a life cycle modelling approach to growth and susceptibility showed that estuarine survival decreased 22 percent following a dose of an organophosphate pesticide (Baldwin et al. 2009). Other studies have taken an intermediate approach and examined how contaminants change the behaviour and susceptibility of juvenile estuarine salmonids to predators. The literature on this topic is extensive and I have given only an overview here.

Low or intermediate levels of turbidity might provide juvenile salmonids with protection from predation by reducing their visibility (R.S. Gregory and Levings 1998). Higher concentrations can affect the health of fish by damaging gill functions (Lake and Hinch 1999) and affecting their respiration rate and ability to swim. Effects can be species-specific, depending on the spacing between gill filaments relative to the size of the suspended particle causing the turbidity (Muraoka, Amano, and Miwa 2011). Behaviour can also be affected by turbidity. When juvenile Chinook salmon in tanks were exposed to high levels of turbidity (a slurry of particles, 50,000 NTU), they delayed seeking cover as a refuge from predation (Korstrom and Birtwell 2006). In estuary mesocosm experiments, juvenile Chinook salmon showed stress effects when exposed to a fungicide used in the forest industry, and were more susceptible to predation by yellowtail rockfish (*Sebastes flavidus*) (Kruzynski and Birtwell 1994). The use of carbaryl, a pesticide used to control burrowing shrimp on estuary mudflats, was a threat because the chemical could affect the neural system of cutthroat trout (Labenia et al. 2007). Laboratory experiments showed that predation by lingcod on the carbaryl-exposed cutthroat trout was higher. Bird predation can be enhanced by degraded water quality. Before recent concern over pollution, sewage and pulp mill outfalls were often located in estuaries, and the warm effluent would lead to fish kills with subsequent predation by gulls (e.g., Tyne River estuary, England [Raven and Coulson 2001]).

Changes in light levels have been seen as a concern for predation on salmonids in disrupted salmonid estuaries. Artificial lighting on bridges could cause juvenile sockeye salmon in a river to seek cover at night, avoiding cottid predation but possibly affecting migration behaviour (Tabor, Brown, and Luiting 2004). Attraction to light can be detrimental. Lights on a bridge at the Puntledge River estuary in British Columbia attracted juvenile Chinook salmon and made them more susceptible to seal predation (Yurk and Trites 2000). Sound pollution can affect behaviour and physiology in salmonids. In laboratory tests, juvenile Chinook salmon showed a flight or avoidance response to a low-frequency (10 Hz) sound to which they did not habituate (F.R. Knudsen et al. 1997; Sand et al. 2001). Fleeing behaviour is a well-known stressor (e.g., Johnsson, Höjesjö, and Fleming 2001). Sound levels similar to those generated from underwater noise from seismic testing caused temporary changes in the blood vessels of Atlantic salmon (Sverdrup et al. 1994).

In summary, the effects of habitat change clearly have the potential to enhance predation on salmonids. For channels, deepening may increase the number of potential marine fish predators, and changes in the adjacent shoreline can result in increased bird predation. For beaches and vegetation, refuge is likely the key function lost. Juvenile salmonids may use the shallow water, possibility at the expense of increased vulnerability to avian and mammalian predators. It is therefore difficult to infer low predation pressure on juvenile salmonids from observations that there are few piscivorous fishes in shallow estuarine habitats (Sheaves 2001). The cumulative effects of multiple predators found in the disrupted estuary (aquatic, avian, and mammalian) considered together, but with each exploiting salmonids in different habitats, has not been considered, but could be important (see Carey and Wahl 2010).

Predation and Community Change

Salmonids live with a variety of predatory fish species in the estuary, some of which they have coevolved with, but increasingly they are required to live with predators they have not experienced in evolutionary time. These could be invasive nonsalmonid fish species or, in some estuaries, other salmonids that are potentially invasive predators. As well, the increased number of large hatchery-reared salmonids released into salmonid estuaries needs to be considered as they are potentially more vulnerable to general predation than wild fish. At the same time, if they are piscivorous, the hatchery fish potentially increase the predation pressure on wild fish. Adding to the complexity, hatchery-reared salmonids have the potential to decrease predation by possibly shielding co-migrating wild salmonids as they migrate through the estuary (pied-piper effect [Weber and Fausch 2003]). These are some of the interactions encountered in efforts to understand how estuarine fish community changes affect predation on wild salmonids.

Predation on or by domesticated salmonids

Domesticated salmonids use estuarine habitats in different ways than wild salmonids, and their ecology may make them more susceptible to predation. In estuaries where larger hatchery-reared fish are the most common type of juveniles (often smolts), they dominate riverine channels and generally are not abundant on shallow beaches or vegetated habitat. In the Columbia River estuary, over 90 percent of the Chinook salmon smolts in riverine channels were hatchery-reared fish (Weitkamp, Bentley, and Litz 2012).

Depth and flow in the channels are particularly important for reducing predation on hatchery-reared fish, because the fish are released in batches of thousands, resulting in very high pulsed densities of smolts that are on directed migration through the river and estuary to the ocean. These aggregations of salmonids are vulnerable to natural predators in the river, such as northern pikeminnow (Shively, Poe, and Sauter 1996), especially in modified habitats such as reservoirs. Because hatchery-reared fish are fed from the surface in the hatchery, they are surface-oriented in feeding and may therefore be more vulnerable to bird predators (Pacific Northwest Hatchery Scientific Review Group 2009). In the highly modified channels of the Columbia River estuary, hatchery-reared wild Chinook salmon smolts were more vulnerable to Caspian tern predation than wild Chinook salmon, but hatchery-reared and wild steelhead smolts were preyed upon in similar proportions (Ryan et al. 2003). Under chemically simulated predation risk, farmed masu were more willing to leave cover and feed than wild fish, indicating reduced predator avoidance in the farmed fish (Yamamoto and Reinhardt 2003). Hatchery-reared Atlantic salmon smolts show lower gill Na^+-K^+–ATPase activity and lower growth hormone and plasma chloride levels than wild smolts. These three factors indicate that they experience more stress and hence are more susceptible to predation (Handeland, Arnesen, and Stefansson 2003).

Several studies have investigated the hypothesis that hatchery-reared salmonids are significant predators on smaller salmonids, but none were in an estuary (Pacific Northwest Hatchery Scientific Review Group 2009). Studies of time and size at release involving some Pacific salmonids showed larger smolts have better survival, a strategy that has been adopted widely (e.g., coho salmon [Irvine et al. 2013]). Hatchery-reared salmonids in estuaries are therefore often larger than wild salmonids and are potential predators on wild fry or subyearlings. However, because the hatchery-reared smolts move through the estuary more quickly than rearing fry (e.g., Chinook salmon at the Campbell River estuary [Levings, McAllister, and Chang 1986]), the interaction between smolts and fry may be minimized. This is another hypothesis where context is important, because channel flow rates can determine the length of time the hatchery and wild fish encounter one another. As well, depending on how long the hatchery-reared fish have been living in a natural environment, they may not have begun to feed on natural prey (C. Brown and Laland 2001). The stomach contents of several hundred hatchery-reared Chinook salmon smolts were examined in the Campbell River estuary, but no wild Chinook salmon fry were found (Kask, Brown, and McAllister 1988).

The benign effects of hatchery-reared salmonids are not universal. Hatchery-reared cutthroat trout may be significant predators on several species of salmon fry in the estuary (Duffy and Beauchamp 2008). There have also been instances in rivers where larger hatchery-reared salmonids have fed on smaller wild salmonids (e.g., large steelhead preying on coho fry [Naman and Sharpe 2012]).

Few data are available on the possible predatory effects of escaped farmed salmonids. Adult Atlantic salmon that escaped from aquaculture operations in Chile were not feeding in the fjords – most fish reported by Soto, Jara, and Moreno (2001) showed empty stomachs or feed pellets. Data from the King River and Gordon River estuaries in Australia showed no evidence of predation by escaped adult farmed Atlantic salmon on other salmonids (e.g., sea trout) in the estuary (Abrantes et al. 2011). Schools of escaped farmed salmonids may attract bird and mammalian predators of wild salmonids (Butler and Watt 2003), although this hypothesis remains to be tested.

In general, results show that hatchery-reared juvenile salmonids may be more vulnerable to estuarine predators than wild fish. Their directed and rapid migration through the estuary may reduce the time when they are available to predators, but their surface orientation due to hatchery feeding practices may increase their susceptibility. The evidence to date indicates that the hatchery-reared fish and escaped farmed salmonids are not significant predators on smaller wild salmonids, with some exceptions. It is possible, however, that this conclusion cannot be extended to the other increasingly common changes in the estuarine community, such as greater numbers of invasive predators.

Invasive predatory fish

Predation is often thought of as the greatest threat to salmonids in estuaries disrupted by invasive fish species. Forecasting their impacts requires detailed data on consumption rates, prey population, and predator population. The northern snakehead (*Channa argus*), a predatory fish species from northeast Asia, is an example. The northern snakehead is thought to be a serious threat to native estuarine salmonids where it invades, even if it is present in small numbers. It is tolerant of brackish water and has been found in several rivers and estuaries outside its natural range, including the Razdol'naya River estuary in Russia, where there is concern about the impacts of this species on the endemic fish fauna, including chum salmon fry (Kolpakov, Barabanshchikov, and Chepurnoi 2010). In North America, the species has established a population in the Potomac River, Maryland (Odenkirk and Owens 2007), and could spread downriver to the upper part of the river's estuary where anadromous brook trout are found (Eastern Brook Trout Joint Venture 2014). However, the possible migration of the northern snakehead into the Potomac River estuary is conjecture at this time, and application of risk models may help predict the seriousness of the situation (see Copp et al. 2009). Numerical models have also been used to help forecast the effects of introduced species. Striped bass (*Morone saxatilis*) are an invasive predator on juvenile Chinook salmon in the riverine channels of the Sacramento–San Joaquin River estuary. A population model for Chinook salmon survival that incorporated predation showed that, if striped bass predation could be eliminated completely, the probability of quasi-extinction for the endangered winter Chinook salmon population would decline from 55 percent to 23 percent, and the probability of recovery within fifty years would rise to 14 percent from 3.8 percent (Lindley and Mohr 2003). A variety of tools are available to help understand the impact of invasive nonsalmonid predatory fish on salmonids in estuaries.

In some cases, where an invasive salmonid has been introduced into an ecosystem where a different genus of native salmonids is found, adaptive change in the native species may diminish the impacts of the predatory invader and potentially promote coexistence between the invader and the native species (Strauss, Lau, and Carroll 2006). However, there are examples where the invasive species is clearly a threat. Brown trout are a known predator on masu and sockeye salmon fry in Japanese rivers (Kitano 2004). There is clearly a potential for increased predation risk from brown trout that may use the freshwater tidal zone and not migrate to the ocean (see Figure 13). As well, the sea trout has been known to prey on Atlantic salmon smolts in estuaries (Thurow 1966), and thus is a potential threat to other species of salmonids. The effects on fish other than salmonids could be more significant, especially in the Southern Hemisphere estuaries where the native fish fauna have never been exposed to salmonids – for example, predation by sea trout on the native mote sculpin (*Normanichthys crockeri*) and Patagonian blennie (*Eleginops maclovinus*) (Zama 1987) at the Relconcavi Fjord in Chile. Our understanding of the predation impact of invasive salmonids in Southern Hemisphere estuaries is particularly poor (McDowall, Allibone, and Chadderton 2001).

As with other impacts of changes in biotic interactions due to disruption, predation by invasive fish on salmonids in the estuary has to be looked at in the context of the data. Recent findings from invasion ecology suggest that predatory invaders across ecosystems and biomes have larger detrimental impacts than nonpredatory invaders (M.S. Thomsen et al. 2011 and references therein). The presence of invasive predators also results in the establishment of new and unprecedented links in the estuarine food web. The implications of new predators in these hybrid food webs are unknown, but unique and unexpected food links have been found in some systems (e.g., cannibalism in invasive brook trout in lakes [Biro, Beckmann, and Ridgway 2008]).

Table 7 summarizes the various types of estuarine disruption in terms of effects on salmonid predation. It is not clear in most industrialized estuaries which of the following has been the most deleterious to salmonid survival: predation rate change due to habitat loss, to community alteration, or to both operating together. A significant body of evidence suggests that channel disruption, either by shoreline change, which enhances bird predators, or possibly deepening, which increases proximity to large fish predators, may be important. For beaches and vegetation, direct loss of habitat, and hence refuge function, is probably key to predation rate, but needs to be placed in the context of species and region. In addition, the effects of predation cannot be easily separated from competition when studying the disrupted salmonid estuary.

Competition and Habitat Loss
The relationships between habitat, food, growth, and carrying capacity are important in the context of exploitative competition in the salmonid estuary. Estuarine habitat loss can affect competition for food among salmonids and other fish species in the estuary.

Beaches
An early study at the Nanaimo River estuary, an estuary where log storage affected about 50 percent of the sandflats and mudflats, showed some evidence of food overlap between juvenile salmonids and non-salmonid fishes on the disrupted beaches (Bell and Kallman 1976). Sibert and Kask (1978) found statistically significant diet overlaps with juvenile Chinook salmon and Pacific herring, three-spined stickleback, and shiner perch, but not with juvenile chum and coho salmon, staghorn sculpin, prickly sculpin (*Cottus asper*), or Pacific sand lance (*Ammodytes hexapterus*). Chum salmon fry caught on sandflats or mudflats also showed diet

Table 7

Summary of possible effects of habitat loss or community change on predation and competition

Habitat or disruption category	Representative type of change or disruption	Possible effect on predation	Possible effect on competition
Beaches	Direct loss by infilling for industrial development or other impairment, e.g., port development.	Loss of refuge function, as substrate habitat is typically in shallow water where large predators are not found.	Can lead to "overloading," decrease in living space, and possible food limitation.
Channels	Reduction in the number of channels used in migratory routes.	May lead to concentrations of smolts, which may increase their vulnerability to bird and fish predation.	Can lead to "overloading," decrease in living space, and possible food limitation.
Vegetation	Direct loss by infilling for industrial development or other impairment, e.g., log storage.	Loss of refuge function, as vegetation can provide cover.	Can lead to "overloading," decrease in living space, and possible food limitation.
Artificial habitats and invasive ecosystem engineers	Presence of artificial habitat, indicating that natural habitat has been replaced.	Utilization of deeper water, where larger predators are found, by juvenile salmonids that normally occupy shallow-water estuarine habitat.	When constructed or developed, can lead to decrease in living space and possible food limitation.
Decreased water quality	Decreased dissolved oxygen, increased temperature, presence of lethal or sublethal contaminants.	Greater susceptibility to predation of salmonids stressed by contaminants.	Species that are more tolerant of poor water quality outcompeting salmonids if diets are similar.
Invasive species	Colonization following introduction via accidental or intentional vector.	Significant threat to juvenile salmonid in estuaries posed by predatory invasive fish.	Invasive fish with high growth rates becoming exploitative competitors with salmonids.
Domesticated salmonids	Overloading resulting from hatchery salmonids using channels.	Some evidence of predation on wild fish.	Possible effects on carrying capacity.

overlap with Pacific herring, three-spined stickleback, and shiner perch, but not with staghorn sculpin or prickly sculpin.

Channels
Riverine channels are probably where intra- or interspecific effects of "overloading" in the estuary is most intense. In an undisrupted estuary, feeding salmonids have opportunities to disperse into a multichannelled system. Physical changes that fill in channels, reduce connectivity, or constrict water flow may exacerbate feeding competition. Intraspecific competition may be more intense in riverine channels with dams if the structures impede upstream migration, forcing the salmonids to live in the lower reaches (e.g., whitespotted char in rivers on Hokkaido, Japan [Morita, Yamamoto, and Hoshino 2000]).

Vegetation
In brackish marshes at the Nanaimo River estuary disrupted by log storage, juvenile Chinook salmon diet overlapped with juvenile coho salmon diet but not with the food habits of chum salmon fry, three-spined stickleback, shiner perch, or prickly sculpin (Sibert and Kask 1978), and thus there may have been no competition between juvenile Chinook salmon and the latter fishes. Chum salmon fry diet overlapped with three-spined stickleback diet but not with that of shiner perch or prickly sculpin. In contrast to sandflats or mudflats at this estuary, Pacific herring, staghorn sculpin, and Pacific sand lance were not part of the fish community (Sibert and Kask 1978) and did not feature as competitors. In the restored but not completely recovered brackish marsh and high-salt marshes at the small (1.8 km^2) Salmon River estuary in Oregon, intraspecific competition for food in juvenile Chinook salmon was suggested by the fact that daily ration was directly related to observed stomach fullness, which in turn was weakly but significantly negatively correlated with the density of juvenile Chinook salmon in the tidal marsh (Bieber 2005; Gray 2005).

Artificial habitats and habitat from invasive ecosystem engineers
Aspects of enhanced competition in habitat from invasive ecological engineers are included in the discussion of the effects of community change below.

Water quality
Competition can be increased by degradation in water quality, which can cause changes in species composition. The species balance of the

fish community in the Kyronjoki River estuary in Finland shifted from Atlantic salmon – which is considered a species sensitive to habitat and pH change – to more tolerant and possibly competitive fish, such as whitefish and sea trout, when acid run-off was a problem (Hildén and Rapport 1993).

To summarize, we know very little about how habitat change affects intra- and interspecific competition for salmonids in the estuary. Interactions between habitat change, fish species community composition, and food productivity need to be considered. In the typically modified single channel, competition for food may be exacerbated, but as data are not available this remains speculative. The data on juvenile Chinook salmon feeding on beaches and on vegetation at the disrupted Nanaimo River estuary are illustrative of those two habitats. Juvenile Chinook salmon diet overlapped with three out of eight possible fish species on disrupted beaches, but with only one species out of five possible fish species on disrupted vegetation. Whether this apparent "relaxation" of competition was related to the number of fish species potentially competing for food or to the relatively lower degradation of the vegetated habitat was not determined. The species composition of the salmonid and non-salmonid fish community is also changing rapidly due to intended releases of cultured salmonids as well as to the presence of invasive species, and these changes likely have effects on competition, as I discuss next.

Competition and Community Change
In the estuary, salmonids live with a variety of potentially competitive fish species that they have coevolved with. As in the case of predation, however, they are increasingly required to live with potential competitors that they have not experienced in evolutionary time.

Competition with hatchery-reared salmonids
Numerous studies have identified exploitative competition between cultured and wild salmonids in estuaries as an issue for Pacific salmonids (e.g., Zaporozhets and Zaporozhets 2012; Pacific Northwest Hatchery Scientific Review Group 2009) and for Atlantic salmon in the Baltic Sea (Jonsson and Jonsson 2006). There was some evidence for density-dependent growth at the Campbell River estuary, where large biomasses of hatchery-reared salmonids shared the estuary with wild Chinook salmon fry (Figure 25). Food availability may have been a limiting factor, although data on diet overlap were not available (Korman, Bravender, and Levings 1997). In some cases, the timing of hatchery releases and the size of fish can reduce competition, as shown by earlier studies at

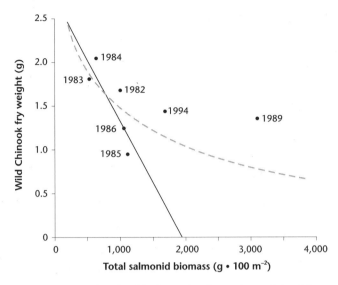

Figure 25 Weight of wild Chinook salmon fry on May 15 of the sampled year at the Campbell River estuary, British Columbia, in relation to total biomass of salmonid juveniles (hatchery-reared and wild) in the estuary. —— line fitted 1982–86; --- line fitted 1982–94.
Note: The data could reflect wild fry growth over the assumed residency period of about sixty days.
Source: From Korman, Bravender, and Levings 1997.

the Yaquina River estuary in Oregon, where batches of coho salmon smolts were released several times during the season and there was little overlap in their use of the estuarine space (Myers and Horton 1982). The possible displacement of smaller wild fish (fry) from habitats where they are normally found (e.g., shallow riverine channel shorelines) has been observed in some systems dominated by hatchery-reared fish (e.g., Chinook salmon in the Columbia River estuary [Weitkamp, Bentley, and Litz 2012]). This may result in a type of intraspecific asymmetric competition characterized by a reduction in "subniche" overlap caused by individuals of the dominant (generally larger) conspecifics. However, the term "asymmetric intraspecific competition" is usually reserved for the phenomenon where a dominant species displaces individuals of the subordinate species from a mutually preferred habitat (K.A. Young 2004). Hatchery-reared fish are often larger than their wild counterparts and thus eat larger prey once they have switched to a natural diet. This relationship may reduce food competition. For example, the smaller Atlantic salmon post-smolts in the northern Baltic Sea ate surface insects, whereas hatchery-reared post-smolts fed on Atlantic herring (Salminen, Erkamo, and Salmi 2001). The situation varies from species to species and may be acute in areas where cultured salmonids greatly outnumber wild salmonids. For example, millions of chum salmon fry are released annually from Japanese hatcheries into rivers on Hokkaido Island (a total of about 1 billion in the year 2000) (Morita et al. 2006; Kaeriyama and Edpalina 2008). Most of the rivers do not have runs of wild chum salmon, but in the systems that do competition with the cultured fish may be exacerbated in the estuary. Some studies with chum salmon fry in Japan suggest that exploitative competition is higher outside of the estuary. A negative relationship between salinity and stomach fullness was found for hatchery chum salmon fry caught in the coastal waters of Hokkaido (Fukuwaka and Suzuki 2000). These findings suggested that food intake was limited as the chum salmon fry migrated into coastal habitats and out of the estuary. In the outer Taku River estuary and Taku Inlet in Alaska, little evidence was found for competition between wild and hatchery-reared chum salmon, even during peak chum salmon densities. Instead, prey categories and communities were partitioned according to size-related foraging behaviours and salmon development (Sturdevant et al. 2012).

To summarize, in some parts of the world, wild salmonid abundance is generally low at present; most estuaries are presumably below carrying capacity for wild salmonids, currently reducing the prospect of competition within the domesticated (hatchery) forms, but this hypothesis needs testing, and urgently. In the northwest Pacific, millions of artificially reared chum and pink salmon are transiting estuaries with unexplored consequences for other salmonids (Ruggerone and Connors 2015). Hatchery-reared Atlantic salmon smolts were dominant in five out of eight salmonid estuaries in Europe and North America listed by Thorstad and colleagues (2012). Thus, in some estuaries, hatchery-reared salmonids now account for most of the salmonid biomass with high food demand, a very different situation from the natural ecosystem. A caveat is that a shift to natural food may not have occurred, depending on how long the hatchery-reared fish have been in the river-estuary system, which in turn is partially dependent on how far upriver the hatchery is located.

Invasive nonsalmonid species
Invasive species, including plants, invertebrates, and nonsalmonid fishes, add a new layer of complexity to the study of exploitative competition in salmonid estuarine ecosystems, and their role in this biotic

interaction is poorly understood (Sanderson, Barnas, and Wargo Rub 2009). An example is the appearance of invasive molluscs in the stomachs of juvenile salmonids in the Columbia River estuary (e.g., the Asian clam, *Corbicula manilensis*, in steelhead smolts [McCabe et al. 1983] and the New Zealand mudsnail, *Potamopyrgus antipodarum*, in subyearling Chinook salmon [Bersine et al. 2008]). Whether this augmented food supply will benefit salmonids by increasing growth, as was found for the invasive gammarid amphipod *Gammarus pulex* and brown trout in an Irish river (Kelly and Dick 2005), has yet to be determined. Is it possible that these represent alternative food supplies for the salmonids, thus alleviating competition for the native food supply?

Loss of native species due to dominance of invasive species may lead to changes in potential competition for food. In the Sacramento–San Joaquin River estuary, juvenile Chinook salmon feed in patches of the invasive ecosystem engineer water hyacinth plant (Toft et al. 2003), where the invasive amphipod *Crangonyx floridanus* was abundant. *C. floridanus* was not eaten by juvenile Chinook salmon but was consumed to some extent by potential competitors such as bluegill (*Lepomis macrochirus*). This was in contrast to the native amphipod (*Hyallela azteca*), which was more abundant in the native pennywort plant (*Hydrocotyle umbellata*). *H. azteca* was relatively heavily consumed by both juvenile Chinook salmon (Index of Relative Importance [IRI] 15 percent) and bluegill (IRI 28 percent) (Toft et al. 2003). Whether or not the increase in consumption of *C. floridanus* was significant enough to alleviate any potential trophic competition cannot be determined without further data, but the example is illustrative of the interactions with invasive species. As well, the Chinook salmon also ate alternative food, such as cladocerans and insects (Toft 2000). Thus, any potential competition for the native food may have been alleviated by the switch to other components of the food web.

Diet overlap has been used as a metric to infer competition with invasive nonsalmonid fish. The yellowfin goby (*Acanthogobius flavimanus*) in the Sacramento–San Joaquin River estuary is an example of an invasive species using disrupted, flow-controlled riverine channels. Food overlap between juvenile Chinook salmon and steelhead and the yellowfin goby was identified as a possible mechanism that could lead to competition (Workman and Merz 2007).

Relative abundance of an invasive species is a possible indicator of potential competition. In restored brackish marsh at the Columbia River estuary, the banded killifish (*Fundulus diaphanus*), an invasive species from the northeast Atlantic, was as abundant as chum and coho salmon fry and Chinook salmon subyearlings (Roegner et al. 2010). Given that a related killifish, California killifish (*F. parvipinnis*), showed a very high production and consumption rate of small crustaceans (Kwak and Zedler 1997), basically the same diet as the salmonids, it is possible the banded killifish at the Columbia River estuary was a competitor for food.

Effects of invasive salmonids on competition
Experiments to establish salmonids in different parts of the world may provide some insight on the effects of invasive salmonids on competition. Harache (1992) concluded that the relative lack of success of Pacific salmonids transplanted to the northwest Atlantic may be due to ocean and estuarine temperatures, but noted that competitive interactions with native nonsalmonid fish species may also be important. More subtle effects, such as competition between native and introduced salmonids, or with species in salmonid ecological guilds, are also possible.

Invading anadromous salmonids initially transplanted into rivers present interesting examples of how possible competition with native salmonids in estuaries may be preventing the invaders' establishment, although the competitive interactions in fresh water and the ocean cannot be excluded. Competition with Pacific salmonids may have prevented Atlantic salmon from establishing in British Columbia, in spite of transplant efforts early in the twentieth century (MacCrimmon and Gots 1979). Competition in fresh water may be reduced now that native salmonid abundance has decreased in the region (McPhail 2007). Some invasive salmonids may use the upper estuary, such as brown trout, but the native salmonids may interfere with their use of habitats as the sea trout form, further seaward. Thus, estuarine exploitative competition may interfere with "leapfrogging" (successive colonization of adjacent estuaries and their rivers) as the sea trout anadromous form, between estuaries.

The successive colonization between adjacent Northern Hemisphere river-estuary systems by invasive salmonids varies without a specific pattern and appears to be somewhat limited, based on the small amount of data available. Most of the invasive salmonids have been transplanted into fresh water (e.g., brown trout [polytype sea trout] from Europe to Newfoundland and Labrador [Westley, Ings, and Fleming 2011]) or to streams in other parts of the same northern continent (e.g., brook trout from Québec, eastern Canada, to British Columbia, western Canada [Karas 2002]). Although brown trout and brook trout are present in coastal rivers in British Columbia, Washington, and California, they have not been found in estuaries in these regions (McPhail 2007; Dr. Casey Rice, Northwest Fisheries Science Center, Seattle, personal communication, October 28, 2010), and apparently have not been able to spread between systems via their anadromous forms. For example, sea trout were found in tributaries of the lower Columbia River, Washington, over thirty years ago (Bisson et al. 1986) but have not been recorded in the thoroughly sampled Columbia River estuary. Estuaries in these regions usually support at least five or six species of native estuarine salmonids. In Hokkaido, where only three species of native estuarine salmonids are found, the sea trout has been discovered in four estuaries, but has been tracked into coastal waters and may be spreading to other estuaries (Honda et al. 2012). Sea trout are found along the south coast of Newfoundland and Labrador (Westley and Fleming 2011; Westley, Ings, and Fleming 2011), where apparently the only native salmonids using estuaries are Atlantic salmon and Arctic char. Sea trout have been purposely introduced into the Manasquan River in New Jersey (Mitchell 2003) but their spread into other systems has not been documented. Sea trout and steelhead have also spread to only a few estuaries in Québec (Dumont et al. 1988; Dumont and Mongeau 1990), where anadromous brook trout, Arctic char, and Atlantic salmon are found. Brook trout have not dispersed between estuaries in the Baltic Sea, where Atlantic salmon and sea trout are endemic (Dr. Kai Korsu, personal communication, February 18, 2011).

The situation seems to be different in the Southern Hemisphere. Invasive salmonids appear to have spread more successfully in Southern Hemisphere estuaries, where exploitative competition within the salmonid ecotrophic guild may not be a factor. However, this is only an hypothesis and mechanisms that enable invasion of organisms into this hemisphere are poorly known in general (see Rodríguez 2001 concerning South America). Sea trout, Chinook salmon, steelhead, and sea-run brook trout are now found in numerous South American, Australian, and New Zealand estuaries. Sea trout and Atlantic salmon have colonized a few systems on an Antarctic Ocean island. The few reports on abundance or density indicate that when the invasives were found in an antipodean estuary, they were a minor component of the fish community. For example, in the Waimakariri River estuary in New Zealand, juvenile Chinook salmon accounted for less than 1 percent of the individuals caught. The Chinook salmon were not thought to be competing for food resources with endemic fishes (Eldon and Kelly 1985). Other Northern Hemisphere salmonids have not established in the Southern Hemisphere (e.g., chum and pink salmon). The failure of

Atlantic salmon to colonize Southern Hemisphere watersheds extensively is particularly interesting given the considerable propagule pressure possible when farmed fish escape. The possible lack of competition may also be related to the absence of the nonsalmonid families with which anadromous salmonids have evolved in the Northern Hemisphere (i.e., predator release), but once again this is only a hypothesis. Abiotic factors such as temperature and habitat availability may play a role, in addition to genetics of the originally transplanted populations.

Competition for spawning beach habitat is another aspect of invasive salmonid ecology in the estuary – perhaps this is an example of possible interference competition. The unprecedented abundance of invasive pink salmon – recent population estimates of spawners reach up to 300,000 to 500,000 fish in the rivers and estuaries along the White Sea in Russia – has given rise to concern about "suppressing" native species in this region (Matishov and Berestovskii 2010). Pink salmon can spawn on estuarine beaches. Interference competition for spawning sites with Atlantic salmon seems to be minimal, as pink salmon were observed spawning in the Neiden River in Norway just above the tidal area of the lower river, at sites not used by Atlantic salmon (Bjerknes and Vaag 1980).

To summarize, a number of possible links to survival can be made between changes in competition in the disrupted estuary relative to the natural setting (Table 7), but our knowledge is very weak in this regard. Increased use of the estuary by domesticated fish is probably the most important factor. The continuation of major ocean ranching programs for several species will require relatively small populations of wild salmonids to share estuarine habitats with large numbers of cultured fish, with possible density-dependent growth effects resulting from exploitative competition.

General Conclusion Regarding Effects of Habitat Loss and Community Change on Fitness Components

Our knowledge of how habitat and community changes in the salmonid estuary has affected the three fitness components for survival (osmoregulatory adjustment, growth, and biotic interactions) has increased dramatically in the past few decades. To help organize thinking or conclusions about the three components, they might be arrayed as the three axes of a ternary plot or ecological triangle, such as that shown in Figure 26. How the factors increase or decrease along the three axes and their rate of change are unlikely to be monotonic. If they were, a nomogram approach might be possible (see Raubenheimer 2011). The values of

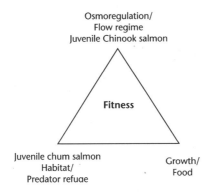

Figure 26 A ternary plot to array fitness elements thought to support survival of juvenile salmonids: osmoregulation, growth, and biotic interactions.
Notes: Each element may in turn be conditioned by river flow regime, food, and habitat/ predator refuge, respectively. Juvenile Chinook salmon and chum salmon are shown on apices as example species.

the three state indicators in a particular salmonid estuary relative to a benchmark could be used, but the functional relationships need to be discovered. For example, the quantitative relationship between wetland area (ha) and how much predator refuge is provided by a given area would be a very useful metric. In the interim, to develop hypotheses and guide practitioners as to management strategies, species might be aligned on the three corners according to their estuarine adaptations. For example, estuary-reared Chinook salmon might be placed in the upper osmoregulation corner and predation-prone chum salmon in the lower left predation refuge corner. In turn, osmoregulation is conditioned by river flow (sufficient fresh water must be provided into the estuary) and predation refuge is conditioned by the areal extent (adequate refuge habitat, such as vegetation, must be available).Thus, although our understanding is far from complete and is uneven around the estuarine "salmonid world," there are sufficient data in each of the three components to provide managers with guidance for conservation. On a cautionary note, the context of any cause-and-effect relationships should be examined. As well, in systems where the databases are limited, only weak inferences may be possible.

Looking inward, however, perhaps we have yet to advance our capability to assess, and deal with, another human-induced survival effect in the estuary: preying on salmonids for our own purposes. This is a survival effect that can be additive independent of habitat or biotic interaction effects. The survival components of fitness therefore also have to include another major anthropogenic effect: fishing. Removal of adult fish by commercial, recreational, or subsistence fishing – in other words, predation by humans on salmonids in estuaries – also requires consideration, as discussed in the next chapter in the context of ecosystem services provided by the estuary.

15
Harvesting and Production of Salmonids and Other Ecosystem Services Provided by the Estuary

Ecosystem services are the many conditions and processes associated with natural ecosystems that confer some benefit on humanity (Fisher, Turner, and Morling 2009). There is a vast literature on ecosystem services provided by estuaries; fisheries harvesting and production is a well-recognized aspect that is generally considered a cultural service (e.g., Holmlund and Hammer 2004). The estuarine ecosystem also provides services such as habitat, which in turn supports salmonid growth and survival, so the services complement one another. Some ecosystem services provided by salmonids in estuaries that are sometimes overlooked include those associated with spiritual values, cultural identity, social cohesion, and heritage values (Chan, Satterfield, and Goldstein 2012). This chapter gives an overview of how the various salmonid species contribute to ecosystem services, and comments on the current status of the service provision and the salmonids' conservation status in selected estuarine areas.

Commercial and Subsistence Harvesting
All species of adult salmonids aggregate in estuaries before ascending their natal rivers to spawn, and it is at this point that a major ecosystem service factor affecting salmonid survival begins – harvest by humans. Fishers have been removing salmonids for human consumption from estuaries for millennia. There is no ambiguity or weak data on the links between this activity and salmonid survival – the individuals are removed and die.

All eighteen species of estuarine salmonids are harvested as adults for food, either in commercial (industrial) fisheries or subsistence fisheries to provide important local sources of nutrition, especially for Native peoples (see Appendix Table 1.7 at http://hdl.handle.net/2429/57062).

Harvest data for salmonids from commercial and subsistence fisheries around the world cannot be summarized easily for estuaries because international statistical data are typically given only by jurisdiction or statistical area (e.g., Food and Agriculture Organization of the United Nations 2014). In this chapter, I give a general description of the estuarine region where fishing occurs, and comment on conservation problems related to estuarine harvesting.

Riverine channels are the migratory routes for adult salmonids moving upstream, and these are typically where salmonids are harvested in the estuary. Gillnets, box net fisheries, or other passive gear types are used as the salmonids are vulnerable to fishing gear while in narrow riverine channels. Beach seines are also used in some estuaries (see Appendix Table 1.3 at http://hdl.handle.net/2429/57062). Haedrich (1983) pointed out that in boreal estuaries salmonids account for the greatest numbers and biomass of fishes, so they are a logical target for fishers in this region and elsewhere in the temperate and Arctic regions. As well, if climate change warming increases the suitability of the Arctic for fisheries, there may be greater opportunities for expanded harvest in these regions (Zeller et al. 2011). If the estuarine fisheries can be managed carefully, harvesting techniques can offer some conservation options.

Harvesting salmonids in the estuary has the advantage of avoiding the problems of a mixed-stock fishery, because some salmonid runs are carefully timed and exploitation rates can be adjusted according to population size and timing. In this way, small or weak populations or life stages can be protected – this has been labelled the "bycatch problem." An example is the protection of the endangered coho salmon and steelhead populations in the Thompson River, a tributary of the Fraser River in British Columbia, from the gillnet fishery of the Fraser River estuary. These populations co-migrate with the much larger chum salmon runs, and as a conservation tactic the chum salmon gillnet fishery is timed to avoid exploiting the endangered species (Pacific Salmon Commission 2008). Experiments have also been conducted in an attempt to recover coho salmon from the stress of being captured in the chum salmon fishery (Buchanan et al. 2002).

Another strategy is to try to reduce bycatch by changing gear types or fishing locations. An example is the capture of threatened coho salmon and steelhead stocks in the gillnet fishery for sockeye salmon in the Skeena River estuary in British Columbia. The coho salmon catches have been reduced by changes in gillnet mesh and fishing methods (Department of Fisheries and Oceans 2001). The gillnet fishers experimented with the use of alternative mesh sizes and "weed lines" that allow a space between the net floats and the top mesh of the gillnet in efforts to reduce catch rates on steelhead that show a greater tendency to swim near the surface (C.C. Wood 2008). Selective gear such as fish wheels or large beach seines can also be deployed, and endangered species separated from those not threatened by fishing. In England, bycatch of Atlantic salmon and sea trout in gillnet fisheries for other species, such as European seabass (*Dicentrarchus labrax*), mullet (*Mugil* spp.), or European flounder (*Platichthys flesus*), can be reduced by setting nets parallel to the shore or in deeper water, as well as seasonal shifts (Potter and Pawson 1991). Bycatch within a species can also be a problem. When salmonids that are below legal size are caught and released, sometimes with fatal injuries or because of minimum size restrictions in some estuaries, this results in the undesirable loss of important life stages, such as older age classes of sea-run migrants or returning veterans (e.g., Arctic char at the Sylvia Grinnell River estuary, Nunavut [Gallagher and Dick 2010]). As a further example, Atlantic salmon kelts were a bycatch in Atlantic salmon fisheries in the middle sections of the Teno River in Finland (Niemelä et al. 2000). The bycatch problem for co-migrating salmonids in the coastal ocean can also be resolved by diversion of fisheries effort to estuaries and rivers, a strategy that has been used in many fisheries around the world. As but one example, in the European Union there is a drive to decrease Atlantic salmon catches in coastal mixed-stock fisheries (W.W. Crozier et al. 2004). Estuarine catches in Northern Ireland increased relative to ocean catches between 2004 and 2014 (International Council for the Exploration of the Sea 2015; Figure 27). Over and above the bycatch problem, however, is the universal problem of overexploitation and efforts to manage contemporary salmonid fisheries so that the resource is sustainable.

First Nations and Aboriginal peoples were able to conduct selective subsistence fisheries in estuaries. For example, traditional methods for capturing salmonids in estuaries on the northwest coast of North America typically included stone traps that caught fish at low tide, stake traps built with wood, or spears (D. Kennedy and Bouchard 1983). Similar methods were used in Arctic estuaries (Balikci 1980). These nonmechanical gears would enable selection of species and impart low catchability on the salmonids.

Many of the estuarine salmonid fisheries around the world are now considered to have low abundance relative to the preindustrial era. In the quest to restore and manage the fisheries effectively, local, regional, national, and international scientific and management bodies have been established. A premise is that survival can be controlled by regulating

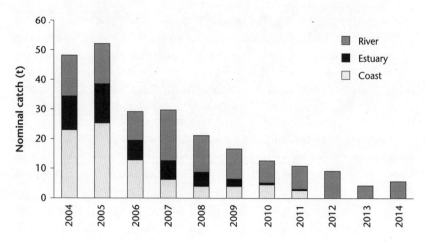

Figure 27 Shift in Atlantic salmon catches from coastal to estuarine fisheries in Northern Ireland.
Source: Modified from International Council for the Exploration of the Sea 2015, with permission.

(reducing) fishing effort. Fishing effort can be measured but is sometimes difficult to manage. This a major field of research, and many scientific journals, books, and policy documents are available. Besides efforts to regulate legal salmonid fisheries, authorities also have to deal with illegal practices. Because river mouths are often areas of major human populations, illegal fishing or poaching in estuaries is a major conservation issue in some areas and for particular species. Examples range from several species of estuarine salmonids in the Russian Far East (Zaporozhets and Zaporozhets 2012) to sea trout in the estuaries of the Caspian Sea (Niksirat and Abdoli 2009).

Recreational Fishing

Recreational or sport fishing is an increasingly important aspect of estuarine salmonid harvesting. Sport fishing for salmonids is now an international pursuit (see *The Salmon Atlas*, http://www.salmonatlas.com). Because of the aesthetic value of catching salmonids, almost all of the species subject to commercial and subsistence fisheries are also caught by sport fishers (see Appendix Table 1.7 at http://hdl.handle.net/2429/57062). Recreational fisheries management is also a major field of research with an extensive literature. Because many species are endangered, especially the iteroparous forms with a long lifespan and complex life history (e.g., Sakhalin taimen [Fukushima et al. 2011]), overexploitation by recreational fishers is an issue. In many regions, depending on abundance and local conditions, sport fishing regulations require salmonids to be released after they are caught (e.g., Dolly Varden in some estuaries in British Columbia [British Columbia Ministry of Forests, Lands, and Natural Resource Operations 2015] and sea trout in the Rio Grande River estuary, Argentina [O'Neal and Stanford 2011]).

Production by Ocean Ranching

Ocean ranching is a form of salmonid stock enhancement using hatcheries. Again, this is a major topic with specialized journals and books that provide support and guidance (e.g., Lorenzen 2008). The process involves mass releases of hatchery-reared juvenile salmonids (fry, smolts, or post-smolts) into rivers or sometimes from pens in the estuary. The fish feed and grow on natural prey in estuarine and marine environments and are subsequently recaptured when they return as adults, adding biomass to the fishery (Salvanes 2001). The growth and survival of the artificially reared juveniles while in the estuary is dependent on the ecosystem services provided by the estuarine habitat (e.g., temperature, food supply, currents to enable migration, etc.). Ocean ranching programs are conducted for at least ten of the eighteen estuarine salmonid species (see Appendix Table 1.7 at http://hdl.handle.net/2429/57062), indicating the importance of the programs for efforts to maintain salmonid harvesting.

Production by Aquaculture

The salmonid farming industry has grown dramatically in the past few decades, and production of farmed salmonids in the coastal zone now exceeds that of wild salmonids. Farm production of Atlantic salmon around the world was in the order of 1.4 million tonnes in 2010, approximately 700 times the production of wild Atlantic salmon (Food and Agriculture Organization of the United Nations 2014). Seven of the eighteen estuarine salmonid species are cultured in sea pens in various countries around the world (see Appendix Table 1.7 at http://hdl.handle.net/2429/57062). Most salmonid farms are not located in the estuary proper, because the low salinities reduce growth rates (Pennell and Barton 1996). However, the maintenance of estuarine circulation patterns (e.g., countercurrent flow) is important to maintain water quality for the farms, including the flushing of waste material. Farming operations are therefore drawing on the ecosystem services provided by the estuary (e.g., Galland and McDaniels 2008). There are also aquaculture-related "population restoration methods," such as development of broodstock reconditioning. Kelt reconditioning is being proposed as a stock enhancement measure in sea trout, and a program in Ireland involved an estuary as a holding location (Poole et al. 2002).

Social Value of Salmonids in Estuaries

Greater interest in conservation of salmonids in natural ecosystems has put a premium on locations where people can view them. Large estuaries with major fisheries are areas that attract public interest in salmonids, because the fishing fleets attract attention. Even small estuaries located in heavily urbanized areas in major cities can be used to demonstrate salmonid ecology to schoolchildren. This is an example of the importance of salmonids without any commercial value in the usual economic sense (Mina 1991). Citizen participation is possible in conservation estuaries with small populations of salmonids (see Appendix 3 at http://hdl.handle.net/2429/57062). In many countries, salmonids are an iconic species that have spiritual values for Native people. Salmonids were important for Native people living in estuaries on the northeast Pacific, as the fish provided an easily harvested source of protein and had great spiritual and social value. Socially oriented ecosystem services provided by salmonids in estuaries are difficult to quantify (Chan, Satterfield, and Goldstein 2012), but are important components of any forward-looking civic planning.

Conclusion

Harvesting and salmonid production are a very important ecosystem service provided to humans by the estuary. Salmonid harvest was one of the reasons why cities and towns were built on estuaries, and in regions where salmonids are an important subsistence food, the fisheries are still the basis for the existence of settlements at river mouths. The proximity of estuarine sport fisheries for salmonids is one reason why many people find it desirable to live near an estuary, and there is increasing appreciation of the social value of salmonids in estuaries. Definitive data are lacking, but the production of wild salmonids from estuaries as a source of food may be generally declining worldwide. If this is so, the estuary is increasingly providing ecosystem services to grow artificially reared salmonids. Ocean ranching and salmon farming have increased the number of artificially reared salmonids in estuaries and the near-ocean zone in many regions. These practices have implications for disease and parasitism issues (Håstein and Lindstad 1991). An overview of this rapidly changing field in the next chapter will complete our discussion of specific factors affecting salmonid survival in estuaries.

16
Health of Salmonids in Estuaries

This chapter gives an overview of the role of disease and parasites, two very important factors that affect the health and survival of salmonids in estuaries. Disease is influenced by the concurrent interaction of the host and pathogen with the environment, and this complex of interactions is called the web of causation (Arkoosh et al. 2004) or the epidemiological triangle (host, parasite, and environment [P.D. Harris, Bachmann, and Bakke 2011]). Parasites can be defined as organisms that live in close association with a host organism, and it is reasonable to assume that the host bears some cost (Ebert 2005). Parasites do not necessarily kill their hosts, but they may render them more susceptible to predation. As part of the web of causation, parasites can also frequently cause disease by overwhelming the immune defences of the host. Investigation of the web of causation requires special techniques as the organisms involved are often microscopic and disease responses are often measured biochemically.

Determining the significance of parasite-induced diseases for salmonid survival in the field relative to other ecological factors is challenging, because the weakened or dead fish are usually consumed by predators. Unlike some ecological processes, however, parasites and their transmission pathways can be studied in the laboratory under controlled conditions (e.g., Urawa 1993), facilitating an understanding of pathway effects. In the field, diseases can be diagnosed using manuals and guidebooks (e.g., Noga 2010; Roberts 2012), but treatment of diseases in wild salmonids is impractical. With cultured salmonids, on the other hand, veterinarians can provide medication to alleviate diseases or parasitic infections. Disease is an indirect but growing concern for the health and productivity of salmonids in estuaries. Production of some salmonid food organisms in estuaries may decrease because of higher parasite loads due to global warming (e.g., amphipods [Poulin and Mouritsen 2006]), with implications for food supply and growth. In addition, parasites can quickly spread around the world via a wide range of vectors, including ballast water, invasive species, and aquaculture activities (Harvell et al. 2004). The study of microbial and parasitic infections in salmonids is a specialized area of research and, because of its importance to fish culture, has resulted in a vast literature.

Microbial Diseases
Most of the pathogens or microbes that cause disease in salmonids have very complex life histories, and their prevalence is not necessarily related to estuarine conditions. A wide variety of microorganisms are involved, including both viruses (Crane and Hyatt 2011) and bacteria (Kinne 1984). Salmonids weakened by contaminant stress may be common in estuaries since pollutants from both upstream and local sources end up there. Stress can increase infection rates. Juvenile Chinook salmon in contaminated estuaries in Puget Sound, Washington, were found to be susceptible to vibriosis, a bacterial disease caused by *Vibrio anguillarum,* when exposed to chlorinated and aromatic compounds (Arkoosh et al. 2001). Diseases that kill adult salmonids in the estuary or river before they spawn are important, as they result in losses of individuals that are poised to ensure reproduction of the population. Research dealing with pre-spawning mortality is growing rapidly, using new genetic techniques that are expanding our knowledge of salmonid diseases in the estuary. Transcriptomic techniques that quantify changes in pooled mRNA have been used to investigate pre-spawning mortality of sockeye salmon in the Fraser River estuary in British Columbia. Genes highly correlated with both cellular immune responses typical of exposure to viruses or intracellular parasites, and humoral immune responses triggered by bacteria or extracellular parasites, were among those significantly switched on or upregulated in gill tissue collected from fish in fresh water, more so than tissues collected from fish in the ocean (T.G. Evans et al. 2011). These findings support the hypothesis raised by K.M. Miller and colleagues (2011) that a genomic signal associated with elevated mortality was a response to a virus infecting fish before their entry into the river. It would be useful to determine whether genetic changes occurred specifically in the estuary, possibly in response to osmoregulatory stress or contaminants.

Parasites
The health effects of a parasitic copepod, the salmon louse (*Lepeophtherius*

salmonis: Copepoda, Caligidae) is a representative case history for parasites, although numerous additional taxa are parasitic, ranging from protozoans to lamprey. The salmon louse has been investigated on several species of wild salmonids in estuaries, especially in salmon farming regions, where the salmon louse has developed on pen-reared salmonids. The data on the salmon louse show the importance of the estuarine environment for infection rates and mortality of the salmonid host.

The effect of the salmon louse on adult sockeye salmon that were held up by low flows and high river temperatures in the Somass River estuary in British Columbia is another example of the importance of interactions between estuarine habitat and water quality, the wild salmonid hosts, and the parasite. In this situation, the sockeye salmon were exposed to an extended period of crowding, increasing the number of hosts available to the louse, with warm water temperatures allowing the parasite population to increase rapidly. The result was a high infection rate resulting in fish suffering from skin lesions and other injuries, and higher prespawning mortality than in normal years, when the fish could migrate relatively quickly through the estuary (S.C. Johnson et al. 1996).

There is a very large and growing body of literature dealing with the spread of the salmon louse from farmed Atlantic salmon to wild salmonids, and the review by P.D. Harris, Bachmann, and Bakke (2011) covers many of the important findings to that date. Numerous papers have been published on the effects of sea lice on other salmonids, including sea trout, chum and pink salmon, and masu (see Nagasawa 2004; Costello 2006). In British Columbia fjords, but not in their estuarine regions per se, pink salmon fry mortality and reduced populations of pink salmon adults have been attributed by some authors to sea lice infections from Atlantic salmon fish farms (Krkošek 2010), but this cause-and-effect relationship has been disputed by others (Brooks and Jones 2008; Marty, Saksida, and Quinn 2010). Because of the complexity involved in determining the effects of the salmon louse on pink salmon population dynamics, modelling has been used extensively by the various investigators (e.g., Krkošek 2010). A Royal Commission report from British Columbia gives further details on this issue with respect to sea lice as a mortality agent for juvenile sockeye salmon (Kent 2011). The conclusions of the Royal Commission highlighted the possible importance of sea lice effects, but evidence linking the parasite directly to inter-annual differences in sockeye salmon survival was equivocal (Cohen Commission 2012). In Norway and Ireland, large-scale experimental releases of tagged, infected Atlantic salmon smolts treated with a sea lice disinfectant have been more definitive. Infected Atlantic salmon that were not treated with the disinfectant suffered a 39 percent increase in smolt-to-adult mortality (Krkošek et al. 2013).

In some situations, estuaries are refuges from sea lice and infestation that occurs outside of the estuary in higher-salinity water, possibly on the seaward edge of the outer estuary. The dense concentrations of sea lice in Atlantic salmon farms that spread to wild sea trout has been cited as the reason for the premature return of infected sea trout to estuaries and fresh water in Norway (Gjelland et al. 2014). In brackish water, the additional stress of osmoregulation is reduced. As well, the survival of larval stages of the sea lice is lower when salinity is less than 29 psu (Bricknell et al. 2006). Explanations of the effects of microbial and parasitic infections on salmonid survival are almost always context-specific.

The presence of invasive species, seasonality, and source of intermediate hosts are some of the factors influencing salmon mortality from parasite-related infections. Infestation of sea lice on native salmonids (e.g., pink salmon in British Columbia) may be a case of parasite spillback. The term "parasite spillback" refers to an indirect food web interaction in which an invasive species (e.g., Atlantic salmon outside its natural range) supplies an alternative host for a native parasite (the salmon louse), thereby increasing the abundance and spread of the parasite and its impact on native hosts (Kelly et al. 2009). In Norway, biologists have observed seasonal and species-specific variations in sea lice infestation. In the Altafjord, Norway, spring migrating Atlantic salmon post-smolts were not affected, whereas summer migrating sea trout and Arctic char were (B. Finstad et al. 2011).

An example of a disease agent that can spread via brackish water is the disease-causing parasite *Gyrodactylus salaris*, which lives on the skin of freshwater fish and which was responsible for decimating wild Atlantic salmon smolt populations in forty-five Norwegian rivers (Johnsen and Jensen 1991). Conditions in estuaries are conducive to the spread of this parasite, because it can tolerate low salinities and its salmonid host can carry it from one river system to another via the brackish water dispersed along the coast (Høgåsen and Brun 2003), using the estuary as a portal. Native to the Baltic Sea, *G. salaris* is also illustrative of an invasive parasite capable of between-coasts transfer. The infectious phase of the organism could be transferred to other areas of Europe via ship ballast water discharged in estuaries (Peeler and Thrush 2004). The development of ballast water dispersal models could assist forecasting if discharged ballast water containing the parasite would spread along a coastline with estuaries (Larson et al. 2003). Models could also help estimate dispersal rates between estuaries while *G. salaris* is living on the host. In Oslofjord, Norway, the risk of the parasite spreading was estimated for three neighbouring rivers situated at different distances from an index river. Smolts moving along the coast that have not completed the smoltification process may ascend a non-natal river via its estuary. For the nearest river, which shared an overlapping estuarine zone with the index river, the model estimated an annual risk of 31 percent that at least one infected Atlantic salmon smolt would ascend this river. The results of the simulation were highly sensitive to salinity along the migration route (Høgåsen and Brun 2003).

Salmonid infection with parasites that have an invertebrate intermediate host endemic to the estuary has not been documented, but an example may be the parasite *Parvicapsula minibicornis*, a kidney parasite associated with pre-spawning mortality of sockeye salmon in the Fraser River (Crossin et al. 2008; Bradford et al. 2010). This parasite has a freshwater polychaete worm (*Manayunkia speciosa*) as an intermediate host in rivers (e.g., Klamath River, California [Bartholomew, Atkinson, and Hallett 2006]), but this host is not found in freshwater reaches of the Fraser River estuary (Reynoldson et al. 2005). However, a closely related species, *M. aestuarina*, is found in the very low salinity parts of the Fraser River estuary (Otte and Levings 1975), and it is possible that this is the intermediate host involved. These are only a few examples of the many unknown factors and interactions that need to be investigated to increase our knowledge of how salmonid viruses, bacteria, and parasites in estuaries affect fitness.

Conclusion

Although microbes and parasites are well recognized as likely sources of significant mortality for wild juvenile and adult salmonids in estuaries, most of the direct information on the effects of these agents has been obtained in studies of cultured or captive populations. Hatchery-reared fish can pose a disease threat to fish or natural origin both before and after their release from the hatcheries, and studies on their health have focused attention on this problem (Pacific Northwest Hatchery Scientific Review Group 2009). Data from the research are valuable and can help reduce risk by ensuring that the fish are as disease-free as possible before release. Field data on parasites after the release of hatchery-reared fish or on wild fish are also important. The study of microbes and parasites in juvenile Chinook salmon and coho salmon in twelve estuaries in Washington and Oregon (Arkoosh et al. 2004) is a good example of the type of synoptic work that can form a baseline for further studies. It is critical, however, to recognize that disease prevalence in estuaries may be linked with estuarine water quality conditions, and that the two factors can act in tandem to decrease the survival of estuarine salmonids.

Looking Ahead to a Broader Perspective on Adaptations and Conservation

As a general summary at this point in the book, the discussions of estuarine changes in osmoregulatory ability, food supply, and growth and mortality from biotic interactions, harvest, and health give a detailed and focused perspective on factors that are currently affecting the distribution, abundance, and survival of estuarine salmonids. In the following chapters, I broaden the discussion to provide a possible framework for long and short evolutionary factors, both abiotic and biotic, that shape an estuary so that salmonids can exploit it. I then translate this framework in the context of indicators of the suitability of an estuary for salmonids, which leads to a discussion of future considerations for the conservation of salmonids in estuaries.

17
What Shapes an Estuary for Salmonids?

What key factors determine whether salmonids around the world can use an estuary or not? Chapter 3 discussed the main physicochemical factors that salmonids in estuaries are adapted to; Chapters 8–11 discussed the three important interacting biological variables influencing survival (success of osmoregulation adjustment, growth and food, and biotic interactions); and Chapter 14 discussed how human-induced habitat loss and change has affected the variables. This chapter draws the literature together from previous chapters and provides additional information to summarize the physicochemical and biological factors to develop a portrait of the idealized salmonid estuary in its natural state. It also discusses a conceptual model that may also determine whether an estuary is usable for the generic anadromous salmonid. The model is partially based on aspects of the estuarine ecology of salmonids moved to the Southern Hemisphere by humans, because these transfers were mostly to relatively pristine estuaries in the southern latitudes. Although probably similar to Northern Hemisphere estuaries in terms of habitat characteristics, the Southern estuaries were not natural to the salmonids. The introduced salmonids were required to live with a totally different community of fishes that they had not coevolved with. In a sense, this simulates the dispersal of salmonids into Northern Hemisphere estuaries in geological time, as the salmonids would also have been invading fish communities with new biotic interactions, especially as they moved into warmer regions.

Key Attributes of Salmonid Estuaries
The ecological filter concept used by Jackson, Peres-Neto, and Olden (2001) for freshwater fishes is a possible multiple-stage scheme that helps explain attributes that enable or restrict the dispersal of salmonids into

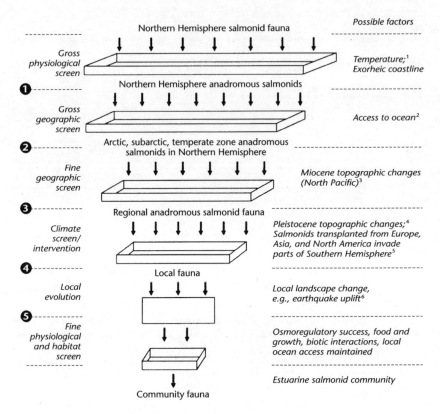

Figure 28 Possible multiple-stage scheme that helps explain factors that enable or restrict the dispersal of salmonids into estuaries by affecting survival and hence individual fitness.

Notes: Filters 1–4 are the longer-term evolutionary processes, or distal fitness elements. Filter 5 includes the proximal fitness elements in the fine-scale physiological and habitat screens and may be related to short-term evolution.

Sources: Adapted from Jackson, Peres-Neto, and Olden 2001. Superscripts indicate references for filters: (1) Jackson, Peres-Neto, and Olden 2001; (2),(3), and (4) D.R. Montgomery 2000; (5) B.S. Dyer 2000; (6) Atwater et al. 2001.

estuaries in an evolutionary sense, as well as in real time from human dispersal. Jackson, Peres-Neto, and Olden (2001) used the scheme primarily in the long-term evolutionary sense, but I also include short-term evolution (genotype × environment), as well as human-caused dispersal. Following the sequence in Figure 28, the possible factors are as follows.

Primary Environmental and Abiotic Factors: The Distal Filters

Temperature and the presence of an exorheic coast are possibly the master variables influencing the broad distribution of salmonids into estuaries. Together with key macrohabitat features, such as dispersal mechanisms and the geological setting, these factors might be considered the distal filters.

Temperature
Temperature is probably the first ecological filter for the distribution of fish. Temperature affects osmoregulatory adjustment rate and other physiological variables. Temperature likely controls how far south salmonids range in the Northern Hemisphere and how far north in the Southern Hemisphere. The range of upper lethal temperature for some species of salmonids is about 2° (24°C–26°C) and the lower lethal temperature is close to 0°C. Whether the river or estuarine water temperature is the limiting factor is not known, but estuarine water temperature could determine if juvenile salmonids moving along the coast could colonize a watershed-estuary-ocean system.

Exorheic coast
Exorheic coasts are margins of the continents with rivers that actually drain into the ocean. They are an obvious requirement for anadromous salmonids, which need fresh water for spawning and brackish water for estuarine life. River discharges along an exorheic coast will create low-salinity water masses along the continental margins. The perimeter length of exorheic coast potentially colonizable by salmonids has varied in geological time with climate change and continental drift. The contemporary world total of exorheic coast is 387,103 km (Dürr et al. 2011), but this includes warm and cold areas that are not suitable for salmonids, so this figure is an overestimate of available coastline.

Geologic setting for access to the ocean, and estuarine habitat geomorphology
A primary requirement for anadromous salmonids is the presence and maintenance of a topographic setting that enables a river to flow into the ocean, on a gradient that spawning ground-bound salmonids can ascend. In long-term evolutionary time, at least in the north Pacific region, it is likely the topographic upheaval that created diverse rivers and estuaries in the Miocene era were factors that established the topography for the salmonid estuary. The role of the estuary as a migratory route to the ocean for juveniles is as important as its role as a rearing habitat (McDowall 1995), and the interaction between juvenile migration and rearing also needs to be considered. The seasonal and interannual differences in breaching of the sand and gravel barriers in some lagoon

estuaries are an example. If the barriers are not breached, migration to the ocean is not possible. Geological dynamics can also facilitate habitat development by creating beaches, channels, and platforms for vegetation growth (e.g., change from earthquakes; see Atwater et al. 2001). The range of estuary types used by salmonids to access the ocean around the world varies dramatically. As an example of the complexity, in Europe the Atlantic salmon is distributed over 30° of latitude, from the Kara River estuary in Russia to the Douro River estuary in Portugal (MacCrimmon and Gots 1979). This 5,000 km reach of coastline includes multiple estuary types (see Figure 5), ranging from fjords to lagoons (Barnes 1991).

Dispersal by humans
The very rapid global dispersal of salmonids by humans in the past two centuries is an evolutionary anomaly that nevertheless must fit into the contemporary ecological filters. Human dispersal of salmonids provides a possible model for natural spread into new regions. Salmonids were transplanted into Southern Hemisphere watersheds and began to use estuaries when their genetics enabled anadromy. Estuarine salmonids have also spread into the Southern Hemisphere estuaries by escaping from aquaculture operations in this part of the world.

Osmoregulatory Success, Growth, and Biotic Interactions: The Proximal Filters

Osmoregulatory success, growth, and biotic interactions likely operate together to provide the final filter that determines the salmonid species composition of the estuary. These three proximal factors are also possibly important for short-term evolution and genotype/environment interactions.

Osmoregulatory success and salinity at the estuary as it relates to flows
Salinity is the main toxic factor that juvenile anadromous salmonids must deal with as they move into the estuary from the river. There are major variations in the shape and size of the salt wedge and freshwater lens in salmonid estuaries, and thus specific reaches of the exorheic coast need to be considered. At the macroscale, the fact that precipitation exceeds evaporation in the northeast Pacific Ocean (Tully and Barber 1960) and that there is little inflow from other parts of the north Pacific (Warren 1983) means that the entire region has an estuarine character (surface salinity <32.8 psu, especially in coastal areas where major rivers discharge). This may have led to widespread dispersal and colonization

by anadromous salmonids (D.R. Montgomery 2000). On the other hand, even estuaries in high-salinity coastal regions can be used as long as freshwater flows from the catchment basin are maintained. At the Deseado River estuary in southern Argentina, an estuary used by sea trout (Baigún and Ferriz 2003), salinity shows a nominal range and reflects the relatively low discharge rate of the river (mean annual discharge of 5 m^3 • s^{-1}) (Depetris et al. 2005) and the high salinity of the south Atlantic Ocean at the river mouth (32.8–33.5 psu) (Piccolo and Perillo 1999).

Dispersal of sea trout in the Southern Hemisphere provides a model for subsequent spread from one estuary to another. An adequate flow of fresh water into an estuary is necessary to attract spawning veterans that disperse along a coastline, seeking additional rivers for spawning, as was observed with invasive sea trout in the Kerguelen Islands, District of the French Southern and Antarctic Lands (Launey et al. 2010). A sufficient supply of fresh water stored in the watershed and released to the estuary at freshet is thus an important abiotic factor enabling the spread of salmonids.

Growth and a food supply appropriate for residency and life support
An estuary should be able to provide food to sustain smolts, sea-run migrants, or kelts while migrating, energy to recover from any stress from osmoregulatory adjustment, and, in certain species and populations, sustenance to enable fry and parr to grow to a size that enables ocean entry and osmoregulation in full ocean salinity. It is important to note that the estuary can obtain energy subsidies, and so in situ secondary production of salmonid food can be supplemented.

Biotic interactions: role of habitat in predation and competition
There is a range of habitats and subhabitats in estuaries. Although there is partial consensus that some provide a predation refuge function for juvenile salmonids, the advantages appear to be species- and context-specific. Brackish marsh vegetation likely provides refuges for residing juvenile Chinook salmon by providing cover, but interaction with other factors, such as turbidity, may be involved. Estuarine hydrodynamics and water properties also play a role. For Atlantic salmon smolts, river flows are likely important to move them through estuaries to avoid possible bird and fish predation. Some estuaries that only have channels and beaches and lack vegetation for predator refuge are used by salmonids. In these settings, the shallow water is likely important as a refuge.

Salmonids have adapted to Southern Hemisphere estuaries and seem to coexist with a wide range of potentially competitive fish species as well as potential predators. Conversely, and only as a hypothesis, competitive interactions with native salmonids may be inhibiting the dispersal of invasive salmonids between or within continents in the Northern Hemisphere.

In summary, a series of distal factors – temperature (climate-controlled), presence of rivers, access (geologically and oceanographically controlled), dispersal rates (natural and human-controlled) – and proximal factors (osmoregulatory success, food, and biotic interactions) are likely important in shaping the contemporary estuary for salmonids. The scheme of distal and proximal factors is provisional and requires comprehensive research for verification. In the interim, conservationists and managers have a range of methods available for assessing the condition or health of an estuary for salmonids.

Ecological Indicators

The assessment of the condition of estuarine salmonid habitat requires the collection of physical, chemical, and biological data, and knowledge about how these different components interact (Breine et al. 2007). Managers usually require such information to be translated into a simple metric or index, usually known as an ecological indicator, which can be used to evaluate the current state of the salmonid estuary relative to a pristine one. Ecological indicators are often preferred over more classic methods that rely on measurements of numerous physical and chemical variables, since bioassessment methods provide an opportunity to evaluate the condition of the environment without having to capture the full complexity of the entire ecosystem. Below, I provide a description of ecological indicators and review selected metrics that may help identify the condition or state of an estuary, or part of an estuary, and its capability to support salmonids. Habitat mapping can be an integral part of the process of developing indicators, and I also give an overview of some contemporary estuarine mapping methods.

Mapping of Salmonid Habitat and the Landscape Approach: Importance for Indicator Development

Mapping of salmonids for habitat management and conservation is a rapidly evolving research area, especially as remote sensing technology becomes more sophisticated. A decade or so ago, low-level aerial photography and habitat surveys to ground truth images using GPS at low tide were the common methodologies. In salmonid estuaries on the west coast of Canada, methods had to be able to map habitat patches with a minimum area of 100 m^2 for management purposes by visual means (Levings et al. 1999). More recent methodologies utilize satellite imagery and thematic mapping (D.R. Green and King 2005), but boundary definition, especially for riparian vegetation habitat that grades into truly terrestrial habitat at upper tide levels, can still be an issue (Levings and Jamieson 2001; Garono and Robinson 2003). At the broad-scale survey level, probability-based sampling methods now provide more objective techniques for habitat mapping (e.g., Environmental Monitoring and Assessment Program [V.D. Engle and Summers 1999]). New methodology also recognizes the importance of mapping dynamic processes such as sediment migration along shorelines, which moves sand and gravel into and out of drift cells. Creeks with low run-off have small estuaries along beaches, and shoreline drift processes can overwhelm the hydraulic forces of the freshwater flow, which is the intrinsic forcing function for river-borne sediment. An example is the mapping of drift cells for nearshore assessment in Puget Sound (Simenstad et al. 2006).

Considering the effects of river flow forces, the recognition of the estuary landform as a dynamic feature, affected especially by fluvial and sedimentary forces, has required reference to the classic topographic description of specific areas where different soil types erode and deposit in relationship to the direction of water flow. Milne (1936) coined the term "catena" from the Latin for chain, for these areas with their contrasting vegetative cover. While the term has been used by botanists and soil ecologists for decades, it has only recently been applied to river floodplain and estuary salmonid habitat (e.g., Stanford, Lorang, and Hauer 2005). The integration of salmonid habitat mapping and estuarine geomorphological data is necessary for management using the modern-day landscape approach.

A true landscape or comprehensive approach includes biological, physical, and socio-economic patterns and processes that function across space and time (Rieman et al. 2015). Conservation and restoration require better integration of estuary managers, improving their ability to effect landscape change and involving them in more inclusive, interdisciplinary science (Kareiva et al. 2007). Some of these initiatives are described in Chapter 18, with the following discussion of some aspects of physical landscape scale measurements needed in salmonid estuaries providing context.

The physical landscape arrangement and connectivity of estuarine habitats are important to juvenile salmonids. Individual juvenile salmonids continually adjust their position as tidal fluctuations alter the distribution of wetted areas, depths, velocities, and chemical gradients. Salmonids interact dynamically with this changing mosaic of mesohabitats and microhabitats along the entire estuarine gradient (Bottom et al. 2005b). Their response to the organization of patches, corridors, and the matrix of habitats through which some estuarine salmonids move in a rearing migration to the ocean is part of the trophic relay of food availability (Kneib 1997).

An example of the practical value of such allometric measurements of salmonid estuarine habitat is the demonstration of relationships between juvenile Chinook salmon food supply and width, depth, area, perimeter, and length of tidal channels at the Chehalis River estuary in Washington (Hood 2002). The biomass of insects exported to the main riverine channel from the tidal channel increased with tidal channel perimeter. The allometric relationship showed that perimeter was directly related to tidal channel area – thus, a very sinuous, winding slough would be expected to export an increased supply of insects, known to be important food to support the growth of juvenile Chinook salmon (Hood 2002). A straight side channel would export fewer insects.

On the small spatial scales used by many practitioners, estuaries are often viewed as "planar" landscapes, organized into beach, channel, and vegetation patches that are connected at high tide by the overlying estuary water. However, managing the patches as isolated units is clearly not appropriate when allometric measurements are made and their interconnectedness is recognized. Complex interactions can also occur with other estuarine animals that can modify estuarine vegetation and even develop pools in tidal swamp-forest subhabitats used by juvenile salmonids at low tide (e.g., beaver, *Castor* spp., dams at the Skagit River estuary [Hood 2012]). A physical landscape approach can therefore help promote more realistic habitat management methods.

A physical landscape approach that systematically accounts for multiple habitats could also facilitate investigations of the relationship between salmonid habitat preference and salmonid habitat availability in estuaries. Because sampling methods are often tailored to specific habitats, it is sometimes problematic to factor habitat availability into habitat occupancy and salmonid density. Radio telemetry tag data are a source of relatively unbiased data that do not require capture techniques, with their habitat-specific effectiveness. Radio telemetry tag data on juvenile Chinook salmon from riverine channels in the Columbia River estuary, Washington-Oregon, and lower Willamette River, Oregon, showed that while 35 percent of recoveries were from sand-mud, accounting for the majority of recoveries in the study, sand-mud was also most abundant (35 percent of the total assessed area). Riprap accounted for 18 percent of the area, and 21 percent of the recoveries were made there (Friesen, Vile, and Pribyl 2007). The finding that juvenile Chinook salmon abundance in riprap ranked second in the survey, even though this subhabitat was relatively scarce in the study area, was unexpected. Thus, using a standardized or calibrated sampling method may give us a different perspective on how salmonids exploit the estuary landscape.

Habitat and ecosystem mapping systems and physical landscape conceptual approaches are becoming more sophisticated. These advances are aiding the development of estuarine indicators for assessing the condition of salmonid estuaries.

Ecological Indicators Based on Biological and Habitat Data

Ecological indicators using biological and habitat data have a long history of use in terrestrial and freshwater habitat, but have been applied to a lesser extent to salmonids in estuaries. First established by botanists, biological indication can be defined as "making use of the specific reactions of organisms to their environment ... many plants rather precisely reflect the values of environmental factors" (Diekmann 2003, 493). In a more general sense, ecological indicators have been defined as measurable characteristics of the structure (e.g., genetic, population, habitat, and landscape pattern), composition (e.g., genes, species, populations, communities, and landscape types), or function (e.g., genetic, demographic/ life history, ecosystem, and landscape disturbance processes) of ecological systems (Niemi and McDonald 2004).

For the interested citizen, the abundance of salmonids in an estuary is an indicator of its status (e.g., sea trout: Celtic Sea Trout Program in Irish and British estuaries [Celtic Sea Trout Project 2014]). The situation is more complex, however, because the anadromous salmonids experience mortality in the river and ocean as well as the estuary. Their low abundance in an estuary is not necessarily an indicator of the estuary's suitability for salmonids. Nevertheless, there may an optimum or maximum density of various species that estuaries can support (i.e., their carrying capacity; Chapter 14). Presence or absence data are also of interest. In Scotland, the number of salmonid species present in an estuary – even though the data used are from bottom trawl data, which may not reflect salmonid presence – is used as an indicator of an estuary's suitability for salmon (Scottish Environment Protection Agency 2014a). Salmonid density is an appealing estuarine indicator for salmonids and can be meaningful when appropriate standardization methods are applied. Perhaps as alternatives to these biological metrics based on fish abundance and distribution, a variety of physical and botanical surrogates have been proposed.

Surrogate indicators for salmonid habitat may be categorized as: (1) pressure indicators, which deal with the stressors affecting the ecosystem – in other words, indicators of deviations from the natural habitat that are considered deleterious; (2) state indicators, which indicate the condition or quality of habitat at a particular time; and (3) quantity of habitat, which is usually an areal measurement. This was the approach taken in the development of the Canadian wild salmon policy (Stalberg et al. 2009). Pressure indicators for wild salmonids in estuaries that were considered included marine vessel traffic, estuary habitat disturbance, and permitted waste management discharges. State indicators that were considered included wetland area, estuary chemistry (e.g., dissolved oxygen), and contaminants. All three indicators require reference to a benchmark, emphasizing the importance of reference or undisturbed estuaries in close proximity to the estuary under study. I next discuss several state indicators for salmonid estuarine habitat and water quality to provide a wider perspective on surrogate metrics.

Wetland area

Wetland area, sometimes used as a surrogate for beach and vegetated habitat, is a possible state indicator of the quantity of habitat available for predator refuge, as well as basal food supply. The key quantity indicator in the Canadian wild salmon policy was estuarine wetland area (riparian vegetation, brackish marsh, eelgrass, and mudflat [Stalberg et al. 2009]). This indicator has been applied in numerous estuaries elsewhere, such as in the Columbia River estuary. This very large estuary was thought to be deficient in macrodetritus, an important basal food originating from wetland plants such as sedges, due to diking and industrial development. The relative mass of carbon in macrodetritus produced in situ from estuarine wetlands relative to carbon in microdetritus derived from freshwater phytoplankton produced in upstream reservoirs showed that microdetritus dominates the contemporary estuary. Before the reservoirs were constructed, wetland plants annually contributed an estimated 19,938 t of carbon to the macrodetritus pool, compared with only an estimated 3,605 t by 1980 (Sherwood et al. 1990). This represented an 82 percent decrease in macrodetritus mass. Another example can be drawn from the Fraser River estuary. In 1985, the Fraser River Estuary Management Program (2014) established one of the earliest estuarine classification systems for managing salmonid habitat. Based on vegetation productivity as an indicator, habitats were classified as Red (high value), Yellow (moderate value), and Green (low value) habitat. The classification system enabled colour-coded maps, giving managers guidance for the preservation of salmonid habitat.

The area of an estuary affected by altered water quality conditions is also a possible state indicator. In some instances, this can be measured by remote sensing techniques, including analysis of satellite imagery. When 120 t of oil spilled from a pipeline on Sakhalin Island, Russia, it was estimated, through the use of satellite imagery, to have affected 580 km^2 of salmonid estuary habitat in the outer estuary of the Amur River (Ivanova et al. 2001).

Growth rate and food
Growth rate is a state indicator of feeding success. It has been used to indicate success of salmonid habitat restoration (Volk et al. 2010) and to compare sea trout growth and food supply in both their endemic region (Northeast Atlantic coast [De Leeuw et al. 2007]) and an invaded estuary (South Atlantic coast [O'Neal and Stanford 2011]). Authors also connect the recommended conservation importance of an estuary, or part of an estuary, to abundance, biomass, or productivity of potential food items for salmonids. Examples include the Columbia River estuary, citing insect data (Lott 2004), and the Sixes River estuary in Oregon (Neilson, Geen, and Bottom 1985) and the Squamish River estuary in British Columbia, citing gammarid amphipod data (Stanhope and Levings 1985). The structure of the food web itself has also been proposed as an indicator. One of the descriptors used to delineate "Good Environmental Status" under the European Union Marine Strategy Framework Directive focuses on food webs. The indicator is recognized as perhaps the most challenging to implement and requires considerable research to identify the factors affecting the functioning and dynamics of the food web to supplement the more easily identified data on food web structure (Rombouts et al. 2013).

River flow
River flow and correlated hydrodynamic factors (Beechie et al. 2006) are also likely ecological indicators for the suitability of an estuary for salmonids. Low flow in the river is of special interest, as low discharge can cause problematic temperature or salinity conditions for smoltification. Stranding of fish may occur if flow drops quickly (e.g., in the Yolo

Bypass in the Sacramento River estuary, California [Sommer, Harrell, and Nobriga 2005]). At the Sacramento River estuary, river flow was one of the factors positively correlated with Chinook salmon smolt survival (Newman and Rice 2002). At the same estuary, a flow-related variable was developed ("X2"): the distance from the mouth of the estuary in San Francisco Bay, California, up the axis of the estuary, where tidally averaged bottom salinity is 2 psu (Jassby et al. 1995). X2 can be considered a surrogate for flow (Kimmerer 2004). Statistical relationships between X2 and fish abundance have been developed and found significant at least some of the time, such as for longfin smelt (*Spirinchus thaleichthys*), Pacific herring, starry flounder, splittail (*Pogonichthys macrolepidotus*), American shad, and striped bass (Kimmerer 2002). Juvenile Chinook salmon in the estuary were not mentioned in Jassby and colleagues' analysis (1995), likely because they were not sampled effectively, but inclusion of salmonids would be worthwhile in future work with X2.

To summarize, three key state indicators could be analogous to the proximate factors in the ecological filter scheme presented earlier in this chapter (Figure 28). They are: (1) wetland area, as a surrogate for a predation refuge function; (2) food supply, as a surrogate for growth; and (3) freshwater flow, as a surrogate for osmoregulatory success. As well, the effects of wastes discharged into the postindustrial salmonid estuary often require different kinds of indicators: standards or benchmarks to estimate how far master variables or prominent nonconservative substances and chemical species have deviated from natural levels.

Water Quality Standards and Indicators

Standards for salmonid estuarine water quality have been developed for many elements, compounds, and contaminants (e.g., United States Environmental Protection Agency 2015), and dissolved oxygen is discussed here only as an example. Standards are based on laboratory toxicity evaluations such as bioassays and physiological effect studies, and are often specific to salmonids. As well, biomarkers, indices of complex chemicals known as threats to salmonids or other fish species, have been developed, based on detailed experimentation. Field- or ecosystem-based indicators have also been applied in salmonid estuaries.

Dissolved oxygen
Depletion of dissolved oxygen can kill fish directly, impose sublethal effects, or change behaviour. Deviation from a standard is therefore a critical indicator of the suitability of an estuary for salmonids. The metric

is used in many water quality management schemes in the salmonid world. For estuaries with salmonid fisheries in the United Kingdom, an early researcher found that the observed minimum dissolved oxygen concentrations in natural estuaries generally showed 95 percent and 50 percent exceedance values of around 7 mg • L^{-1} and 10 mg • L^{-1}, respectively, whereas fisheries in estuaries recovering from reduced dissolved oxygen stress are found in waters with minimum dissolved oxygen concentrations with 95 percent and 50 percent exceedance values of 0 4 mg • L^{-1} and 5–8 mg • L^{-1}, respectively (Hugman, O'Donnell, and Mance 1984). It appears that, at least in a study in the Clyde River estuary in Scotland, the more demanding thresholds may relate to smolt migration, since water quality was generally poorer than during the period of adult migration (J.C. Curran and Henderson 1988). Thus seasonal flows, temperature, and other variables need to be considered when applying dissolved oxygen metrics as indicators.

Biomarkers

Indicators of the subtle effects of the complex mixture of pollutants (the chemical soup) in estuaries are particularly challenging to develop and implement. Biomarkers are biological or biochemical variations in the tissue, body fluids, or cells of an organism that provide evidence of exposure to chemical pollutants. Many biomarkers at the cellular or genetic level have been proposed. This is a particularly important topic because of the possible effects of such pollutants on juvenile salmonids during smolting, when they are sensitive to physiological changes that can lead to health or reproductive problems in older life stages. For example, exposure to the pharmaceuticals 4-nonylphenol and 17β-estradiol during smolting negatively affected growth and expression of growth hormones in Atlantic salmon (Arsenault et al. 2004). There is a large and growing literature on biomarkers of contaminant effects on salmonids in estuaries. In Europe, biomarkers are used to meet the European Union Water Framework Directive's challenges for improved detection of the impacts of chemical compounds on aquatic organisms (i.e., improved link between biological effects observed at the community level and monitored chemical concentrations [Sanchez and Porcher 2009]). To be confident that environmental monitoring based on biomarkers represents a quantitative assessment, it is necessary to know the range of natural variability of assessed parameters in healthy organisms, and this requires long-term data sets. Nine core biomarkers have been used to assess the health and toxicological status of wild fish in environmental monitoring networks, and most of these biomarkers could be applied to salmonids in estuaries. The biomarkers ranged from those that have been used for some time (e.g., changes in detoxification enzymes) to newer methods at the DNA level (see Appendix Table 1.8 at http://hdl.handle.net/2429/57062). Body burdens of contaminants known to have deleterious effects on salmonids have also been used as biomarkers. In the Columbia River estuary, L.L. Johnson and colleagues (2007b) monitored levels of persistent organic pollutants (polycyclic aromatic hydrocarbons, polychlorinated biphenyls, dichlorodiphenyltrichloroethanes, and other organochlorine pesticides) in juvenile fall Chinook salmon. Levels were usually above those known to have adverse effects on the health of Chinook salmon.

Biological Community Structure as an Indicator of Estuarine Quality

In some estuaries, the fish assemblages themselves, sometimes including salmonids, are used as indicators; this approach is used in the European Union Water Framework Directive (e.g., Champ, Kelly, and King 2009). However, much of the literature dealing with estuarine indicators using fish or invertebrate community structure focuses on temperate subtropical estuaries close to or south of the latitudinal range of salmonids (e.g., mid-east coast of the United States; see Hughes et al. 2002). The structure of fish communities using species and their tolerance, or lack of it, to reduced water and habitat quality as indicators was first developed by Karr (1981) for rivers in North America. This widely used metric is known as the Index of Biological Integrity (IBI) and requires sensitive species as reference taxa. Indices of biological integrity have also been used in European salmonid rivers (e.g., Loire River, France [Lasne et al. 2007]). Salmonid estuarine components have usually not been considered, but sea trout and Atlantic salmon were factored into the IBI for part of the Scheldt River estuary, Belgium (Breine et al. 2010), and sea trout were a reference species. Thus, there is definite scope for developing salmonid-specific indices for estuaries.

In the European Union, the Water Framework Directive suggests that each member state identify a typology for its estuaries and/or other water bodies, which is the first step in calculating an Ecological Quality Ratio (Hatton-Ellis 2008). Types are based on fundamental ecological drivers such as altitude and geology, and their ecological quality is then compared with a "type-specific reference condition." A series of generic metrics of environmental quality are identified in the directive, mainly related to taxon richness and abundance, and measures of the biological community structure, which conceivably could include estuarine salmonids. The metrics are then used to grade the ecological quality of each water body by comparing it with a reference condition using an Ecological Quality Ratio. The range of the Ecological Quality Ratio is then divided into five classes, from high to bad ecological status (Hatton-Ellis 2008). To my knowledge, an Ecological Quality Ratio that takes into account salmonid abundance in estuaries has not been calculated.

Parts of estuaries, or in some instances the whole estuary, in the European Union have been designated as Special Areas of Conservation (SACs) under the European Union Habitat Directive, and the literature suggests that indicators have been used in their selection. Use of Atlantic salmon and sea trout is mentioned as a qualifying criterion for a "notable estuarine species assemblage" (e.g., Severn River estuary, England-Wales [Henderson and Bird 2010; Severn Estuary/Môr Hafren European Marine Site 2009]). Linkages between salmonids and other endangered species are often acknowledged. The abundance of sea trout as an estuarine prey species for river lamprey (*Lampetra fluviatilis*), an endangered species under Annex II of the European Union Habitat Directive, was one of the indicators used as part of criteria to designate the Severn River estuary as a SAC.

Conclusions

While simple one-dimensional indicators can have value for identifying important attributes of salmonid estuaries, the complexity of their ecosystems argues for a multiple-parameter approach. Béguer, Beaulaton, and Rochard (2007) showed that the distribution of Atlantic salmon, sea trout, and Arctic char in western European rivers is the result of interactions between four main factors: temperature, surface area, stream power, and productivity. A similar approach could be taken with estuaries, if the data were available. It is also possible to extend the concept of global syndromes, i.e., typical patterns of problematic people/environment interactions that can be found worldwide and can be identified as regional profiles of damage to human society and ecosystems (German Advisory Council on Global Change [GACGC] 2000, cited in Meybeck 2003) as indicators of how the capability of estuaries to support salmonids has been modified by human activity. A number of global syndromes for rivers were identified by Meybeck (2003). Because of the tight connection between river and estuarine function, several of these syndromes are also applicable to salmonid estuaries, such as flow regulation, chemical contamination, and eutrophication. However, other syndromes may be estuary-specific due to ocean influences such as changes in salinity distribution, upwelling of low dissolved oxygen, and acidification. The global syndromes require specific management attention in order to conserve and restore the estuary's capacity to support salmonids. This is a complex task, requiring comprehensive ecological and socio-economic information, as well as community planning and societal acceptance of common goals.

18
Future Considerations for Conservation of Salmonids in Estuaries

The emerging conservation movements, engagement of citizens, and comprehensive management plans being developed in many salmonid estuaries around the world offer optimism regarding conservation. There are also other socio-economic factors that need to be considered in local, regional, and global planning. The literature on how communities, states, and countries could conserve salmonids and their estuaries in the social context is voluminous. Estuaries and salmonids are not manageable without considering people and their institutions, but a detailed review of this topic is beyond the scope of this book.

This chapter provides an overview of some of the future considerations and currently available tools for applied ecologists tasked with conserving salmonids in estuaries. In this role, applied ecologists are by definition working with practitioners. I first provide an overview of an emerging critical area that estuarine salmonid ecologists should be aware of – the short-term evolution and the genotype × environment issue. This is followed by a discussion of some of the possible methods and associated problems for identifying global patterns and prioritization for conservation of salmonid estuaries. Finally, I give an overview of some of the management strategies that might help build a robust framework, including a few institutional considerations, for conservation and restoration of estuaries and their dependent salmonid populations.

Short-Term Evolutionary Considerations

It is important for managers to develop an appreciation of possible short-term evolutionary considerations arising from abiotic and biotic factors, for several reasons: (1) the rate of habitat loss in salmonid estuaries may be increasing; (2) there has been an unprecedented dispersal of invasive salmonids into river-estuary systems around the world in the past two centuries; and (3) there is likely a growing number of domesticated salmonids using estuaries.

There are very few studies on the short-term evolution of salmonid life history characteristics in the context of habitat or biotic change that relates specifically to estuaries. A possible analogy is the reservoir-type Chinook salmon smolts from the Columbia River, Washington-Oregon. The original populations of certain tributaries of the river system migrated downstream as subyearlings and therefore may have relied on estuarine conditions before migrating to the ocean. Now, less than seventy-five years after the construction of multiple dams and reservoirs on the system, conditions appear to favour the overwintering of fry and parr in the reservoirs, where they grow to smolt size and migrate as yearlings (J.G. Williams et al. 2008; Waples, Beechie, and Pess 2009). Given that yearlings move through the estuary faster than subyearlings, this implies a decrease in estuary use. It is not clear whether the increased fitness component for survival is attaining large size in fresh water or decreased mortality in the estuary, or some combination. Another natural experiment of this type was described by Morita, Yamamoto, and Hoshino (2000) with anadromous whitespotted char in Japan. The dams in this situation were only twenty to thirty years old, but fish transplanted from above and below a dam to a fishless above-dam site in another stream showed an interesting result. The transplanted fish displayed very low rates of smolting and emigration to the estuary regardless of whether they came from below or above the original dam, suggesting an environment × genotype response to the reservoir habitat. The freshwater resident life history type was selected for, possibly because the fish grew faster above the dam due to lower fish densities (Morita, Yamamoto, and Hoshino 2000). Changes in both the Chinook salmon and whitespotted char freshwater habitat and subsequent short-term evolution could influence the degree of estuary use.

Biotic factors can also induce changes, and predation is a potentially important factor in shaping salmonid life histories in estuaries in the short term. Harvesting adult salmonids in the estuary where they are highly vulnerable to fishing can exert selective pressure on salmonids. Size selection for sockeye salmon in an estuarine gillnet fishery in Alaska caused heritable changes in body morphology of the species (Hamon et al. 2000). There are other examples that relate to life history traits. A linear function between an environmental value or gradient and the expression of a genotype is known as a *reaction norm* (Hutching 2011). Using a model calibrated with data from a river draining into the St. Lawrence estuary in Québec, Thériault and colleagues (2008) predicted that fishing of anadromous brook trout over the course of 100 years causes evolution in the migration reaction norm, with the probability of migrating to the estuary decreasing as harvest rate increased. In this case, the harvest rate is the environmental value causing the shift in the migration timing.

In summary, the fact that fish life histories display relatively high rates of evolution (about one-fourth of those observed for other traits, such as physiology, morphology, and behaviour [Conover et al. 2006]) is extremely important for developing sustainable conservation strategies for species subjected to biotic and abiotic changes (Healey and Prince 1995). Managers of salmonids in estuaries need to be aware of this fact. However, obtaining data on the heritability of estuarine factors affecting the salmonid phenotype (i.e., is the observed variation induced by environment, genetics, or both?) is a challenge, because common garden experiments or growth-transplant experiments are difficult, if not impossible, to conduct. And if the multigenerational experiments are conducted in the laboratory, the results may not be relevant to wild populations (Carlson and Seamons 2008) and may lead to selection of traits from the artificial environment. The use of very large-scale mesocosms that reduce enclosure effects in small tanks may help in this regard (Devlin, Sundström, and Muir 2006). Studies on species living above and below dams can also provide natural experiments on loss or gain of anadromous life history types (Morita, Yamamoto, and Hoshino 2000; Hendry and Stearns 2004). Until new technology is developed to enable better experiments, it is prudent to ensure that management-oriented studies on short-term evolution in salmonids is well coordinated between freshwater, estuarine, and marine habitats. Because of the ever-increasing awareness of field-based salmonid genetic studies as conservation tools (e.g., Ng et al. 2005), it is opportune to develop comprehensive global information banks on the current physical and ecological status of salmonid estuaries and their populations.

Identifying Global Patterns for Conservation

An international science-based scheme that assists conservation groups with the task of rating the importance of various salmon estuaries around the world would be very useful in ensuring restoration and maintenance. There are several issues related to this option.

Estuary Size and Type

Although estuary size is an attractive metric for rating importance, there are shortcomings with data and there can be methodological challenges.

The sizes of estuaries used by salmonids vary tremendously over their range and are sometimes not measured using a standard method between jurisdictions. Estuaries may have different areal measurements depending on how they are viewed by separate agencies. For example, the Port of London Authority, England, uses its Vessel Traffic Services office to manage safe ship movement over 1,335 km^2 of the lower Thames River and estuary (Port of London Authority 2015). A general estuary guide gives a core area of the same estuary as 47 km^2 (ABPmer and HR Wallingford 2007). If problems arise because of differences in areal measurements, they can usually be worked out in detailed planning sessions and updated as technology for assessing area improves.

Estuary type might be another criterion for conservation purposes if it is found that some types are particularly rare or important as alternative habitat for salmonids in certain regions. For example, estuarine lagoons on the coasts of the Kamchatka Peninsula, Russia, are available as alternate rearing habitat for subyearling sockeye salmon fry in the Kamchatka River (accounting for 15 percent of adult returns from 1985 to 2001 [Bugaev 2004]). In general, however, data are lacking for broad-scale regional syntheses or data banks relating to estuaries as alternative habitats relative to freshwater or marine habitats, as well as the relative importance of various estuary types within the suite of estuary forms.

Salmonid Biological Metrics and Estuary Conditions

There is also the problem of identifying an appropriate biological metric. Should salmonid population size be the key measure, or are others, such as a measure of resilience (e.g., salmonid species diversity or life history diversity, including genetic diversity), more important? Contemporary conservation ecology and ecosystem-based management emphasizes diversity, which can facilitate resilience (Kareiva et al. 2007), as discussed later in this chapter. Even though a river-estuary system may rank highly in terms of number of salmonids caught in commercial and recreational fishing, if it is a monoculture of one species, its resilience may be low. Another metric possibly worthy of investigation is the number of iteroparous species. The presence of more repeat migration species would suggest that the estuary is more heavily used by salmonids.

Estuary size and salmonid population data are therefore important as rating criteria for conservation, but need to be used in the context of socio-economic local values. There are thousands of small estuaries in the temperate regions of the Northern and Southern Hemispheres that support small populations of salmonids and their cumulative populations are important, in some regions perhaps more so than the megastocks in larger estuaries. In Norway, tiny estuaries of a few hundred square metres are used by sea trout, either spawned in small streams or migrants from nearby systems (shown in Appendix 5 at http://hdl.handle.net/2429/57062). Such small stocks of salmonids are of often of major local interest to people around the salmonid world, as evidenced by community-based salmonid conservation movements (see Appendix 3 at http://hdl.handle.net/2429/57062). In British Columbia, almost 60 percent of the estuaries are smaller than 30 ha in area (Ryder et al. 2007), and most of these systems support salmonids with widely varying species composition and populations. Chum salmon, coho salmon, and cutthroat trout frequently use the smaller streams and estuaries, often with populations of fewer than 100 spawners (e.g., Carnation Creek, British Columbia [Tschaplinski 1982]).

At the other end of the range, the huge Amur River estuary in Russia (7,400 km^2) supports major populations of chum and pink salmon runs in the range of 4–6 million fish (Novomodnyi, Sharov, and Zolotukhin 2004). Before industrial fishing, the Chinook salmon runs in the Columbia River could have been up to 4.3 million (estimate for 1900; Chapman 1986), but by 1938 they were depleted to about 1 million (van Hyning 1973). The total area of the estuary in 1938 was approximately 893 km^2 (Sherwood et al. 1990) and the estuary was only partially disrupted, so the role of estuarine habitat loss may have been only one component leading to the drop in production. Multiple small estuaries might be an asset for the stocks rearing in a particular water body. An analogy might be the situation in the Wood River watershed in Alaska, where several spawning grounds and populations are important for maintenance of the sockeye salmon stock for a fishery (portfolio effect; Schindler et al. 2010). In England, the numerous intermediate-sized estuaries a few hundred hectares in area (e.g., the 213 ha Avon River estuary in England [ABPmer and HR Wallingford 2007]) are important for sea trout.

A global summary of estuary sizes and conditions that are pertinent to salmonid conservation requires more data than are currently available. Remote sensing has great potential for measuring estuary sizes or categorizing estuaries by type in regions (Kravtsova and Mit'kinykh 2011) and these metrics could relate to salmonid ecology. However, collection of data on salmonid usage of specific estuaries will require a vast increase in on-the-ground surveys. As well, while estuary variation in size and habitat features may be related to salmonid population variability, this is an untested hypothesis. There are no long-term data on salmonid trends in estuaries, at least for carrying capacity and production at the juvenile stage. Long-term trends in adult salmonid catches in major

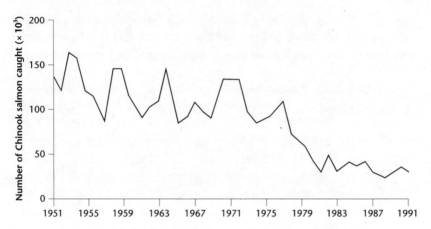

Figure 29 Catch data from the Chinook salmon gillnet fisheries at the Fraser River estuary, British Columbia.
Sources: Data from F.J. Fraser, Starr, and Fedorenko 1982; Department of Fisheries and Oceans 1995.

estuaries are available for some species (e.g., Chinook salmon in the Fraser River; Figure 29), and there is a widespread pattern of reductions in harvest. Whether this is due solely to decreased survival of juveniles in the estuary owing to habitat limitations is not known. Now, with the reduced salmon populations (due to multiple and difficult-to-understand factors), estimates of wild salmonids in relation to the carrying capacity of estuaries are misleading relative to historical conditions. For example, current numbers of North American wild Atlantic salmon are at somewhere around 4 percent of the species' historic abundance (Whoriskey 2009). Estimation of the current carrying capacity of many Atlantic salmon estuaries is problematic, but it is possible that the estuaries' current capability to produce Atlantic salmon is not being exceeded. Unfortunately, there is no way of determining, with contemporary data, whether the capability of the estuary was limiting when Atlantic salmon runs were larger. This dilemma was recognized by early salmonid researchers (e.g., R.J. Miller and Brannon 1982) and obviously is still a problem, exacerbated by current conditions of climate, harvesting, and other factors. The carrying capacity issue might be different for the estuaries that are now used to convey the huge numbers of hatchery-reared fish from rivers to oceans, but the relative value of hatchery-reared fish vis-à-vis wild fish, and hence the concern about "overloading" the estuary, is a societal issue as well as a question of biotic interaction. Therefore, even if it were possible to rank estuaries by size and wild salmonid populations within them, a relationship between these two variables is unlikely. Alternative biological metrics that are supported by available data might be more appropriate.

In terms of productivity and species diversity, estuaries that have (or had in the past) more estuary-dependent salmonid species and life history patterns (i.e., species or forms with a longer residency time in estuaries) might be a possible metric for arraying the importance of particular estuaries around the world. In north Pacific estuaries, Chinook salmon might be the species to focus on initially, since they have life history forms that reside in the estuary longer, and hence are at greater risk from anthropogenic activities. Unfortunately, there has been much apparent loss in life history patterns. In 1916, there were six juvenile Chinook salmon life history types found in the Columbia River estuary throughout the year (fry, four subyearling types with a mixture of freshwater and estuarine rearing histories, and yearlings). In early spring, fry were the most abundant. By 1985, only three were found (fry, one subyearling type, and yearlings), and the subyearling form was dominant, in late spring (Burke 2004; Figure 30). As well, the relatively rapid evolution of life history types within this species makes tracking life history types a challenge. Therefore, further ecological and genetic work is required to implement the life history criterion, especially considering the lack of detailed information from the hundreds of Chinook salmon rivers and estuaries where the species is present (now including estuaries in the Southern Hemisphere). However, the concept of using the criterion is solid, and new genetic techniques can help.

Risk of extirpation is another possible rating scheme. For example, conservation of wild Chinook salmon and steelhead to prevent loss of certain populations is stated as a priority in several regions around the North Pacific Ocean. According to the Wild Salmon Center, an international wild salmonid conservation organization, the rivers and estuaries on the northeast and northwest Pacific Ocean coastal areas rank high in conservation importance for wild Chinook salmon and steelhead (Augerot et al. 2005; Wild Salmon Center 2015). A number of interesting projects have dealt with northwest Pacific rivers and estuaries and their salmonids, and the results support their idea. There are relatively fewer hatchery-reared fish released in this region, and this increases the opportunity for focusing on wild salmonids and their habitat. Examples from the Kamchatka Peninsula are illustrative. Less than 1 percent of salmon harvested on the Kamchatka Peninsula are from hatcheries (Zaporozhets and Zaporozhets 2012). Over 300,000 Chinook salmon returned to spawn annually on the Kamchatka Peninsula rivers between

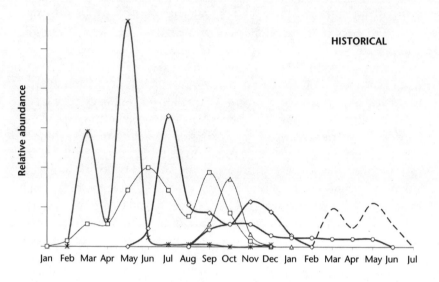
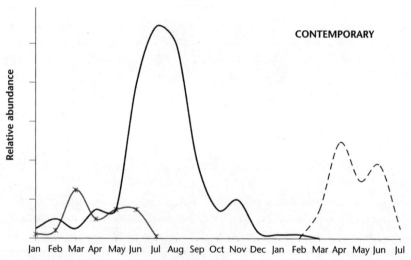

Figure 30 Historical and contemporary life history patterns for juvenile Chinook salmon in the Columbia River estuary, Washington-Oregon, inferred from abundance data and scale information.
Notes: Upper panel – historical life history patterns:
∗ – fry; □, ○, ◇, △ – subyearling types; – – – yearlings.
Lower panel – contemporary life history patterns:
∗ – fry; —— subyearling type; – – – yearlings.
Source: Modified from Burke 2004, with permission.

1951 and 2000 (Rogers 2001). More than 75 percent of Asian Chinook salmon spawn in only two rivers, the Kamchatka River (east Kamchatka) and the Bol'shaya River (west Kamchatka) (Urawa et al. 1998). Although Kamchatka Peninsula Chinook salmon are apparently mainly of the river or yearling ecotype (Vronskiy 1972), whose smolts do not reside in the estuary very long, it is important to note the variation in the populations and estuary conditions in the region. Subyearling Chinook salmon have been observed in the Bol'shaya River population, Russia (Leman and Chebanova 2005) and these fish may possibly use the extensive estuarine habitat found on the west coast of the Kamchatka Peninsula (Gerasimov and Gerasimov 1998). Although detailed mapping of west Kamchatka estuaries is not available, representative east Kamchatka estuaries were characterized by ecologically important vegetation such as eelgrass and macroalgal species (*Laminaria bongardiana* and *Fucus evanescens*) (Maximenkov 2007). The extensive lagoon estuaries of the Kamchatka and Bol'shaya Rivers (223 km^2 and 7.5 km^2, respectively) may be especially important, considering the major populations of Chinook salmon in their watersheds, but this is a hypothesis requiring further investigation.

In summary, given the paucity of available data, a multiple-attribute approach (Prato 2003) is required to conserve estuaries until more comprehensive scientific criteria can be developed to rate their importance. Such an approach would take into account a variety of ecosystem features (e.g., flow, wetland area, and fish communities), and can, in some situations, include economic attributes (Prato 2003). The inclusion of estuaries in wild "salmon stronghold" watersheds (Rahr and Augerot 2006; Aas, Policansky, Einum, and Skurdal 2011b) is clearly also very important, but more data on specifics are required before metrics such as estuary size within a watershed can be considered a criterion in selecting the stronghold. If river size and estuary size can be positively correlated, as Hood (2004a) found in northwest North America, then historically large watersheds with large estuaries might have supported larger salmonid populations. This may have been the case for Chinook salmon, as there was a positive relationship between river discharge and spawner populations, at least for the period 1952 to 1976, which was prior to recent decades of poor Chinook salmon survival. However, Healey (1991) noted that while there was a general correlation with discharge, populations varied considerably throughout the range. Populations in the centre of the range, where estuarine conditions might be expected to be optimal, were not more abundant, but this part of the range (Oregon, Washington, and southern British Columbia) is also where anthropogenic disruption of estuaries is most evident. Further data are required to understand the reasons why. Nevertheless, Healey's observation (1991) supports the premise that populations at the edge of their range may have unique adaptations. As we have poor understanding of the exact factors linking salmonid survival and estuarine habitats, there is strong reason for conserving them using a broad suite of strategies.

Management Strategies That Might Help Build a Robust Framework for Conservation of Salmonids in Estuaries

While fishing for salmonids in estuaries may be more effectively managed today (e.g., poaching may be controlled), industrial developments such as ports may continue to threaten salmonids' estuarine habitats. Further, how a society values each salmonid species for harvesting in an estuary influences management strategies and may or may not align with ecological goals (Carey et al. 2012). Harvesting of salmonids in estuaries, and elsewhere in their life cycle, may be the most important factor affecting the long-term survival of a population. Fishing removes adults, which are the survivors from groups of individual fish that have benefited from the estuary's resources as juveniles, sea-run migrants, or kelts and then returned to the estuary after time in the ocean. Fishing is an ecosystem services issue that can involve finding a balance between provision of food (commercial and subsistence fisheries), recreational services (sport fishing), or other social needs (e.g., First Nations spiritual value). At the same time, destruction of habitat results in loss of other services. For example, dredging of sand and mud flats to provide docks for deep ocean shipping could lead to habitat loss and a reduction in salmonid survival. Sometimes habitat damage can be mitigated by proper site planning or risk analysis (List, Mirchandani, Turnquist, and Zografos 1991) may be applied. Thus estuaries are areas where salmonids are particularly vulnerable to human activity at several stages in their life history. Society has to decide which source of salmonid estuarine mortality is tolerable or, using the best available science, strike a balance.

Conservation managers working in estuaries therefore have to take into account competing needs such as "salmonids *versus* ports" and, working with conservation scientists, seek consensus among diverse groups of people (Kareiva and Marvier 2012). The comprehensive landscape approach is a possible means to that end. A comprehensive landscape or community-based approach to ecosystem-based management incorporates consideration of the human biological system as well as the estuarine ecosystem (Rieman et al. 2015). This approach is viewed as an alternative to centralized approaches to resource management (Gruber 2010) and can include a variety of strategies. The management of the salmonid estuary is an excellent example of the mix of scientific, management, and social challenges involved in developing a comprehensive landscape approach for conserving and restoring ecosystems.

Estuaries as Conservation Areas
Protected-area strategies or similar large-scale concepts of marine spatial planning can be used as conservation tools to maintain the flow of ecosystem services provided by the salmonid estuary. The implementation of a salmonid estuary conservation goal could be considered an element of a Marine Protected Area (MPA) strategy, a field of endeavour that has become fully developed only in the past few decades. MPAs may be generally defined as "areas of the ocean designated to enhance conservation of marine resources" (Lubchenco et al. 2003). Some of the earlier papers on MPAs clearly recognized the importance of estuaries in their fullest landscape sense (Agardy 1994), although without a focus on salmonids. The goal of some estuarine MPAs, particularly those where fishing is prohibited, is to achieve greater local fishery production from previously overfished areas (McKinley et al. 2011). This is a difficult goal for salmonid estuary managers without considering the fisheries in the river and ocean that intercept migrating salmonids on their way through the estuary. As a compromise, estuaries may also be included as components of MPAs (Jamieson and Levings 2001), as in Gilbert Bay, Newfoundland and Labrador. Subsistence fisheries for Atlantic salmon and Arctic char are allowed (landings restricted to personal consumption), but habitats are protected (Wroblewski et al. 2007). In general, the focus of the MPA strategy in salmonid estuaries has been on the importance of their habitats in the provision of ecosystem services that support salmonids.

Countries with endangered or unique anadromous salmonid populations include estuaries as important areas for ecosystem conservation in national habitat cataloguing systems for management (e.g., the *United Kingdom Estuary Guide*, developed for flood and coastal erosion risk [ABPmer and HR Wallingford 2007]). The estuaries in these cataloguing systems may not be listed as MPAs, but the data therein may support fish species other than salmonids with estuarine requirements that are listed as significant in the national or regional policy. Some of these fish species may be "umbrella species" – a species whose conservation is expected to confer protection on a large number of naturally co-occurring species (Roberge and Angelstam 2004). In the Severn River estuary in England, the twaite shad (*Alosa fallax*), an anadromous species with estuarine requirements similar to sea trout and Atlantic salmon, was mentioned as a species whose presence made the estuary a Special Area of Conservation (Severn Estuary/Môr Hafren European Marine Site 2009). In the United States, the National Estuarine Research Reserve System (National Oceanic and Atmospheric Administration 2014b), established in 1972, provides habitat and biological protection for selected estuaries. Some reserves specifically protect salmonid estuarine habitat (e.g., vegetation in Padilla Bay near the Skagit River estuary, Washington [National Oceanic and Atmospheric Administration 2014b]). The Pacific Estuary Conservation Program in British Columbia (Nature Trust of British Columbia 2013) is another example of a focused effort to conserve estuarine habitat, in this case by purchasing wetland habitat. The Norwegian salmon fjord system, which prohibits salmon farming and other development in certain estuaries in order to protect wild Atlantic salmon, also deserves mention (North Atlantic Salmon Conservation Organization 2009). Estuarine habitats of salmonids may be protected with wetland-requiring species other than fishes. For example, habitats of Sakhalin taimen may be afforded protection in the Akkeshi Lagoon (Bekanbeushi River estuary, Japan) because the area is a bird sanctuary under the Ramsar Convention on Wetlands (Ramsar Convention Secretariat 2014). Estuary management to protect salmonids is now being implemented in the Southern Hemisphere. Sea trout habitat in the Manawatu River estuary in New Zealand is also protected under the Ramsar Convention on Wetlands (Ravine 2007), and the Aysén Fjord in Chile is the subject of a coastal management scheme (Marín, Delgado, and Bachmann 2008). These examples provide perspective on some of the MPA-related strategies used to conserve salmonid estuaries.

In summary, designation of natural estuaries or undisrupted parts of estuaries as protected areas, using a variety of strategies tailored to a region, is important for conserving salmonids. Remote or relatively undisturbed estuaries can provide researchers with important insights into how natural estuaries function and are a source of intact genetic resources (Levings 1993). Park systems can contribute to the task; for example, the Noatak National Preserve (within the United States National Parks system, on the Arctic northwest coast of Alaska) prohibits ocean ranch/hatchery operations within its boundaries (Eggers and Clark 2006). International initiatives are also important, such as the protection of sea trout (Black Sea salmon) within the UNESCO Biosphere Reserve on the Danube River estuary in Romania (Navodaru, Staras, and Cernisencu 2001). Because social dimensions need to be considered in different ways in various regions (Klein et al. 2008), it is likely that broad-based management and planning schemes (the comprehensive landscape approach) will be required.

Salmonid Estuary Management Plans Involve a Variety of Ecosystem Services

Estuary management and planning are essentially subsets of coastal zone planning, marine spatial planning, and ecosystem-based management, which are very large topic areas covered in review articles (e.g., Ballinger et al. 2010), websites for specific areas (e.g., the northwest Atlantic coast [Atlantic Coastal Zone Information Steering Committee 2014] and the British Columbia coast [British Columbia Marine Conservation Analysis 2014]), and books (e.g., Moksness, Dahl, and Støttrup 2013). And, as pointed out by the authors of the Fowey Estuary Management Plan in England, "estuary management" is a generic term representing a spectrum of planning and management activity. Factors ranging from ecological concerns to historical aspects to tourism are important (Fowey Estuary Partnership 2014), and to this I would add fishing plans to promote conservation or enhancement of the salmonid resource. The management and policy implications should be supported by science, and so the design of management measures and metrics used also requires attention.

The history of salmonid estuary planning clearly shows that successful ecosystem-based management requires sensitive socio-ecological assessment procedures, tools for evaluating ecological quality, and well-built monitoring programs based upon pertinent indicators (Ducrotoy and Dauvin 2008). Assessment procedures could include socio-ecological aspects such as development of a vision, goals, and objectives; tools could include salmonid survey procedures and methods; and monitoring programs could include structured decision making incorporated into an adaptive management plan (Runge 2011). Planning should be tailored to local governance arrangements as well as appropriate spatial scales. Essentially, institutional arrangements, such as working relationships between planning and management groups and how they interact, should be taken into account. Coastal zone management books and general planning references (e.g., R. Gregory et al. 2012) cover many of these issues. Because of the iconic status of salmonids in estuaries, the role of the citizen in planning is paramount, as are partnerships between agencies, conservation groups, and development-oriented groups such as harbour authorities.

Differences in political systems and citizen attitudes are variables that affect partnership success, factors that are difficult to measure (Dorcey 2010) but important to consider in any ecosystem-based management initiative (Layzer 2012). Institutional and economic understanding can be important in achieving successful estuary management and conserving salmonids. Institutions are the mechanisms that integrate the human and ecological spheres (Hanna 2008). Improved understanding among institutions can be achieved through the involvement of a mix of expertise in salmonid estuary planning groups. For example, the board of directors of the Lower Columbia Estuary Partnership (Lower Columbia Estuary Partnership, 2015) in Portland, Oregon, includes representatives from the agriculture, shipping, property law, and education sectors, as well as planners and estuarine ecologists. All sector representatives work towards restoring salmonid habitat in the estuary, essentially using bargaining and economic trade-offs between land uses when deciding which part of the estuary should be restored for salmonids. The decision-making process (including socio-economic trade-offs) to protect and restore estuarine salmonid habitat can be smoother where the local governance structure recognizes the role of incentives and transaction costs. The containment of transaction costs requires an institutional setting that reflects the properties of both the salmon ecosystem and human systems (Hanna 2008); and these have to be tailored to the local setting. Other socio-ecological factors entering into salmonid estuary management plans are more directly related to salmonid production as an ecosystem service planning task with the fisheries management sector.

An important value-laden trade-off issue is the societal decision in some parts of the world on whether to conserve salmonids through ocean ranching, hatcheries, and other artificial rearing methods or to focus totally on restoration of wild populations (see G.J.A. Kennedy and Crozier 1997). This choice has an important bearing on how to manage an estuary. For example, juvenile Chinook salmon released from hatcheries are usually smolts that have different habitat requirements from wild fish subyearlings. Careful management of shallow-water habitats may not be "cost-effective" if the wild smaller fish are not present. Decisions may have been made to supplant them with hatchery-reared fish. Matching the desired salmonid life history with its estuary should be explicitly considered in conservation plans.

Another ecosystem service issue relates to the increasing domestication of salmonid production using aquaculture, and the challenges involved in locating salmonid farms relative to sensitive habitats of wild fish (Galland and McDaniels 2008; Moksness, Dahl, and Støttrup 2013). This siting problem involves economic issues, such as cost to transport feed to the farms. However, one criterion for siting salmonid farms also relates to distance from estuaries. Although such farms are usually located outside of estuaries in higher-salinity water, where the fish grow faster, there may also be ecological reasons for the separation. A possible rationale for siting relates to issues such as straying of escaped fish into the estuary and, if they are adult salmonids, possible colonization of the stream (Levings et al. 1995). Rainbow trout can travel 25–40 km within a week after escaping from fish farms (Skilbrei and Wennevik 2006). Rainbow trout that escaped from fish farms in Chile interbred with established populations in streams (Consuegra et al. 2011), showing the possible introgression of genes from farmed salmonids to wild stocks. The possible introduction of disease or other health problems from the farmed fish to wild fish is another concern (Galland and McDaniels 2008). For example, salmon lice from infected Atlantic salmon farm cages were found on sea trout up to 31 km distant, providing data for possible application as a separation distance from sea trout estuaries (Middlemas et al. 2013). The escape rates of farmed fish into estuaries vary around the world, depending on cage structures, monitoring procedures, local regulations, and ocean conditions such as storm prevalence. Escaped Atlantic salmon have established in systems in northern Europe (e.g., the Teno River in Finland and Norway [Erkinaro et al. 2010], and rivers in the Faroe Islands [Thorstad et al. 2011]) and their colonization prospects elsewhere are still an open question (Thorstad et al. 2008). Another possible issue for wild estuarine salmonids is the cumulative effects (e.g., nutrient enhancement, dissolved oxygen depletion, and other waste issues) of fish farms in the coastal water body adjacent to the estuary (Galland and McDaniels 2008). In most countries where salmonid farming is practised, additional comprehensive environmental assessments and monitoring are required in order to deal with the waste problem issue (e.g., Scottish Environment Protection Agency 2014b).

Coastal planning and ecosystem-based management are components of the comprehensive landscape approach and offer good strategies for the proper use of resources and for sustaining salmonid estuaries. There is an increasing number of actual ecosystem-based plans in various regions. A decade ago, only one out of nine estuary management plans in British Columbia called for the ecosystem approach (G.L. Williams and Langer 2002). The development of the Pacific North Coast Integrated Management Area in British Columbia, which incorporates watersheds, estuaries, and the adjacent ocean together with coastal community social requirements, is an example of a program that can move this region closer to an ecosystem-based management approach (Pacific North Coast Integrated Management Area 2014). The inclusion of traditional ecological knowledge in discussions with communities can help in this regard (Menzies and Butler 2006). Implementing coastal planning and ecosystem-based management is not easy. To assist, a growing number of decision-making tools aim to include socio-economic considerations in planning and management of natural resources in general (e.g., Natural Capital Project 2014; models relating to climate change [Convention on Biological Diversity 2012]). The number of jurisdictions requiring conservation measures will increase as anadromous salmonids spread into more Southern Hemisphere estuaries and into high-latitude Arctic and Subarctic estuaries in the Northern Hemisphere as rivers and oceans warm with climate change (Nielsen, Ruggerone, and Zimmerman 2013; Cheung et al. 2013).

Looking Upstream and Downstream
The management of salmonid habitat in estuaries is inextricably connected with river and ocean dynamics, so both should be under the joint purview of river, estuary, and ocean managers (Wolanski 2007). Several salient river and ocean factors need to be kept in mind when planning. River flow is crucial to ecosystem functioning in estuaries, and various management strategies have been suggested to maintain adequate discharges. For example, "environmental flow" refers to a variable water flow regime that has been designed and implemented, such as through intentional releases of water from a dam into a downstream reach of a river and then into the estuary, in an effort to support desired ecological conditions and ecosystem services. Environmental flows that maintain the integrity of the estuary – and hence sustain salmonid populations using the estuary – are a function of the state of the river and its watershed (Newson, Sear, and Soulsby 2012) and are especially important. Environmental flows have been raised as a major concern in the conservation of iteroparous species of salmonids, whose adults, sea-run migrants, returning veterans, and kelts sometimes must pass through a dammed river with fishways and its estuary multiple times during their life (Kraabøl et al. 2009). Spills from reservoirs are often not planned:

they are conducted to prevent flooding, to avoid stress on dams, and for other engineering reasons. The challenge of river and estuary management planning therefore no longer relates to an engineering problem (how to control the river and estuary) but is a management role (how to allocate river water among competing uses while still maintaining ecosystems) (Zhou and Tong 2010). An example of an effort to jointly manage the river and estuary is the Thames Estuary Partnership (2015) in England, which brings stakeholders together to work towards the best social, economic, and environmental outcomes for the Thames. The Partnership specifically includes stakeholders concerned with the river and its tributaries (The London Rivers Action Plan 2015). There are three key steps that could help prevent the deleterious effects of dams on estuaries and their watersheds: (1) integrated river basin planning, including stakeholder surveys, options assessment, and dam siting; (2) dam designs, operational plans, and monitoring; and (3) adaptive management (B.D. Richter et al. 2010). Implementation of these steps is essential for the maintenance of estuarine salmonid habitat on regulated rivers. However, integration of ocean ecosystems is also a necessary consideration for all river-estuary systems.

While the whole coastal ocean is basically uncontrollable (except for climate change and greenhouse gas management), there are factors in the proximate ocean that may be. Features such as riverine channel depth – and hence salt wedge penetration – can be influenced by dredging. Incoming tidal currents are controlled by tidal surge barrages. There is scope for further incorporation of structural changes in the estuary. For example, some types of channelization may be beneficial if they result in ocean flows that restore countercurrent flows in estuaries that have become shallow due to filling.

In summary, management strategies need to transcend distinctions between freshwater, estuarine, and marine habitats and consider critical habitat links between them. Many of the stressors affecting fitness elements described in detail in Chapter 14 (water quality, freshwater flow, habitat loss, biotic interaction changes, etc.) can be put in perspective by describing their seaward (downstream) and landward (upstream) provenance (Table 8). Because of this two-pronged-threat scenario, the salmonid estuary is extremely vulnerable to deleterious change, possibly reducing its adaptive capacity to cope with major ecological shifts such as climate change (N.L. Engle 2011).

Although the stressors can be dealt with on a case-to-case basis, the greatest challenge for salmonid estuary science is to understand their cumulative effects on the various salmonids. The problem of cumulative

Table 8

Summary of potential external stressors upstream and downstream of the salmonid estuary

Stressor	From seaward side (downstream)	From landward side (upstream)
Water quality	Climate change, especially increasing temperatures	Watershed activities and wastewater discharge, which can reduce dissolved oxygen and increase contaminant levels
Ocean regime shifts	Reduced upwelling, which can reduce nutrient supply to the estuary	Changes in rain and snowfall, which can modify river discharge
Flow	Barrages affecting tidal circulation	Dams and barriers, which can block river flow
Migration barriers	Storm surge barrages	Dams
Sediment	Coastal drift cells, which can provide sand and gravel may be blocked	Dams may trap sediment essential for deltaic processes
Habitat loss and food web changes	Upwelling and climate change	Urban and industrial development
Changes in estuary volume	Dredging for deep ocean shipping	Major floods (erosion can deepen channels) or excessive erosion in watershed (infilling and excessive sediment can result in decreases in water volume and also decreases sediment capacity)
Harvesting	Coastal interception, which does not allow protection of weak stocks	If upper river stakeholders are allocated more fish, this could result in increased conservation in the estuary
Density effects	Returning adults from ocean ranching can cause bycatch problems	Hatchery releases could cause "overloading"
Invasive fish species	Some species come from the ocean, e.g., shad	Majority of species appear to be arriving from upstream, e.g., cyprinids

Source: Modified from Levings 2000b.

effects is not unique to salmonid estuaries (Crain, Kroeker, and Halpern 2008), but is particularly challenging to understand because the dynamics of the river, estuary, and ocean need to be considered together when survival is assessed (Ross et al. 2013). In addition, as pointed out by Kennish and colleagues (2008) and Breitburg and colleagues (1999), whether stressors act synergistically or antagonistically or are neutral appears to be context-dependent, contingent on the stressor, biological taxon, and background information such as ambient nutrient levels in eutrophicated estuaries. Integrated data banks on river, estuarine, and ocean habitats are required so that the stressors can be examined together. Statistical analysis using an information-theoretic or multi-model approach (Burnham and Anderson 2002) with test river flows and estuarine juvenile Chinook salmon survival data in the Sacramento–San Joaquin River estuary in California showed the value of this approach (Zeug and Cavallo 2013).

Besides integrating river and ocean information in the immediate vicinity (the estuary's watershed and the adjacent ocean), salmonid estuary managers have to be aware of incoming invasive species. These can also arrive from upstream (often secondary dispersal from already established populations) and downstream (primary spread of new inoculants from elsewhere in the world).

Preventing Invasive Fish Species from Establishing in Salmonid Estuaries

Several methods can be deployed to manage invasive species, focusing on controlling or limiting the routes or vectors that fish invaders use in order to disperse. Control programs (such as bounty fishing for invaders) are possible, but may not be effective in estuaries, unlike in rivers and lakes, which are more bounded ecosystems.

Some invasive fish species found in salmonid estuaries come from upstream in the watershed, where they have been accidentally or deliberately released. Others may be escaped salmonids from aquacultural operations in the coastal region. If humans release live fish from sources outside the watershed (e.g., for bait, fish released from aquaria, etc.), then direct observation and removal, regulatory checks, and educational programs can help. The prompt removal of a single northern snakehead fish from an isolated pond not too far from a tributary of the Fraser River estuary is an example (Scott et al. 2013). The spread of fish via ships' ballast water (e.g., the black goby [Wonham et al. 2000]) is an international problem that can be reduced by ensuring that ballast water is not taken up in one estuary and released in another (Di Bacco et al. 2012). There are few management options for invasive species that move into the estuary and river from the ocean (e.g., American shad [Sanderson, Barnas, and Wargo Rub 2009]). Risk assessment tools can also help forecast which species are more likely to spread from one system to another (Copp et al. 2009), and results can then be used to focus regulatory checks. International data banks on the ecological requirements of potentially invasive fish species can help set up risk assessment models and environmental match/mismatch matrices.

Monitoring of the estuarine fish community is required for early detection of invasive species. Regional data banks (including photo banks) provide the basic information on species' presence. At the local level, data banks can enhance vigilance by providing scientific resources to assist citizen scientists (see Appendix 3 at http://hdl.handle.net/2429/57062), thus deploying many additional eyes on the estuary. Invasives other than fishes (e.g., vegetation and invertebrates) are part of the problem, because their growth and development change estuarine salmonid habitat and food webs. Disrupted and early stages of restored estuary habitat are more susceptible to invasive species and procedures may be required to prevent or reduce this threat. Long-term monitoring is therefore also required to successfully implement restoration in salmonid estuaries.

Restoration of Salmonid Estuarine Habitat

Salmonid estuary restoration is a widely used and accepted management methodology that deserves discussion. Ecological restoration is trying to do in a matter of years what takes decades or centuries under natural conditions (Hildebrand, Watts, and Randle 2005). Therefore, it is not surprising that restoring salmonid habitat in estuaries is subject to uncertainty, and although researchers have proposed scientific criteria to measure success or failure (e.g., Simenstad and Cordell 2000), the criteria have not been widely applied. As well, there has been discussion about the importance of balancing restoration of structure, function, and process in restoration projects (L.R. Brown 2003; Simenstad, Reed, and Ford 2006). Hydrological models can help design restoration strategies to ensure that functions such as connectivity are realized (e.g., Diefenderfer and Montgomery 2009). Restoration may not be possible because of economic or social constraints. In highly industrialized estuaries such as the Rhine River estuary in the Netherlands, some habitat changes are recognized as irreversible (bij de Vaate, Breukel, and van der Velde 2006).

As well, there are important contextual issues, such as the ability of the water overlying the restoration projects to support salmonids. Pollution abatement may be required.

Numerous estuarine salmonid habitat restoration programs have been conducted in various regions. A variety of books and review papers are available for guidance on salmonid estuary restoration (e.g., Cattrijsse et al. 2002; C.A. Rice et al. 2005; Ducrotoy 2010; Borja et al. 2010).

Beaches

The purposeful restoration of beach habitat has received very little attention. However, dredging of gravel, sand, and mud in estuaries to maintain shipping channels and develop port facilities is a common practice in industry and has frequently led to the unintended creation of beach habitat. If sand replaced an artificial subhabitat such as riprap that destroyed a natural beach, this could be viewed as restoration. Disposal of the material leads to creation of "spoil" islands, artificial beaches, and sandbanks along channels. Created beaches may mimic natural habitats and juvenile salmonids can use them for feeding and possibly other functions (e.g., Chinook salmon in the Fraser River estuary; Macdonald 1984).

Channels

Recovery of riverine channels re-establishes connectivity between the ocean and the river for salmonids going to the ocean or returning to spawn, and is therefore a very important aspect of ecosystem recovery (Kondolf et al. 2006). Dam breaching can restore natural river flows into the estuary (Pess et al. 2014). Barriers to adult salmonid passage caused by tidal weirs, shipping locks, and blocked or lost riverine channels may be overcome using special channels (fishways) that bypass the blocking structure (e.g., Scheldt River estuary, Belgium [Buysse, Coeck, and Maes 2008]). At the Tuloma River estuary in Russia, fishways were built to facilitate salmonid migration past a hydroelectric dam (Karppinen et al. 2002). Breaching of causeways can quickly restore riverine channel habitat for migrating salmonids (e.g., sea-run brook trout after breaching of a dam on the tidal part of the Petitcodiac River, New Brunswick [Petitcodiac Watershed Alliance 2014]). Tidal channels, especially those leading to higher intertidal zones, can be restored and reconnected to riverine channels by digging trenches through disrupted or filled areas. This was the approach taken to restore tidal channels for coho salmon fry at the Squamish River estuary in British Columbia (Ryall and Levings 1987).

Vegetation
Some early attempts to restore brackish marsh were conducted in the Fraser River estuary. Marsh plants were transplanted into areas that had been disrupted, or onto natural sand flats or artificial sand flats created by dredge spoil material. The viability of replacing natural sand flats with vegetation habitat requires further investigation (Levings 2000a). In some situations, salt marsh has been purposely developed by creating a relatively high elevation platform from dredged material, which has then been colonized naturally by drifting plant fragments (Adams and Williams 2004).

Dike breaching to restore water flow into previously flooded vegetated areas can recover vegetation quickly and does not involve habitat substitution. Using this method, the abundance and productivity of the basal vegetative elements of the food web may be at least partially recovered within a few years, while increase in water volume available to salmonids is immediate as fish move in from adjacent channels. This technique has been used to restore marshes and tidal swamp-forested vegetation in the northeast Pacific (e.g., Nisqually River estuary, Washington [http://hdl.handle.net/2429/57062]) and numerous other estuaries in the region; see Gray et al. 2002 for a representative study). Breaching has also been conducted on the Mersey estuary, England, and although salmonids were not recorded after the breach, other juvenile fish were (Colclough et al. 2005). An assumption with this technique is that the native plant assemblage will reappear, following natural succession, but this does not always occur. Sometimes invasive plant species will dominate the recovered habitat (Tanner et al. 2002). Subsidence of the previously unflooded area can be issue as well, as it can take decades for sedimentation to restore the estuary to an elevation suitable for native plants (Frenkel and Morlan 1991), depending on the local sedimentation regime. A second method for restoration of brackish marsh that does not involve substitution habitat is removal of the fill material to enable recovery of the vegetation. Some plant rhizomes and seeds can survive under fill (Allison 1995) and the plant material may remain viable for years.

Wet meadows are sometimes difficult to restore, because they are often behind dikes designed to protect urban areas from flooding or to develop land for farming agriculture ("polderization"; see Ducrotoy 2010). However, they can be restored under some circumstances. In Europe, an innovative technique used to restore wet meadows is the creation of flood control areas under the influence of a controlled reduced tide

(Jacobs et al. 2009), but it is not clear whether this technique is useful for restoring salmonid habitat.

There is an extensive literature on eelgrass restoration (Paling et al. 2009). Restoration of this plant typically involves the transplanting of shoots (van Katwijk et al. 2009). Thom and colleagues (2012) found that survival of transplanted eelgrass in northeast Pacific estuaries was affected by algal blooms of "unusual magnitude"; thus, potential restoration of this habitat is dependent on the local context. Restoration of algal habitat can be accomplished by establishing hard material (e.g., rocks or riprap) for algal spores to settle on; in some instances, transplanting of sporelings has been conducted (e.g., *Laminaria saccharina* [Russell et al. 1983]). Riparian vegetation restoration to aid the recovery of salmonids has received considerable attention in rivers and streams (Nislow 2010) and methodologies (e.g., transplanting riparian shrubs such as willow) are similar to those for brackish marsh vegetation. Riparian vegetation restoration was integrated into a general landscape model for the Sacramento River estuary in California (Kimmerer, Murphy, and Angermeier 2005).

Artificial habitats and habitat from invasive ecosystem engineers
Restoration from the effects of artificial habitats such as docks, wharves, and other structures can be achieved when ports are abandoned or facilities are removed to another location. Removal or control of invasive ecological engineers such as invasive plants is more difficult, sometimes requiring chemical control (Toft et al. 2003) or annual harvesting.

Water quality improvement
An overarching issue related to habitat restoration is the problem of water quality improvement. Clearly, if water quality is a limiting factor for salmonids in an estuary, then habitat restoration will be unsuccessful. This is an aspect of cumulative effects (Hildén and Rapport 1993) – both water quality and habitat quality/quantity should be improved in an overall program. Water quality has been improved in salmonid estuarine regions such as British Columbia using a variety of treatment methods, especially for pulp mill effluents (Waldichuk 1993), and acid mine drainage problems have been resolved with water treatment plants (O'Hara 2007). Provision of light by specialized systems has been proposed to overcome shading under docks (Ono and Simenstad 2014). Advances in sewage treatment technology and other types of pollution control (see Pokhrel and Viraraghavan 2004) have improved dissolved

oxygen levels in many of the formerly polluted estuaries of the world, enabling safe migrations through rivers and estuaries by salmonids (e.g., Mersey River estuary, England [Jones 2006]; Seine River, France [Perrier et al. 2010]). However, the situation may be less optimistic for legacy contaminants that are buried in estuarine sediment (e.g., Columbia River estuary [Alvarez et al. 2013]) and are a threat to salmonids when the estuary is disrupted by dredging. Water quality and habitat restoration should be able to proceed in unison to restore the various degraded components of the salmonid estuary, but the process can be challenging and requires monitoring and evaluation.

Monitoring and Evaluation of Restoration

Monitoring is a part of the required adaptive management process (Marmorek and Peters 2001) and is applicable to both habitat restoration and water quality. While automated water quality monitoring systems can help track improvements in salmonid estuary water properties (e.g., Glasgow et al. 2004), monitoring of estuarine habitat recovery requires different techniques.

The environmental variability of salmonid estuaries can occur at various spatial (e.g., square metres, hectares) and temporal scales (days, months, years) (Kimmerer et al. 2008), and should be allowed for in the evaluation in any restoration scheme (Moyle et al. 2010). For example, thirteen years after brackish marsh was restored by planting vegetation on artificial intertidal islands at the Campbell River estuary in British Columbia, the vegetation community was still changing (Dawe et al. 2000). Thus, while a number of bioengineering techniques are available to help restore salmonid habitat in estuaries, it is important to consider a number of complex ecological issues, such as spatial and temporal variability and recovery time when planning and evaluating the role of restoration as a management tool (Palmer 2009). Long-term (decade-scale) monitoring of the success of restored habitat has rarely been done, but temporal changes beyond the immediate restoration effect are to be expected (Thom, Zeigler, and Borde 2002). There are also policy implications when habitat restoration methodology is used in mitigation or compensation projects in an attempt to balance industrial and ecological needs in salmonid estuaries, but these are beyond the scope of this book. The following is an overview of some of the complexities involved in restoration evaluation and of scientific needs for evaluation design and the adaptive management scheme.

Functional equivalency is often assumed, but is restored estuarine habitat actually ecologically equivalent to those that had been lost

(Palmer 2009), and is it performing the same functions to support salmonids? In this context, the issue of control or reference habitats when assessing success or failure of habitat restoration is a design issue not often assessed, and usually comparisons are with an undisrupted area of the subject estuary, which can confound statistical analyses. An out-of-estuary reference area is preferable, but this may also be problematic, because in many instances the historical estuarine habitat and fish community is no longer present in an entire region, often due to the very low abundance or extirpation of salmonids. The Gironde River estuary in France is an example. Salmonids are a minor component of the estuarine fish community guilds in contemporary studies (Lobry et al. 2003), even though they were likely more abundant historically. The Tagus River estuary in Portugal is another example. Salmonids were not mentioned in a study of this estuary (Cabral, Costa, and Salgado 2001), but Atlantic salmon were present historically in this river system and hence would have been present in the estuary (Netboy 1980). However, even an approximate reference estuary can be useful. An example of the importance of a reference estuary is the comparative study of the heavily industrialized Seine River estuary and the pseudo-natural Somme River estuary, also in France (Ducrotoy and Dauvin 2008). For planning and evaluation of estuary habitat restoration, the use of hydrodynamic models (Levings and Nishimura 1997; Yang et al. 2010) and mesocosms (Callaway, Zedler, and Ross 1997) can also be useful.

Metrics defining performance of estuarine habitat restoration include measures of benefit to salmonids, such as recovery of food species, diet and foraging success, residency and growth, condition, and life history diversity (Simenstad and Cordell 2000; Gray et al. 2002; Bottom et al. 2005b). These are important metrics, but the most important challenge for technologists involved in salmonid estuary restoration is assessment of success in terms of improved salmonid survival. Actual survival in terms of fry to smolts or smolt to adult could be measured experimentally in restored and reference habitats. Population models can assist in this regard. Leslie matrix modelling, which uses an age-structured model of population growth, was used by Greene and Beechie (2004) to simulate density-dependent interactions at the Skagit River and Duwamish River estuaries in Washington. The authors compared the relative importance of different habitats under three density-dependent scenarios: juvenile density-independence; density-dependent mortality within streams, delta, and nearshore; and density-dependent migration among streams, delta, and nearshore. Their findings indicated that nearshore habitat relationships may play significant roles for salmon populations and that the relative importance of restoring habitat in various areas will depend on the mechanism of density-dependence influencing salmon stocks.

In summary, restoration of degraded salmonid estuaries should be an integral part of any conservation management plan. Estuarine salmonid habitat restoration requires ongoing scientific guidance to develop goals and objectives, to apply proper techniques and monitoring tools, and especially to incorporate ecological processes in the methodology. The life history targeted is an important issue when developing goals. The role of the community and citizen scientists in restoration of salmonid estuarine habitat cannot be overemphasized. There are many projects that local conservation groups can assist with to restore salmonid habitat, ranging from community-based mapping to scoping out of potential areas, to helping with monitoring effectiveness (see Appendix 3 at http://hdl.handle.net/2429/57062). Community involvement in restoration projects can help foster buy-in and expedite the joint engagement of ecological and social sciences needed in the comprehensive landscape approach. Ecological integrity and resilience are more conceptual aspects of the restoration of the physical and biological features of the salmonid estuary that also need to be considered.

Recovery or Maintenance of Ecological Integrity and Resilience as Salmonid Estuary Management Goals

An overarching goal for estuary managers concerned with conservation and survival of wild salmonid populations should be the maintenance and recovery of ecological integrity and resilience. There are various definitions of these terms, but the following captures most of the two concepts. Ecological integrity is said to exist when an ecosystem is deemed characteristic for its natural region, including the composition and abundance of native species and biological communities, rates of change, and supporting processes (Parks Canada Agency 2000). Resilience was earlier defined as the amount of disturbance that an ecosystem can withstand without changing self-organized processes and structures (defined as alternative stable states in Holling 1973). Resilience is now also considered to include responses of the socio-economic system (e.g., Kareiva et al. 2007; Bottom et al. 2009) and fisheries (Healey 2009), but in this book I focus on the narrower ecological aspects.

Recovery of estuarine functions to date has stressed productivity and abundance rather than ecological integrity and resilience, possibly because ecological measures are not usually a component implemented in the management plans. Researchers have documented how estuarine restoration can recover specific habitats used by specialized life histories (e.g., channels and vegetation used by estuarine-rearing coho salmon fry at the Columbia River estuary [B.E. Craig, Simenstad, and Bottom 2014]), so the knowledge base on resilience recovery is growing. A major problem of using resilience as a measure, when it is defined as the capacity to absorb or to recover from a perturbation, is that this is known only for the largest disturbance yet experienced for specific ecosystems. Because of our limited ability to predict the dynamics of an ecosystem under stress, it is difficult in practice to quantify the resilience of an ecosystem (Maltby, Paetzold, and Warren 2010). Nevertheless, a variety of metrics have been proposed in salmonid estuaries for these important ecological variables.

Ecological integrity can be assessed by monitoring community structure and comparing it with reference natural systems or historical conditions using indices such as species lists or measures of species diversity, but ecological processes that maintain the structure also need to be considered. Because invasive species are likely to be constantly moving into salmonid estuaries, their presence and confounding effects make identifying reference areas a challenge. Nevertheless, attempts have been made in some regions to reconstruct the reference fish community in the estuary. The fish-based Index of Biological Integrity developed by Scholle and Schuchardt (2012) is an example; a historical reference represents a benchmark for the fish-based assessment of the ecological status of this index. The current ecological status is assessed through the similarity or dissimilarity to the reference community and classified in the relevant status class according to a five-level system (high, good, moderate, poor, and bad). Atlantic salmon and sea trout were included in a diadromous species guild (i.e., fishes that migrate to the ocean for spawning, such as European eels, as well as anadromous species such as salmonids) in the analyses by Scholle and Schuchardt (2012). Process-related measurements of the biotic interactions and processes, environmental regimes, and landscape structure are more challenging to obtain in the salmonid estuary than species composition, but are essential to consider when establishing goals for maintaining ecological integrity. At the minimum, a conceptual ecological model (e.g., processes to maintain Chinook salmon in the Cosumnes River Protected Area, California [Parrish, Braun, and Unnasch 2003]) should be developed to frame how key processes such as connectivity can be maintained in the salmonid estuary (e.g., avoiding diking projects that block flows between riverine channels and side channels). Thus, a continuing quest for measures of ecological integrity is important and needs to be linked with the question of how to assess resilience.

The number of life histories displayed by a salmonid species is a good measure of diversity, which in turn can reflect resilience in the sense that the species has a number of "bet-hedging" strategies available for coping with changing environments (Einum and Fleming 2004). As with ecological integrity, however, processes and the landscape approach are important. Identification and enumeration of the number of life histories originally shown by a species is one approach, keeping in mind that short-term evolution could have occurred. Other metrics are now being developed that might aid in measuring life history diversity. A complex statistical index that incorporates behaviour and morphological variation in individual juvenile Chinook salmon has been developed for the Columbia River (index of early life history diversity [G.E. Johnson et al. 2014]). Newer genetic tools such as microsatellites (Moran et al. 2013) and transcription analysis (T.G. Evans et al. 2011) are rapidly being applied to life history assessment. Modularity is a process and socioeconomic concept that could also be used to assess resilience and help maintain or reconcile life history diversity. Modularity is the process of maintaining redundant habitats or other entities (e.g., subpopulations, social groups, or economic units) that are connected enough to allow interactions, but not so connected as to experience the same conditions (Rieman et al. 2015). In the context of ecosystem services such as harvesting, greater efforts are needed to detect and preserve the full spectrum of life history strategies until the implications of their loss can be understood.

Resilience should be maintained by sustaining the full range of original life history types of salmonid species, even if they are not detected as adults in fisheries or if they occur as adults in small numbers with very high interannual variability (e.g., an almost extirpated population of Chinook salmon in an upper tributary of the Columbia River [Copeland and Venditti 2009]). Without extensive sampling, the minor contributors may be missed, or they may be forms that are too small for harvesting and pass through nets. Other traditional methods, such as scale analyses and related monitoring programs, may also miss minor life history types. Better identification of minor or specialized life histories, possibly adapted to unique habitats, could help restoration programs, because it is unrealistic to expect the programs to regenerate exact replicas of natural habitat, especially as climate change is modifying the original or historically recorded zoogeographic boundaries of salmonids (Cheung et al. 2012). While local adaption and short-term evolution may be powerful forces in salmonids (D. Fraser et al. 2011), the ability of the genome to cope with a variety of habitats and niche shifts may be present in isolated and as yet unknown populations (Dolloff, Flebbe, and Thorpe 1994). It is sometimes difficult to separate phenotypic and genotypic changes in the trait in question. Nevertheless, the versatility and colonizing ability of salmonid populations may arise from genetic material in marginal populations (Scudder 1989), such as those at boundary ranges or in small, unique habitats that can often be easily disrupted by our industrial and urban activities. As well, a reduction in habitat diversity can result in loss of resilience (Waples, Beechie, and Pess 2009). Preserving the original life history types could maintain the ability of the species to adapt to constructed or restored habitat and to other environmental variations, such as climate change (L.G. Crozier et al. 2008). If the restored habitat fails over time, the species might then be at a competitive disadvantage. On the other hand, if a species adapts to exploit restored or artificial habitat, the species could lose its capacity to benefit from natural areas.

Our knowledge of the specialized salmonid life histories that appear to be specifically adapted to estuaries is steadily advancing. Some examples include early life history forms that use the estuary (see Appendix Table 1.6 at http://hdl.handle.net/2429/57062), such as Chinook salmon migrant fry or subyearlings, nomad or estuary-rearing coho salmon, whitling or finnock sea trout that return to the river after a few months in the ocean, and river- or sea-type sockeye salmon; estuary-spawning chum and pink salmon in Alaska (Helle, Williamson, and Bailey 1964); older age classes of lake trout in the Canadian Arctic (Swanson et al. 2010); and maturing Atlantic salmon in the Koksoak River estuary in Québec (Robitaille, Cote, Schooner, and Hayeur 1986). The chances of missing these specialized forms are great now that wild salmonid populations have been reduced in many regions. It is hoped that DNA surveys and advances in genetics will enable collection of better information in the future.

Recognition of these forms by scientists suggests that resilience may be implicit in the thinking of estuarine salmonid ecologists in many regions around the world. For example, Leman and Chebanova (2005, 326) stated in reference to Chinook salmon (called "king salmon" by the authors) in the Bol'shaya River estuary in Russia: "The population of the Bol'shaya River King salmon is heterogeneous ecologically; the riverine type dominates and the abundance of King salmon migrating downstream (to the estuary and then to the ocean) as subyearlings is low. However, under favourable conditions it may produce strong generations." On the other side of the Pacific, J.A. Miller, Gray, and Merz (2010) reached a similar conclusion when documenting the difference in survival between large hatchery-reared Chinook salmon (parr and smolt) and small wild Chinook salmon (fry) in the Sacramento River in California. The fry contributed 20 percent of the fish that returned as adults, and the fry stage is known from otolith microchemistry to use the estuary for rearing.

Specific life history features and adaptations may enable the reutilization of estuarine habitats, and hence recovery of ecological integrity and resilience. Resource polymorphisms are defined as the occurrence of discrete intraspecific morphs showing differential niche use. Resource polymorphisms are usually expressed through discrete differences in feeding biology and habitat use (T.B. Smith and Skúlason 1996). Restoration of anadromous species that use estuaries but have been blocked by dams or other obstructions can occur with species that have consolidated landlocked forms in reservoirs above the barriers. One example is the return of anadromous sockeye salmon from nonanadromous kokanee populations thought to be landlocked behind a dam nearly ninety years after the dam was built at the Alouette River in British Columbia, a tributary to the Fraser River estuary. DNA data confirmed that the adult sockeye salmon were from the same population as the smolts that migrated downstream earlier through a new water release structure (Godbout et al. 2011). Populations elsewhere, such as Black Sea salmon (sea trout) in the Dnieper River, Ukraine, which are now isolated by reservoirs and are found only as the freshwater form in the upper river (Vasil'eva 2003), may also have the potential to recover should the barriers be removed. Genetic surveys of the species in blocked areas could help forecast such restoration potential. Another example is the potential for anadromy in landlocked Atlantic salmon, as shown by tagging experiments and isotope analysis for this form in a New Brunswick watershed (Carr, Whoriskey, and Courtemanche 2004). Resource polymorphisms, specific to estuary use by salmonids, can be a vital component of resilience recovery.

Conclusion

A number of strategies are available to help conserve estuarine salmonids and their habitats, ranging from spatial planning initiatives to MPAs to restoration. Importantly, strategic planning must look upstream and downstream, not just in the estuary. Perhaps the major challenge in implementing the strategies is the issue of cumulative effects, but evolving statistical methods and modelling may help. Maintenance of resilience and recovery of ecological integrity are two important concepts that managers need to focus on while developing plans for salmonid estuaries. Although many authorities recommend reconnecting social and ecological resilience in salmonid ecosystems in the comprehensive landscape approach, this an anthropocentric issue that requires much social discourse for implementation (Lackey, Lach, and Duncan 2006; R. Gregory et al. 2012). Considering the wide range of societal arrangements involved in salmonid estuary management around the world, it is likely that multiple paths will be taken in the policy and management arenas to achieve conservation goals. As the various policies are chosen, it is prudent to continue focusing on scientific efforts to improve our understanding of salmonid ecology in estuaries.

19
Conclusion

Progress in Our Knowledge of Salmonid Ecology in Estuaries

Our knowledge of salmonid ecology in estuaries, and the application of that knowledge, has increased dramatically in the past few decades, and I salute colleagues who have persevered with their research and studies. Developing hypotheses, designing studies, and executing field studies in the challenging estuarine environment and in the laboratory require dedication. Their efforts have been fruitful. Using a standard citation source (Google Scholar, April 22, 2015) with the keywords "estuarine salmonid" to compare the number of citations in decades thirty years apart, showed that there were 1,450 references from 1972 to 1982, and 10,100 from 1996 to 2006. While this is only a rough approximation of the production of data in this area of research, and the metric is likely correlated with greater scientific effort in general, it is a strong indication of steadily growing interest in estuarine salmonid ecology. Expansion of studies into unexplored geographic areas has led to the successes and accomplishments in our knowledge of the salmonid estuary. Many of the historical data on estuarine salmonid ecology were on the *Oncoryhnchus* genus in the North Pacific Ocean and Atlantic salmon in the North Atlantic Ocean, but we have now expanded information on other *Salmo* species, as well as on *Salvelinus* and *Hucho* species from other parts of the world. Greater global collaboration, evidenced by the numerous international agencies and conservation organizations now concerned with salmonids, has helped in this regard. As well, the concept of a salmonid estuary management plan was in its embryonic stage in the 1970s, but by the beginning of the twenty-first century comprehensive plans were being developed and implemented in many parts of the salmonid world. The scientific advances have in large part pushed the need for conservation, aided of course by changing social needs and the environmental movement. While the simple beach seine and a pair of rubber boots are still among the most valuable devices in the estuarine salmonid ecologist's toolbox, our understanding has benefited from the vastly improved technologies developed in the past few decades.

A blend of laboratory studies, field data, and technological wizardry has enabled estuarine salmonid researchers to contribute to society's need for management information. Progress has been made on all the fitness elements, although somewhat unevenly. Data banks on habitat change and survival effects relating to osmoregulatory adjustment, especially water quality, have advanced substantially, most likely because a mixture of laboratory and field experimentation is possible in these areas of research. Laboratory effects of temperature on smolting can be extrapolated to the field when considering climate change, for example. Obtaining data on the effects of shifts in food availability and their cascading effects on carrying capacity is more of a challenge in the estuary. Forecasting effects from laboratory or mesocosm experiments can require many assumptions. The study of food webs in the salmonid estuary has made remarkable progress, however, in spite of the need to deal with a myriad of invertebrate and fish prey. The application of biogeochemical techniques, such as stable isotope analyses, has led to great strides in this field. Concerning biotic interactions, assessment results of habitat change on fitness elements relating to predation and competition have perhaps accrued more slowly. This area of research requires substantial contextual information on ecosystem elements that are likely to be estuary-specific. Detailed knowledge of species-specific behaviour such as residency and migration, in addition to basic data on salmonid density or abundance, can provide the needed context. The development of electronic tagging techniques has enhanced our ability to assess biotic interactions, especially predation. International databases have increased our understanding of invasive species in salmonid estuaries (including salmonids themselves in some regions), as has sharing of regional information on the growing number of domesticated (hatchery-reared) salmonids. The remarkable advances in genetic techniques, including analyses at the genome level, are rapidly increasing our understanding of habitat effects on all three fitness elements. In addition, the burgeoning science of estuarine oceanography and geology dealing with hydrology, sediment transport, and estuarine water mass dynamics has provided data to help us understand the processes involved in maintaining the habitat complexity of the estuary. Exploring the frontiers of estuarine salmonid ecosystems has involved a multidisciplinary approach, ranging from molecular biology to basic field biology.

Suggestions for Priority Research
In general, the emphasis in salmonid research for many years has been on gathering data in rivers and oceans. It is time to focus on estuaries, where there is still much to learn about salmonids. The following (not in rank order) are some overarching ideas, challenges, and data needs for future research and management.

Inventory Lost, Damaged, and Vulnerable Salmonid Estuaries
An inventory of salmonid estuaries lost or damaged because of anthropogenic activity is required. Estuaries and coasts are generally acknowledged to be among the most disrupted aquatic ecosystems on the planet, but we need better information specifically on the state of salmonid estuaries around the world. An inventory could help prioritize estuaries for further study. A data bank on vulnerable salmonid estuaries, developed at the local, regional, and international levels, would also be useful.

Conduct Comparative Studies on a Variety of Estuary Types
Our concept of the classic salmonid estuary needs to be broadened, as there is remarkable variation in shapes and configurations in estuary geomorphology and types. Much of the literature is from gently sloping estuaries with sand and mud beaches and vascular plants. We need more data from estuaries dominated by different beach, channel, and vegetation subhabitat types. There is a need for specific studies on intact estuaries in the Northern Hemisphere, possibly with a comparative approach to identify limiting factors across the north Pacific or north Atlantic Ocean or another region. Better data from remote and undisturbed estuaries will provide more insight into how fitness components are related to adaptation. Long-term monitoring of the salmonid estuarine ecosystem will be necessary.

Improve Metrics: Estuary Size, Juvenile Abundance, and Survival
Regional or international forums for developing standardized methods for assessing estuary size are required. At the same time, improvements in standard methods for sampling the abundance and distribution of juvenile salmonids in various estuary types around the world are critical to enable construction of databases to correlate abundance with estuary conditions and survival to later life stages. Greater standardization would facilitate efforts to compare the salmonid density data available in the literature. Continuing efforts to measure juvenile salmonid survival across the full spectrum of life stages, especially within the estuary and

from the estuary to the oceans, would also be helpful for a variety of species. Further miniaturization of electronic tags would greatly aid such efforts.

Investigate the Genetics and Pace of Short-Term Evolution of Salmonid Estuarine Life Histories

There is great scope for expanding our understanding of the evolutionary plasticity of salmonids with respect to their adaptations to the estuary. When global data are assembled on a seasonal basis for the eighteen taxa discussed in this book (see Appendix Table 1.9 at http://hdl.handle.net/2429/57062), it is clear that salmonids of various life stages are present in an estuary in any month of the year. We currently know very little about the specific adaptations that enable this widespread use. New data will help integrate short-term evolution considerations for management.

Enhance Our Understanding of Cumulative Effects

Researchers and managers recognize that salmonid estuary stressors act together, but methods to sum up effects could be improved. Types of stress vary around the world, but port development, urbanization, and water pollution are widespread and they are rarely considered together. Harvesting of salmonids in estuaries is an almost universal stressor – globalization of commercial and recreational fishing has left very few predator refuges in the world's oceans and it is likely that the same situation prevails in salmonid estuaries. All these effects need to be viewed together when assessing the health of salmonid estuaries.

Improve Methods of Assessing Restoration

Estuarine habitat restoration to remove stressors will be increasingly important to help maintain salmonids in estuaries. Restoration methods should focus on processes and recovery of connectivity so salmonids can access a variety of habitats, thus aiding in the recovery of resilience. Indicators for success are needed, and long-term monitoring should be undertaken to track the trajectory of the recovering habitats. An adaptive management plan should be in place.

Improve Our Understanding of How Hybrid Food Webs Change Trophic Relationships and Biotic Interactions, and How They Are Formed

Inevitably, hybrid food webs with invasive species (plants, invertebrates, fish, birds, and possibly even mammals) in them will become more common in salmonid estuaries. The emergence of these types of food webs is a critical problem, complicating our understanding of bottom-up ecosystem processes. A crucial issue in this regard is whether or not the invasive prey fed upon by salmonids are as important for fish growth and survival as the native food species. Among top-down effects, the linkage between predation and estuarine habitat deserves focus. For example, it is not clear whether invasive vegetation is as important as native vegetation in the refuge function. This issue of course includes invasive salmonids, and we need better information on the genetic basis that enables or inhibits their dispersal between estuaries and watersheds. For example, invasion along a coast by sea trout would be more predictable if the ancestral history of brown trout–stocked rivers were known.

Investigate Our Natural Food Web Paradigm Further

Food limitation and its effects on salmonid survival in the estuary is an understudied area that merits further research. There is still much to be learned about food selection from various habitats, prey vulnerability, and other aspects of salmonid feeding in estuaries. A general theory of fish feeding has been sought by many researchers, but there are still problems identifying what salmonids will eat from the suite of potential prey in an estuary. An unresolved issue for studies of food relationships in relation to carrying capacity remains: which methods measure actual food availability for salmonids in estuaries? There has been considerable research on salmonid food webs at the secondary and tertiary levels in the various channel habitats. However, there is a need to investigate the relative importance of autotrophic *versus* heterotrophic production at the basal level, especially in large freshwater tidal channels with phytoplankton production in deeper water and detrital production in shallow areas.

Expand Behavioural Studies

Our knowledge of how salmonids behave in estuaries could be improved in a number of areas. Further field studies are required to develop detailed data on schooling of salmonids in estuaries in relation to predation. Adaptations to the actual saltwater/freshwater interface, which all estuaries have, are probably unappreciated fitness elements that deserve attention. What specific behaviours (e.g., salinity preferences at the microhabitat scale) are shown at the saltwater/freshwater interface? In situ observations using scuba or other remote cameras would yield interesting information. At the other end of the estuary, in the freshwater tidal zone, improved data on territorial behaviour of salmonids before they smolt would be valuable, especially observations in relation to habitats as refuges.

Look North and South in the Context of Climate Change and Development

Increased internationally coordinated monitoring of salmonid estuaries and their fish communities is required to track change related to climate. Climate change and global warming, which have latitudinal gradients, are clearly overarching issues. While the direct signal of global warming is usually temperature, with its pervasive influence on individual salmonid estuarine physiology and ecology, numerous collateral community effects could to be measured, such as increased number of nonsalmonid invasive species, changes in migration patterns, colonization of new river-estuary systems, and abandonment of others. These effects will result in changes in the species composition of the estuarine salmonid community itself, most likely in the Arctic Ocean and possibly in the Antarctic Ocean.

Management challenges are being met in many temperate salmonid estuaries, but the rate of development in the polar regions is increasing and greater vigilance is required. Invasive salmonids are spreading further into polar regions at the same time that humans are increasing their exploitation of other resources, such as oil and gas, which will require port development and lead to greater shipping traffic. Global trends in salmonid estuary management systems over the equator-pole gradient could be synthesized, adding value to the work being done in various regions of the estuarine salmonid world.

Embrace Process

Future work needs to embrace estuarine ecological processes in study design. Processes maintaining salmonid estuarine habitat (flow, connectivity, sedimentation, vegetation growth, etc.) should be measured and accounted for in management schemes. Ecosystem simulation models can help scope out experiments to assess multiple effects of process change on salmonid survival in estuaries. There is now a proliferation of end-to-end models, also known as whole-of-system models in the literature. End-to-end ecosystem modelling approaches differ from earlier models by attempting to represent the entire ecological system (including human components) and the associated abiotic environment (extending to climate impacts), integrating physical and biological processes at different scales and allowing for dynamic two-way coupling (or interactions) between ecosystem components.

Look Upstream and Downstream
Researchers, conservationists, and applied ecologists working in estuaries should always look upstream and downstream when studying limiting factors for salmonids. Further development of linked river-estuary-ocean models is required. The estuary should be considered when developing environmental flows for the river. Greater recognition that the estuary is a sink for contaminants introduced into the river is also required. Conservation of anadromous salmonids requires recognition that estuaries are part of the river-estuary-ocean habitat continuum. This is a research area where limnologists, estuarine scientists, and oceanographers could work together to advantage.

Assess and Preserve Specialized Life History Forms and Species to Maintain Resilience in the Salmonid Community
There is strong consensus that the resilience of the salmonid estuarine ecosystem may be a function of life history diversity. An inventory of specialized life history forms is needed before they are possibly lost from the gene pool. Forms that rear in the estuary or pass through and return to the estuary after a short time in the ocean or the estuary might be priorities for preservation, and there is clearly very great scope for further research on how the various fitness components of survival in the estuary are related to genotypic and phenotypic variation.

Understand the Implications of Increasing Domestication
Given the increasing use of the estuary as a conduit for hatchery-reared fish en route to being ocean-ranched, it is important to understand the ramifications of the likely rise in the throughput of smolts. As the years pass, future generations of salmonids may be almost entirely domesticated in some systems, with attributes not as finely tuned to the estuary as their wild ancestors were. Better understanding of their competitive ability and predator effects vis-à-vis wild fish is required. In addition, the danger of escaped farmed salmonids moving into estuaries is always a possibility. In many parts of the world, wild salmonid abundance is generally low at present; most estuaries are presumably below carrying capacity for wild salmonids, currently reducing the prospect of competition within the wild forms, but this hypothesis should be tested. On the

other hand, from the viewpoint of the ocean rancher or the aquaculturist, maintenance of a flow of ecosystem services from the estuary (e.g., flushing, suitable temperature, and dissolved oxygen) is important.

Summing Up

Over thirty years ago, I recommended that researchers continue to test the null hypothesis that "juvenile salmonids are not dependent on estuaries" (Levings 1984). Even though the foregoing list of suggested research is lengthy, remarkable progress has clearly been made towards rejecting this null hypothesis. We now have a much better understanding of how salmonids are adapted to the estuary through fitness elements relating to osmoregulatory adjustment, growth, and biotic interactions. The consensus is that salmonids *do* depend on estuaries, but they also depend on watersheds and oceans. We have arrived at that conclusion through multiple routes and using many approaches. These have ranged from rather gross and uncontrolled approaches such as juvenile-to-adult survival transfer experiments, to sophisticated genetic research, to detailed structure-oriented and process-oriented ecological studies. In our attempts to parse out fitness elements and their survival benefits, we may have been too demanding in our thirst to document specific habitat usage for species with complex life histories and residency in multiple habitats. Along the way, studies of ecological processes, the implications of short- and long-term multiple-habitat use, and cumulative effects may have lagged. However, the combination of structural and functional approaches has helped conservationists and managers to slow or halt salmonid estuary loss, and in some cases to recover estuaries from disruption. While proof of the estuary's importance is a fundamental need, an understanding of the ecological processes and performance of habitats responsible for salmonid survival is the way to move forward.

Furthermore, no scientific formula exists that prescribes a single course of action that is both necessary and sufficient to conserve salmonids (Nicholas 1997). There is no single management strategy or universal framework that fits all salmonid estuaries, so conservation and restoration efforts should be tailored to the specific ecosystem service needs of society and the policy arrived at by local institutions. We tend to deal with stressors on fitness elements one at a time, but it is extremely important to recognize that the stressors act together. The effects of some stressors are likely accelerating in this century, driven by globalization, human encroachment on the coast, human population growth, and climate change. They include increases in fisheries harvest, port development, coastal urbanization (including pollution effects), increases in water temperature, and changes in river discharge. Understanding these cumulative effects on estuarine salmonid fitness elements is the grand challenge for today's estuarine salmonid researcher. In the meantime, the conservation and restoration components of management plans will be increasingly important to help maintain salmonids in estuaries.

Glossary

Acclimatization experiments – term sometimes used to describe transplanting of salmonids from their endemic or native region to an area where they did not exist before.

Adaptive management – identification of uncertainties and establishment of methodologies to test hypotheses concerning those uncertainties.

Allometric relationship – can be determined by study of habitat scaling, suggesting that landscape form can be viewed as a system of related rates of change between various geomorphic features of the physical landscape.

Anadromous – describes fish that migrate from salt water to spawn in fresh water.

Aquaculture – cultivation of aquatic animals and plants, especially fish, shellfish, and seaweed, in natural or controlled marine or freshwater environments.

Autotrophy – a food web pathway that involves photosynthesis.

Bayesian network – a graphic model that includes probabilistic relationships among variables.

Bedload – sand, gravel, or rocks transported along the bottom of a river.

Benthivorous – obtaining food from the bottom of the estuary or water body.

Bet-hedging – an evolutionary strategy to avoid risk by showing variable fitness in one life stage or the other.

Bioassay – determination of the strength or biological activity of a chemical, by comparing its effects with those of a standard preparation on a test organism.

Biodiversity – the variability among living organisms from all sources, including terrestrial and marine and other aquatic ecosystems and the ecological complexes of which they are a part.

Biofilm – aggregates of small benthic algae, largely comprising diatoms, protozoans, and bacteria.

Biomarker – a biochemical, genetic, or molecular characteristic or substance that is an indicator of a particular biological condition or process. Often related to effects of contaminants in organisms.

Bottom-up effect – refers to controls on the abundance and/or community structure of organisms that derive from supply of resources.

Brackish – refers to a mixture of salt and fresh water.

Bycatch – the portion of a commercial fishing catch that consists of animals caught unintentionally.

Catchability – susceptibility to fishing gear; the factors responsible for generating fishing mortality other than fishing effort per se.

Catchment basin – watershed or river basin; the entire geographical area drained by a river and its tributaries.

Chernobyl – the site of a nuclear power plant in Ukraine where a catastrophic accident occurred in 1986 and dispersed radioactivity around the world.

Coevolved – describes two organisms each of which has adaptations that are specifically linked to the biology of the other.

Cohort – a group of fish that have shared specific habitats together during a particular time span, e.g., salmonids that have hatched at the same time.

Colonization – the process where a species moves into an area that it has not lived in before and establishes a self-reproducing population.

Common garden experiment – a classical approach to quantifying genetically based phenotypic differentiation among populations. Typically, observations are made when organisms in the seed or egg stage are moved from one environment to another environment.

Community – a group of organisms occurring in a particular environment, presumably interacting with each other and with the environment and separable from other groups by means of an ecological survey.

Compensation – a habitat management policy concept wherein replacement of a destroyed habitat unit by development of another similar unit nearby is considered to have achieved "no net loss" of habitat.

Conservation list – a list of species that are of concern to global, national, or local agencies due to very low population levels, and that require measures to avoid extinction.

Conservative properties of seawater – properties that are changed only by mixing.

Conspecifics – members of the same species.

Context – concept wherein variables characterizing the sampling scheme, the spatial scale, and the types of organisms or habitats involved affect the results of ecological observations.

Cortisol – the main steroid hormone involved in stress responses, produced in the interrenal cells of the kidney in fishes.

Daily caloric expenditure – amount of energy used in a 24-hour period for vital processes such as movement, energy consumption, digestion, etc.

Daily ration – amount of food taken in over a 24-hour period.

Data storage tags – use small computers that contain a real-time clock, various sensors, and internal memory for data storage; sometimes called archival tags or data logging tags.

Demersal – living near the bottom of an estuary or other water body.
Detoxification enzyme – an enzyme that removes toxic substances from a living organism.
Diadromous – describes fish that spend portions of their life cycles partially in fresh water and partially in salt water.
Diel – involving a 24-hour period that usually includes a day and the adjoining night.
Differential habitat use – data inferring that a species is more abundant in one habitat than another by preference or opportunism.
Distal factor – relating to higher-level or overlying critical attributes that affect whether or not salmonids dispersing along a coastline can use an estuary (e.g., presence or absence of a river mouth).
Drift cell – a specific segment of a beach where sediment is moved along the shoreline by wave action, typically between points or other landforms.
Ecological filter – a set of hierarchical factors that influence the species composition of a local fauna.
Ecological integrity – the condition when an ecosystem is deemed characteristic for its natural region, including the composition and abundance of native species and biological communities, rates of change, and supporting processes.
Ecological niche – the functional position of an organism in its environment, comprising the habitat in which the organism lives, the periods of time during which it occurs and is active there, and the resources it obtains there.
Ecology – the study of the relationships among living organisms and between those organisms and their environment.
Ecophysiology – science of the interrelationships between the physiology of organisms and their environment.
Ecosystem service – a contribution of the ecosystem to production of the final good or service; the benefits of nature to households, communities, and economies.
Ecotone – zones of transition between adjacent ecological systems.
Ecotype – a locally adapted population of a widespread species. Such populations show minor changes of morphology and/or physiology that are related to habitat and are genetically induced.
Electrophoretic enzyme study – separation of enzymes representing genetic material based on their differential migration in an electromagnetic field.
Endocrine disruptors – chemicals that may interfere with an organism's endocrine system and produce adverse developmental, reproductive, neurological, and immune effects.
Environmental flow – the quantity and quality of water from a river required for ecosystem conservation and resource protection.
Environmental suitability models – models that propose to predict the suitability of habitat for a target species based on its known affinities with environmental variables.

Estuary – a semi-enclosed and coastal body of water, with free communication to the ocean and within which ocean water is diluted by fresh water derived from land.

Euryhaline – describes an organism that can tolerate a wide range of salinities.

Eutrophication – condition where nutrient concentrations, often nitrates, are higher than natural estuarine levels, causing excessive plant growth and subsequent low dissolved oxygen levels when the plants decompose.

Exceedance – the amount by which something, especially a pollutant, exceeds a standard or permissible measurement.

Exploitative competition – occurs when consumption of a limiting resource by one species makes that resource unavailable for consumption by another.

Extirpation – extinction of a species within a portion of its range.

Family – in the hierarchical Linnaean system of classifying organisms, a family is more precise than an order but less precise than a genus.

Fecundity – the potential reproductive capacity of an individual; measured in salmonids by number of eggs deposited on the spawning ground.

Fishway – a structure built to facilitate salmonid migration around barriers (such as dams and locks).

Fitness – describes a key concept of natural selection, wherein individuals that survive will make a genetic contribution to future generations, and hence are more fit than those that die before reproducing.

Food conversion efficiency – the amount of body weight gained per unit of food consumed.

Food ration – the consumption of a particular food type over a specific time period, often 24 hours.

Food web – describes a set of different pathways by which energy, nutrients, and other materials move to individual organisms and species.

Genome – full set of chromosomes; all the inheritable traits of an organism.

Genotype – the inherited instructions an organism carries within its genetic code.

Genus – in the hierarchical Linnaean system of classifying organisms, the next level above species.

Globalization – processes that increase worldwide exchanges of national and cultural resources.

Guild – a group of species that exploit the same resources, often in related ways.

Habitat – place where an organism or a biological population normally lives or occurs; the home of a particular organism, where the species is as adaptive as possible to that particular environment.

Heterotrophy – indicates a food web pathway that does not involve photosynthesis.

Humoral immune responses – an immune response mediated through a body fluid.
Hydroacoustic tags – sound-emitting devices that permit remote tracking of fish in estuaries and the ocean.
Hypersalinity – salinity that is much higher than the salinity for average ocean water, which is about 35 psu.
Imago – the penultimate stage an insect attains during its metamorphosis to an adult.
Immunotoxicology – the study of immune dysfunction resulting from exposure of an organism to a contaminant.
Incipient lethal temperature – the upper and lower incipient lethal temperatures represent the temperature values above or below which 50 percent of the population can no longer live over an indefinite period of time.
Industrial fishery – large-scale commercial fishing.
Inferential approach – use of inferential statistics to make judgments of the probability that an observed difference between groups is a dependable one or one that might have occurred by chance.
Information-theoretic – a statistical approach to selecting the most appropriate model from a suite of models.
Insect larva – a specialized feeding stage of immature insects; follows the egg stage.
Interference competition – occurs directly between individuals when the individuals interfere with foraging, survival, and reproduction of others, or directly prevent their physical establishment in a portion of the habitat.
Intermediate host – an organism in or on which a parasite develops to an adult but not sexually mature stage.
Introgression – infiltration of the genes of one species or life history type into the gene pool of another.
Invasibility – the susceptibility of an environment to the colonization and establishment of individuals from species not currently part of the resident community.
Isoform – an enzyme protein that has the same function as another protein that is encoded by a different gene.
Isotonic – when the solution in a cell or other structure surrounded by a membrane has the same salt concentration as the surrounding environment.
Iteroparous – a type of reproduction in which adults reproduce repeatedly.
Kelt reconditioning – the process of culturing kelts in a captive environment until they are able to reinitiate feeding and growth, and again develop mature gonads.
Landscape genetics – describes how landscape features such as river basins, estuary, and coastal features influence spatial-temporal processes of population genetics and the ultimate distribution of genetic variation.

Leapfrogging – situation in which a terminal population at the end of a local range was colonized through long-distance dispersal from another similar but remote population via routes that bypass barriers; successive colonization of estuaries and their landscapes.

Logistic equation – a simple model of population growth in conditions where there are limited resources; an asymptotic or maximum population size is reached when the limit is reached.

Macrohabitat – coarse-grained level of habitat description, often at scales of >100 m discrimination.

Macrodetritus – large particles of a mixture of vascular plant debris and bacteria and other organic material; an operational definition.

Management plan – the framework or blueprint derived from government policy on an estuary; often prescribes multiple use. The plan may describe the obligations that various institutions have to implement to enact the policy.

Master variable – a sole regulator of one or more processes that species depend on.

Mesocosm – a medium-sized experimental chamber or tank that brings a small part of the natural environment under controlled conditions.

Mesohabitat – visually distinct units of habitat with apparent physical uniformity, often with a scale between 1 and 100 m.

Microbenthic algae – small single-celled algae; usually include blue-green algae and diatoms.

Microdetritus – small particles of a mixture of phytoplankton or algae mixed with bacteria and other organic material; an operational definition.

Microhabitat – fine-grained physical requirements of a particular organism's habitat. For salmonids, the physical measurements (e.g., temperature, depth, velocity) are typically made within a few centimetres of a fish's nose, in situ.

Migratory pulse – the peak in number of migrating salmonids over a short time period.

Miocene – geological period extending from about 23.03 to 5.33 million years ago.

Mitigation – a habitat management policy concept that involves taking steps to avoid or minimize negative impacts; can include habitat restoration in some jurisdictions.

Monotonic relationship – functions that tend to move in only one direction along the ordinate (y-axis) as values on the abscissa (x-axis) increase.

mRNA – messenger ribonucleic acids that convey genetic information (often denoted using the letters G, A, U, and C for the nucleotides guanine, adenine, uracil, and cytosine) and direct synthesis of specific proteins within cells.

Multivariate analysis – a generic term for any statistical technique used to analyze data from more than one variable.

Na^+-K^+-ATPase – a key enzyme controlling the sodium and potassium levels in vertebrate cells.

Nomogram – a graphic representation that consists of several lines marked off to scale and arranged in such a way that by using a straight edge to connect known values on two lines an unknown value can be read at the point of intersection with another line.

Nonconservative properties – properties that are changed by processes other than mixing (e.g., changes owing to biological processes such as photosynthesis).

Nonlinear approaches – required when the relationship between variables does not form a straight line.

Non-natal estuary – an estuary used by salmonids that were not spawned in the river (or one of its tributaries) draining into the estuary.

Nonparametric statistics – statistical analyses that do not assume that data have come from a type of probability distribution.

Organochlorine pesticides – pesticides containing compounds made up of carbon, chlorine, and hydrogen.

Osmoregulation – control of the levels of water and mineral salts in blood.

Otolith – hard calcium carbonate structures located directly behind the brain of teleost fish; also known as ear bones.

Parametric statistics – statistical analyses that assume that data have come from a type of probability distribution, usually the normal distribution.

Parasite – an organism that lives on or in a host and gets its food from or at the expense of its host.

Phenotype – an organism's observable characteristics or traits (morphology, behaviour). Phenotypes result from the expression of an organism's genes as well as the influence of environmental factors and the interactions between the two.

Pheromones – a chemical produced by an animal that changes the behaviour of another animal of the same species.

Piscivorous – preying on fish as a feeding habit.

PIT tags – Passive Integrated Transponder (PIT) tags are very small radio-frequency devices that do not require a battery. When hit by a signal, the tag responds by transmitting a unique code that is picked up by an antenna and transmitted to a digital recorder.

Planktivorous – feeding on plankton.

Pleistocene – the geological period that lasted from about 2,588,000 to 11,700 years ago.

Polytypic species – species that vary markedly in their colour, life history, and ecology.

Porphyropsin – a photosensitive protein pigment found in the retinas of salmonids.

Progeny – a genetic descendant or offspring.

Propagule pressure – the number of individuals (e.g., eggs or other life history stages) in an introduction event, and the frequency of these events.

Proximate factor – an event that is closest to or immediately responsible for a change in short-term evolution in salmonids after they have dispersed into an estuary.

Radio tags – sound-emitting devices that permit remote tracking of fish in fresh water.

Ramsar Convention on Wetlands – an international treaty for the conservation and sustainable utilization of wetlands, with a focus on birds; named after a city in Iran.

Reaction norm – linear functions between an environmental value or gradient and the expression of a genotype.

Resilience – the amount of disturbance that an ecosystem and associated socio-economic system can withstand without changing self-organized processes.

Resource polymorphism – the occurrence of discrete intraspecific morphs showing differential niche use, usually through discrete differences in feeding biology and habitat use.

Restoration – the practice of renewing and restoring degraded, damaged, or destroyed habitats.

Rheotaxis – oriented movement of a fish in response to a water current.

Rhodopsin – a red photosensitive protein in the retina that is important for vision in dim light.

Riprap – rock deposited on a shoreline to provide a foundation and protect a channel or riverbank from erosion.

River basin – the entire geographical area drained by a river and its tributaries.

Run – migration of a salmon population into an estuary from the ocean, at a specific time.

Salmonid stronghold – a watershed providing hospitable habitat conditions for wild salmonid populations, representing the productivity and diversity for salmonid populations of specific management interests.

Salmonine – a member of the subfamily Salmoninae, family Salmonidae.

Scale resorption – process wherein the outer perimeter of fish scales dissolve and are assimilated in the inner part of the scale.

Seawater challenge – a test to determine whether salmonid smolts can survive in full-strength seawater after rearing in fresh water.

Selective force – any phenomenon that alters the behaviour and fitness of living organisms within a given environment; driving force of evolution.

Self-sustaining – in the context of estuary restoration, requiring minimal maintenance or management, or no maintenance at all.

Semelparous – a type of reproduction in which adults reproduce once, then die.

Short-term (rapid) evolution – a genetic change occurring rapidly enough to have a measurable impact on an adaptive feature.

Smoltification – physiological changes that anadromous salmonids undergo in fresh water or the estuary while migrating towards salt water.
Sonar – abbreviation for "sound navigation and ranging." Sound waves are transmitted into water, and the reflected energy can detect and enumerate fish and other organisms.
Spate – a sudden flood in a river.
Spatially explicit models – represent a heterogeneous space that is continuous or discrete.
Species – a fundamental category of taxonomic classification, in the Linnaean system ranking below a genus or subgenus and consisting of related organisms capable of interbreeding.
Speciose – describes the condition of a taxonomic unit that contains many species.
Sporeling – a young algal plant produced by a germinated spore, similar to a seedling derived from a germinated seed.
Stock – in fish population dynamics terminology, a population of fish that occupies the same area; the individuals in the stock are all part of the same reproductive process.
Stress (syndrome) – manifestation of the response of an organism to any demand placed on it such that it causes an extension of a physiological state beyond the normal resting state.
Sublethal behaviour – behaviour resulting from a stress or contaminant effect that does not lead to death.
Subsistence fishery – fishing for personal consumption; the fish harvested are important to feed a social unit such as a family or village.
Supplementation – using hatchery-reared fish to provide increased harvest opportunities and to mitigate reductions in natural populations.
Taxon – a taxonomic category or group, such as a phylum, order, family, genus, or species.
Taxocene – a taxonomically related set of species within a community.
Tectonic – resulting from structural deformation of the earth's crust.
Ternary plot – a plot on three variables that sum to a constant; this type of coordinate system assumes that the location of a point is specified as the centre of mass, or barycentre, of masses placed at its vertices (e.g., three corners of a triangle).
Thalweg – the line defining the deepest points along the length of a river bed or estuary channel.
Transaction cost – a cost incurred in making an economic exchange.
Threshold quantitative trait – a trait that has an underlying quantitative distribution but appears only if a threshold is crossed; does not follow basic Mendelian rules.
Thyroxine – hormone produced by the thyroid glands to regulate metabolism by controlling the rate of oxidation in cells; in salmonids, the

thyroxine-producing cells are diffuse and distributed throughout the connective tissue of the pharyngeal area.

Top-down effect – control of the structure/population dynamics of the ecosystem by a top predator.

Training – the building of wood, rock, or steel walls to constrain the flow in a dredged river channel in an estuary. Wing dams or groins (structures built in the channel at angles from the shore) can also be used for training.

Transcriptomic techniques – methods to completely profile all the information that appears in the RNA pool (i.e., all expressed genes within a cell, body fluid, or tissue). Cell function is mediated through gene expression involving the production of cellular mRNA (see above).

Trophic – referring to food or feeding.

Trophic subsidy – flows of biologically fixed energy and nutrients from one ecosystem to another.

Typology – the study or systematic classification of types of areas such as water bodies or estuaries that have characteristics or traits in common.

Umbrella species – a species whose conservation is expected to confer protection on a large number of naturally co-occurring species.

Upregulation (of genes) – occurs within a cell triggered by a signal (originating inside or outside the cell) and results in increased expression of one or more genes; thus, the protein(s) encoded by those genes increase in concentration.

Vascular plant – plant having a specialized system for conducting nutrients and water.

Wash load – fine sediment in near-permanent suspension in a river.

Water mass – body of estuarine or ocean water with a distinctive narrow range of temperature and salinity.

Web of causation – in the context of diseases and parasites, the concurrent interaction of the host and pathogen with the environment.

Weight-of-evidence approach – a quantitative method for combining evidence in support of a hypothesis.

Wetland – land area that is saturated with water, either tidally, permanently, or seasonally, such that it takes on the characteristics of a distinct ecosystem.

Wrack line – the line of debris found in the high tide zone; often includes wood chips and sometimes fragmented or whole macroalgae.

Yolk sac – a membranous sac containing yolk attached to recently hatched salmonid fry.

References

Aas, Ø., S. Einum, A. Klemetsen, and J. Skurdal, eds. 2011a. *Atlantic Salmon Ecology.* Oxford: Wiley-Blackwell Publishing.

Aas, Ø., D. Policansky, S. Einum, and J. Skurdal. 2011b. "Salmon Ecological Research and Conservation." In *Atlantic Salmon Ecology*, ed. Ø. Aas, S. Einum, A. Klemetsen, and J. Skurdal, 445–56. London: Wiley-Blackwell Publishing.

Aas-Hansen, Ø., M.M. Vijayan, H.K. Johnsen, C. Cameron, and E.H. Jørgensen. 2005. "Resmoltification in Wild, Anadromous Arctic Char (*Salvelinus alpinus*): A Survey of Osmoregulatory, Metabolic, and Endocrine Changes Preceding Annual Seawater Migration." *Canadian Journal of Fisheries and Aquatic Sciences* 62 (1): 195–204. http://dx.doi.org/10.1139/f04-186.

Able, K.W., T.M. Grothues, and I.M. Kemp. 2013. "Fine-Scale Distribution of Pelagic Fishes Relative to a Large Urban Pier." *Marine Ecology Progress Series* 476: 185–98. http://dx.doi.org/10.3354/meps10151.

ABPmer and HR Wallingford. 2007. "The Estuary Guide: A Website Based Overview of How to Identify and Predict Morphological Change within Estuaries." Prepared for the joint Defra/EA Flood and Coastal Erosion Risk Management R&D Programme, UK. http://www.estuary-guide.net/index.asp.

Abrahams, M.V., and M.C. Healey. 1993. "A Comparison of the Willingness of Four Species of Pacific Salmon to Risk Exposure to a Predator." *Oikos* 66 (3): 439–46. http://dx.doi.org/10.2307/3544938.

Abrams, P.A. 2000. "The Evolution of Predator-Prey Interactions: Theory and Evidence." *Annual Review of Ecology and Systematics* 31 (1): 79–105. http://dx.doi.org/10.1146/annurev.ecolsys.31.1.79.

Abrantes, K.G., J.M. Lyle, P.D. Nichols, and J.M. Semmens. 2011. "Do Exotic Salmonids Feed on Native Fauna after Escaping from Aquaculture Cages in Tasmania, Australia?" *Canadian Journal of Fisheries and Aquatic Sciences* 68 (9): 1539–51. http://dx.doi.org/10.1139/f2011-057.

Acolas, M.L., J. Labonne, J.L. Baglinière, and J.M. Roussel. 2012. "The Role of Body Size versus Growth on the Decision to Migrate: A Case Study with *Salmo trutta*." *Naturwissenschaften* 99 (1): 11–21. http://dx.doi.org/10.1007/s00114-011-0861-5.

Adams, M.A., and G.L. Williams. 2004. "Tidal Marshes of the Fraser River Estuary: Composition, Structure and a History of Marsh Creation Efforts to 1997." In

Fraser River Delta, British Columbia: Issues of an Urban Estuary, vol. 546, ed. B.J. Groulx, D.C. Mosher, J.L. Luternauer, and D.E. Bilderback, 147–71. Geological Survey of Canada Bulletin. Ottawa: Natural Resources Canada, Geological Survey of Canada.

Agardy, M.T. 1994. "Advances in Marine Conservation: The Role of Marine Protected Areas." *Trends in Ecology and Evolution* 9 (7): 267–70. http://dx.doi.org/10.1016/0169-5347(94)90297-6.

Airoldi, L., and M.W. Beck. 2007. "Loss, Status and Trends for Coastal Marine Habitats of Europe." In *Oceanography and Marine Biology: An Annual Review*, vol. 45, ed. R.N. Gibson, R.J.A. Atkinson, and J.D.M. Gordon, 345–405. Ottawa: CRC Press.

Alderdice, D.F. 1972. "Factor Combinations: Responses of Marine Poikilotherms to Environmental Factors Acting in Concert." In *Marine Ecology: A Comprehensive, Integrated Treatise on Life in Oceans and Coastal Waters. Vol. 1: Environmental Factors*, ed. O. Kinne, 1659–1716. New York: Wiley and Sons.

Alexander, G., R. Sweeting, and B. McKeown. 1994. "The Shift in Visual Pigment Dominance in the Retinae of Juvenile Coho Salmon (*Oncorhynchus kisutch*): An Indicator of Smolt Status." *Journal of Experimental Biology* 195 (1): 185–97.

Alexander, W.B., B.A. Southgate, and R. Bassindale. 1936. "Survey of the River Tees, Part II: The Estuary, Chemical and Biological." *Journal of the Marine Biological Association of the United Kingdom* 20 (3): 717–24. http://dx.doi.org/10.1017/S0025315400058276.

Allan, I.H.R., and J.A. Ritter. 1977. "Salmonid Terminology." *Journal du Conseil international pour l'exploration de la mer* 37 (3): 293–99. http://dx.doi.org/10.1093/icesjms/37.3.293.

Allison, S.K. 1995. "Recovery from Small-Scale Anthropogenic Disturbances by Northern California Salt Marsh Plant Assemblages." *Ecological Applications* 5 (3): 693–702. http://dx.doi.org/10.2307/1941978.

Alvarez, D., S. Perkins, E. Nilsen, and J. Morace. 2013. "Spatial and Temporal Trends in Occurrence of Emerging and Legacy Contaminants in the Lower Columbia River 2008–2010." *Science of the Total Environment*. http://dx.doi.org/10.1016/j.scitotenv.2013.07.128.

Andreassen, P.M.R., M.B. Martinussen, N.A. Hvidsten, and S.O. Stefansson. 2001. "Feeding and Prey-Selection of Wild Atlantic Salmon Post-Smolts." *Journal of Fish Biology* 58 (6): 1667–79. http://dx.doi.org/10.1111/j.1095-8649.2001.tb02321.x.

Annett, B., G. Gerlach, T.L. King, and A.R. Whiteley. 2012. "Conservation Genetics of Remnant Coastal Brook Trout Populations at the Southern Limit of Their Distribution: Population Structure and Effects of Stocking." *Transactions of the American Fisheries Society* 141 (5): 1399–1410. http://dx.doi.org/10.1080/00028487.2012.694831.

Araki, H., B.A. Berejikian, M.J. Ford, and M.S. Blouin. 2008. "Fitness of Hatchery Reared Salmonids in the Wild." *Evolutionary Applications* 1 (2): 342–55. http://dx.doi.org/10.1111/j.1752-4571.2008.00026.x.

Araújo, M.B., and M. Luoto. 2007. "The Importance of Biotic Interactions for Modelling Species Distributions under Climate Change." *Global Ecology and Biogeography* 16 (6): 743–53. http://dx.doi.org/10.1111/j.1466-8238.2007.00359.x.

Arismendi, I., J. Sanzana, and D. Soto. 2011. "Seasonal Age Distributions and Maturity Stage in a Naturalized Rainbow Trout (*Oncorhynchus mykiss* Walbaum) Population in Southern Chile Reveal an Adfluvial Life History." *Annales de limnologie / International Journal of Limnology* 47 (2): 133–40. http://dx.doi.org/10.1051/limn/2011012.

Arkoosh, M.R., E. Clemons, P. Huffman, A.N. Kagley, E. Casillas, N. Adams, Herb R. Sanborn, Tracy K. Collier, and John E. Stein. 2001. "Increased Susceptibility of Juvenile Chinook Salmon to Vibriosis after Exposure to Chlorinated and Aromatic Compounds Found in Contaminated Urban Estuaries." *Journal of Aquatic Animal Health* 13 (3): 257–68. http://dx.doi.org/10.1577/1548-8667(2001)013<0257:ISOJCS>2.0.CO;2.

Arkoosh, M.R., E. Clemons, A.N. Kagley, C. Stafford, A.C. Glass, K. Jacobson, P. Reno, M.S. Myers, E. Casillas, F. Loge, et al. 2004. "Survey of Pathogens in Juvenile Salmon *Oncorhynchus spp*. Migrating through Pacific Northwest Estuaries." *Journal of Aquatic Animal Health* 16 (4): 186–96. http://dx.doi.org/10.1577/H03-071.1.

Arpe, K., and S.A. Leroy. 2007. "The Caspian Sea Level Forced by the Atmospheric Circulation, as Observed and Modelled." *Quaternary International* 173–74: 144–52. http://dx.doi.org/10.1016/j.quaint.2007.03.008.

Arsenault, J.T.M., W.L. Fairchild, D.L. MacLatchy, L. Burridge, K. Haya, and S.B. Brown. 2004. "Effects of Water-Borne 4-Nonylphenol and 17-Beta-Estradiol Exposures during Parr-Smolt Transformation on Growth and Plasma IGF-I of Atlantic Salmon (*Salmo salar* L.)." *Aquatic Toxicology* (Amsterdam) 66 (3): 255–65. http://dx.doi.org/10.1016/j.aquatox.2003.09.005.

Atlantic Coastal Zone Information Steering Committee. 2014. "Atlantic Coastal Zone Information." http://coinatlantic.ca/.

Attrill, M.J., and S.D. Rundle. 2002. "Ecotone or Ecocline: Ecological Boundaries in Estuaries." *Estuarine, Coastal and Shelf Science* 55 (6): 929–36. http://dx.doi.org/10.1006/ecss.2002.1036.

Atwater, B.F., D.K. Yamaguchi, S. Bondevik, W.A. Barnhardt, L.J. Amidon, B.E. Benson, Gudrun Skjerdal, John A. Shulene, and Futoshi Nanayama. 2001. "Rapid Resetting of an Estuarine Recorder of the 1964 Alaska Earthquake." *Geological Society of America Bulletin* 113 (9): 1193–204. http://dx.doi.org/10.1130/0016-7606(2001)113<1193:RROAER>2.0.CO;2.

Aubé, C.I., A. Locke, and G.J. Klassen. 2005. "Zooplankton Communities of a Dammed Estuary in the Bay of Fundy, Canada." *Hydrobiologia* 548 (1): 127–39. http://dx.doi.org/10.1007/s10750-005-4730-0.

Augerot, X., D.N. Foley, C. Steinback, A. Fuller, N. Fobes, and K. Spencer. 2005. *Atlas of Pacific Salmon: The First Map-Based Status Assessment of Salmon in the North Pacific*. Berkeley: University of California Press.

Ayllon, F., P. Davaine, E. Beall, J.L. Martinez, and E. Garcia-Vasquez. 2004. "Bottlenecks and Genetic Changes in Atlantic Salmon (*Salmo salar* L.) Stocks Introduced in the Subantarctic Kerguelen Islands." *Aquaculture* (Amsterdam) 237 (1–4): 103–16. http://dx.doi.org/10.1016/j.aquaculture.2004.04.014.

Ayllon, F., P. Davaine, E. Beall, and E. Garcia Vazquez. 2006. "Dispersal and Rapid Evolution in Brown Trout Colonizing Virgin Subantarctic Ecosystems." *Journal of Evolutionary Biology* 19 (4): 1352–58. http://dx.doi.org/10.1111/j.1420-9101.2005.01075.x.

Baigún, C., and R. Ferriz. 2003. "Distribution Patterns of Native Freshwater Fishes in Patagonia (Argentina)." *Organisms, Diversity and Evolution* 3 (2): 151–59. http://dx.doi.org/10.1078/1439-6092-00075.

Baker, P.F., F.K. Ligon, and T.P. Speed. 1995. "Estimating the Influence of Temperature on the Survival of Chinook Salmon Smolts (*Oncorhynchus tshawytscha*) Migrating through the Sacramento–San Joaquin River Delta of California." *Canadian Journal of Fisheries and Aquatic Sciences* 52 (4): 855–63. http://dx.doi.org/10.1139/f95-085.

Baldwin, D.H., J.A. Spromberg, T.K. Collier, and N.L. Scholz. 2009. "A Fish of Many Scales: Extrapolating Sublethal Pesticide Exposures to the Productivity of Wild Salmon Populations." *Ecological Applications* 19 (8): 2004–15. http://dx.doi.org/10.1890/08-1891.1.

Balikci, A. 1980. "Charr Fishing among the Arviligjuarmuit." In *Charrs: Salmonid Fishes of the Genus* Salvelinus, vol. 1, ed. E.K. Balon, 141–203. The Hague: Dr. W. Junk Publishers.

Ballinger, R., A. Pickaver, G. Lymbery, and M. Ferreria. 2010. "An Evaluation of the Implementation of the European ICZM Principles." *Ocean and Coastal Management* 53 (12): 738–49. http://dx.doi.org/10.1016/j.ocecoaman.2010.10.013.

Barnes, R.S.K. 1991. "European Estuaries and Lagoons: A Personal Overview of Problems and Possibilities for Conservation and Management." *Aquatic Conservation: Marine and Freshwater Ecosystems* 1 (1): 79–87. http://dx.doi.org/10.1002/aqc.3270010107.

Barry, K.L., J.A. Grout, C.D. Levings, B.H. Nidle, and G.E. Piercey. 2000. "Impacts of Acid Mine Drainage on Juvenile Salmonids in an Estuary near Britannia Beach in Howe Sound, British Columbia." *Canadian Journal of Fisheries and Aquatic Sciences* 57 (10): 2032–43. http://dx.doi.org/10.1139/f00-157.

Bartholomew, J.L., S.D. Atkinson, and S.L. Hallett. 2006. "Involvement of *Manayunkia speciosa* (Annelida: Polychaeta: Sabellidae) in the Life Cycle of *Parvicapsula minibicornis*, a Myxozoan Parasite of Pacific Salmon." *Journal of Parasitology* 92 (4): 742–48. http://dx.doi.org/10.1645/GE-781R.1.

Barton, B.A. 1996. "General Biology of Salmonids." In *Principles of Salmonid Culture*, ed. W. Pennell and B.A. Barton, 29–95. Amsterdam: Elsevier. http://dx.doi.org/10.1016/S0167-9309(96)80005-6.

Bascom, W. 1964. *Waves and Beaches: The Dynamics of the Ocean Surface*. New York: Anchor Books.

Batchelder, H., and M. Kashiwai. 2007. "Ecosystem Modeling with NEMURO within the PICES Climate Change and Carrying Capacity Program." *Ecological Modelling* 202 (1–2): 7–11. http://dx.doi.org/10.1016/j.ecolmodel.2006.05.037.

Bax, N.J., E.O. Salo, B.P. Snyder, C.A. Simenstad, and W.J. Kinney. 1980. "Salmon Outmigration Studies in Hood Canal: A Summary – 1977." In *Salmonid Ecosystems of the North Pacific*, ed. W.J. McNeil and D.C. Himsworth, 171–202. Corvallis: Oregon State University Press.

Bayer, R.D. 1981. "Shallow-Water Intertidal Ichthyofauna of the Yaquina Estuary, Oregon." *Northwest Science* 55 (3): 182–93.

Beale, H. 1991. "Relative Rise in Sea-Level during the Past 5,000 Years at Six Salt Marshes in Northern Puget Sound, Washington." MS thesis, Western Washington University, Bellingham.

Beamer, E., A. McBride, C. Greene, R. Henderson, G. Hood, K. Wolf, et al. 2005. *Delta and Nearshore Restoration for the Recovery of Wild Skagit River Chinook Salmon: Linking Estuary Restoration to Wild Chinook Salmon Populations.* Supplement to Skagit Chinook Recovery Plan. La Conner, WA: Skagit River System Cooperative. http://www.skagitcoop.org/documents/Appendix%20D%20 Estuary.pdf.

Beamish, R.J., and C.D. Levings. 1991. "Abundance and Freshwater Migrations of the Anadromous Parasitic Lamprey, *Lampetra tridentate*, in a Tributary of the Fraser River, British Columbia." *Canadian Journal of Fisheries and Aquatic Sciences* 48 (7): 1250–63. http://dx.doi.org/10.1139/f91-151.

Beamish, R.J., and C. Mahnken. 2001. "A Critical Size and Period Hypothesis to Explain Natural Regulation of Salmon Abundance and the Linkage to Climate and Climate Change." *Progress in Oceanography* 49 (1–4): 423–37. http://dx.doi.org/10.1016/S0079-6611(01)00034-9.

Beamish, R.J., I.A. Pearsall, and M.C. Healey. 2003. *A History of the Research on the Early Marine Life of Pacific Salmon Off Canada's Pacific Coast.* North Pacific Anadromous Fish Commission Bulletin No. 3. Vancouver: North Pacific Anadromous Fish Commission.

Beamish, R.J., B.L. Thomson, and G.A. McFarlane. 1992. "Spiny Dogfish Predation on Chinook and Coho Salmon and the Potential Effects on Hatchery-Produced Salmon." *Transactions of the American Fisheries Society* 121 (4): 444–55. http://dx.doi.org/10.1577/1548-8659(1992)121<0444:SDPOCA>2.3.CO;2.

Becker, L.A., M.A. Pascual, and N.G. Basso. 2007. "Colonization of the Southern Patagonia Ocean by Exotic Chinook Salmon." *Conservation Biology* 21 (5): 1347–52. http://dx.doi.org/10.1111/j.1523-1739.2007.00761.x.

Beechie, T.J., B.D. Collins, and G.R. Pess. 2001. "Holocene and Recent Geomorphic Processes, Land Use, and Salmonid Habitat in Two North Puget Sound River Basins." In *Geomorphic Processes and Riverine Habitat*, ed. J.M. Dorava, D.R. Montgomery, B.B. Palcsak, and F.A. Fitzpatrick, 37–54. Washington, DC: American Geophysical Union. http://dx.doi.org/10.1029/WS004p0037.

Beechie, T.J., E. Buhle, M. Ruckelshaus, A. Fullerton, and L. Holsinger. 2006. "Hydrologic Regime and the Conservation of Salmon Life History Diversity." *Biological Conservation* 130 (4): 560–72. http://dx.doi.org/10.1016/j.biocon.2006.01.019.

Beechie, T.J., D.A. Sear, J.D. Olden, G.R. Pess, J.M. Buffington, H. Moir, Philip Roni, and Michael M. Pollock. 2010. "Process-Based Principles for Restoring River Ecosystems." *BioScience* 60 (3): 209–22. http://dx.doi.org/10.1525/bio.2010.60.3.7.

Begon, M., C.R. Townsend, and J.L. Harper. 2005. *Ecology: From Individuals to Ecosystems.* 4th ed. Oxford: Wiley-Blackwell Publishing.

Bégout Anras, M.L., E.C. Gyselman, J.K. Jorgenson, A.H. Kristofferson, and L. Anras. 1999. "Habitat Preferences and Residence Time for the Freshwater to Ocean Transition Stage in Arctic Charr." *Journal of the Marine Biological Association of the United Kingdom* 79 (1): 153–60. http://dx.doi.org/10.1017/S0025315498000174.

Béguer, M., L. Beaulaton, and E. Rochard. 2007. "Distribution and Richness of Diadromous Fish Assemblages in Western Europe: Large-Scale Explanatory

Factors." *Ecology of Freshwater Fish* 16 (2): 221–37. http://dx.doi.org/10.1111/j.1600-0633.2006.00214.x.

Bell, L.M., and R.J. Kallman. 1976. *The Nanaimo River Estuary: Status of Environmental Knowledge to 1976: Report of the Estuary Working Group, Department of the Environment, Regional Board Pacific Region (No. 5)*. West Vancouver: Environment Canada. http://www.dfo-mpo.gc.ca/library/16769.pdf.

Bendall, L.B., A. Moore, and V. Quayle. 2005. "The Post-Spawning Movements of Migratory Brown Trout *Salmo trutta*." *Journal of Fish Biology* 67 (3): 809–22. http://dx.doi.org/10.1111/j.0022-1112.2005.00786.x.

Benigno, G.M., and T.R. Sommer. 2008. "Just Add Water: Sources of Chironomid Drift in a Large River Floodplain." *Hydrobiologia* 600 (1): 297–305. http://dx.doi.org/10.1007/s10750-007-9239-2.

Benke, A.C., and C.E. Cushing, eds. 2005. *Rivers of North America*. Burlington, MA: Elsevier Academic Press.

Bennett, W.A., W.J. Kimmerer, and J.R. Burau. 2002. "Plasticity in Vertical Migration by Native and Exotic Estuarine Fishes in a Dynamic Low-Salinity Zone." *Limnology and Oceanography* 47 (5): 1496–1507. http://dx.doi.org/10.4319/lo.2002.47.5.1496.

Berg, L., and T.G. Northcote. 1985. "Changes in Territorial, Gill-Flaring, and Feeding Behavior in Juvenile Coho Salmon (*Oncorhynchus kisutch*) Following Short-Term Pulses of Suspended Sediment." *Canadian Journal of Fisheries and Aquatic Sciences* 42 (8): 1410–17. http://dx.doi.org/10.1139/f85-176.

Berg, O.K., and M. Berg. 1989. "The Duration of Sea and Freshwater Residence of the Sea Trout, *Salmo trutta*, from the Vardnes River in Northern Norway." *Environmental Biology of Fishes* 24 (1): 23–32. http://dx.doi.org/10.1007/BF00001607.

Bersine, K., V.E.F. Brenneis, R.C. Draheim, A.M. Wargo Rub, J.E. Zamon, R.K. Litton, S.A. Hinton, et al. 2008. "Distribution of the Invasive New Zealand Mudsnail (*Potamopyrgus antipodarum*) in the Columbia River Estuary and Its First Recorded Occurrence in the Diet of Juvenile Chinook Salmon (*Oncorhynchus tshawytscha*)." *Biological Invasions* 10 (8): 1381–88. http://dx.doi.org/10.1007/s10530-007-9213-y.

Bieber, A.J. 2005. "Variability in Juvenile Chinook Foraging and Growth Potential in Oregon Estuaries: Implications for Habitat Restoration." MS thesis, University of Washington, Seattle.

bij de Vaate, A., R. Breukel, and G. van der Velde. 2006. "Long-Term Developments in Ecological Rehabilitation of the Main Distributaries in the Rhine Delta: Fish and Macroinvertebrates." *Hydrobiologia* 565 (1): 229–42. http://dx.doi.org/10.1007/s10750-005-1916-4.

bij de Vaate, A., A.W. Breukelaar, T. Vries, G. de Laak, and C. Dijkers. 2003. "Sea Trout Migration in the Rhine Delta." *Journal of Fish Biology* 63 (4): 892–908. http://dx.doi.org/10.1046/j.1095-8649.2003.00198.x.

Billen, G., H. Decamps, J. Garnier, P. Boet, M. Meybeck, and P. Servais. 2006. "Atlantic River Systems of Europe (France, Belgium, the Netherlands)." In *River and Stream Ecosystems of the World*, ed. C.E. Cushing, K.W. Cummins, and G.W. Minshall, 389–418. Los Angeles: University of California Press.

Biro, P.A., C. Beckmann, and M.S. Ridgway. 2008. "Early Microhabitat Use by Age 0 Year Brook Charr *Salvelinus fontinalis* in Lakes." *Journal of Fish Biology* 73 (1): 226–40. http://dx.doi.org/10.1111/j.1095-8649.2008.01930.x.

Birtwell, I.K., and R.M. Harbo. 1980. "Pulp Mill Impact Studies at Port Alberni and Port Mellon, British Columbia." *Transactions of the Technical Section Canadian Pulp and Paper Association* 81: 85–88.

Birtwell, I.K., and G.M. Kruzynski. 1989. "*In Situ* and Laboratory Studies on the Behaviour and Survival of Pacific Salmon (Genus *Oncorhynchus*)." *Hydrobiologia* 188–89 (1): 543–60. http://dx.doi.org/10.1007/BF00027822.

Birtwell, I.K., M.D. Nassichuk, and H. Beune. 1987. "Underyearling Sockeye Salmon (*Oncorhynchus nerka*) in the Estuary of the Fraser River." In *Sockeye Salmon* (Oncorhynchus nerka) *Population Biology and Future Management*, ed. H.D.L. Smith, L. Margolis, and C.C. Wood, 25–35. Ottawa: Fisheries and Oceans Canada.

Birtwell, I.K., R. Fink, D. Brand, R. Alexander, and C.D. McAllister. 1999. "Survival of Pink Salmon (*Oncorhynchus gorbuscha*) Fry to Adulthood Following a 10-day Exposure to the Aromatic Hydrocarbon Water-Soluble Fraction of Crude Oil and Release to the Pacific Ocean." *Canadian Journal of Fisheries and Aquatic Sciences* 56 (11): 2087–98.

Bisson, P.A., J.L. Nielsen, M.W. Chilcote, B. Crawford, and S.A. Leider. 1986. "Occurrence of Anadromous Brown Trout in Two Lower Columbia River Tributaries." *North American Journal of Fisheries Management* 6 (2): 290–92.

Bjerknes, V., and A.B. Vaag. 1980. "Migration and Capture of Pink Salmon, *Oncorhynchus gorbuscha* Walbaum in Finnmark, North Norway." *Journal of Fish Biology* 16 (3): 291–97. http://dx.doi.org/10.1111/j.1095-8649.1980.tb03706.x.

Blackburn, J., and W.C. Clarke. 1987. *Revised Procedure for the 24 Hour Seawater Challenge Test to Measure Seawater Adaptability of Juvenile Salmonids*. Canadian Technical Report of Fisheries and Aquatic Sciences, No. 1515. Nanaimo, BC: Department of Fisheries and Oceans, Fisheries Research Branch.

Boeuf, G. 1993. "Salmonid Smolting: A Pre-Adaptation to the Oceanic Environment." In *Fish Ecophysiology*, ed. J.C. Rankin and F.B. Jensen, 105–35. London: Chapman and Hall. http://dx.doi.org/10.1007/978-94-011-2304-4_4.

Bogutskaya, N.G., A.M. Naseka, S.V. Shedko, E.D. Vasileva, and I.A. Chereschnev. 2008. "The Fishes of the Amur River: Updated Check-List and Zoogeography." *Ichthyological Exploration of Freshwaters* 19 (4): 301–66.

Bond, M.H., S.A. Hayes, C.V. Hanson, and R.B. MacFarlane. 2008. "Marine Survival of Steelhead (*Oncorhynchus mykiss*) Enhanced by a Seasonally Closed Estuary." *Canadian Journal of Fisheries and Aquatic Sciences* 65 (10): 2242–52. http://dx.doi.org/10.1139/F08-131.

Borja, Á., D.M. Dauer, M. Elliott, and C.A. Simenstad. 2010. "Medium- and Long-Term Recovery of Estuarine and Coastal Ecosystems: Patterns, Rates and Restoration Effectiveness." *Estuaries and Coasts* 33 (6): 1249–60. http://dx.doi.org/10.1007/s12237-010-9347-5.

Bottom, D.L. 1997. "To Till the Water: A History of Ideas in Fisheries Conservation." In *Pacific Salmon and Their Ecosystem: Status and Future Options*, ed. D.J. Stouder, P.A. Bisson, and R.J. Naiman, 569–97. New York: Chapman and Hall. http://dx.doi.org/10.1007/978-1-4615-6375-4_31.

Bottom, D.L., K.K. Jones, T.J. Cornwell, A. Gray, and C.A. Simenstad. 2005a. "Patterns of Chinook Salmon Migration and Residency in the Salmon River Estuary (Oregon)." *Estuarine, Coastal and Shelf Science* 64 (1): 79–93. http://dx.doi.org/10.1016/j.ecss.2005.02.008.

Bottom, D.L., C.A. Simenstad, J. Burke, A.M. Baptista, D.A. Jay, K.K. Jones, et al. 2005b. *Salmon at River's End: The Role of the Estuary in the Decline and Recovery of Columbia River Salmon.* NOAA Technical Memorandum, NMFS-NWFSC-68. Washington, DC: US Department of Commerce.

Bottom, D.L., K.K. Jones, C.A. Simenstad, and C.L. Smith. 2009. "Reconnecting Social and Ecological Resilience in Salmon Ecosystems." *Ecology and Society* 14 (1): 5–23.

Boubee, J.A.T., and F.J. Ward. 1997. "Mouth Gape, Food Size, and Diet of the Common Smelt *Retropinna retropinna* (Richardson) in the Waikato River System, North Island, New Zealand." *New Zealand Journal of Marine and Freshwater Research* 31 (2): 147–54. http://dx.doi.org/10.1080/00288330.1997.9516753.

Boula, D., V. Castric, L. Bernatchez, and C. Audet. 2002. "Physiological, Endocrine, and Genetic Bases of Anadromy in the Brook Charr, *Salvelinus fontinalis,* of the Laval River (Québec, Canada)." *Environmental Biology of Fishes* 64 (1/3): 229–42. http://dx.doi.org/10.1023/A:1016054119783.

Boulding, E.G., M. Culling, B. Glebe, P.R. Berg, S. Lien, and T. Moen. 2008. "Conservation Genomics of Atlantic Salmon: SNPs Associated with QTLs for Adaptive Traits in Parr from Four Trans-Atlantic Backcrosses." *Heredity* 101 (4): 381–91. http://dx.doi.org/10.1038/hdy.2008.67.

Boyd, F.C. 1989. "Report of the Working Group on Permissible Levels of Inference in Habitat Management of Salmonids." In *Proceedings of the National Workshop on Effects of Habitat Alteration on Salmonid Stocks,* ed. C.D. Levings, L.B. Holtby, and M.A. Henderson, 194–96. Ottawa: Fisheries and Oceans Canada.

Bradford, M.J., J. Lovy, D.A. Patterson, D.J. Speare, W.R. Bennett, A.R. Stobbart, and C.P. Tovey. 2010. "*Parvicapsula minibicornis* Infections in Gill and Kidney and the Premature Mortality of Adult Sockeye Salmon (*Oncorhynchus nerka*) from Cultus Lake, British Columbia." *Canadian Journal of Fisheries and Aquatic Sciences* 67 (4): 673–83. http://dx.doi.org/10.1139/F10-017.

Brandes, P.L., and J.S. McLain. 2001. "Juvenile Chinook Salmon Abundance, Distribution, and Survival in the Sacramento–San Joaquin Estuary." In *Contributions to the Biology of Central Valley Salmonids. Vol. 2, Fish Bulletin 179,* ed. R.L. Brown, 39–136. Sacramento: California Department of Fish and Game.

Brannon, E.L., M.S. Powell, T.P. Quinn, and A. Talbot. 2004. "Population Structure of Columbia River Basin Chinook Salmon and Steelhead Trout." *Reviews in Fisheries Science* 12 (2–3): 99–232. http://dx.doi.org/10.1080/10641260490280313.

Bravo, S., M. Perroni, E. Torres, and M.T. Silva. 2006. "Report of *Caligus rogercresseyi* in the Anadromous Brown Trout (*Salmo trutta*) in the Río Gallegos Estuary, Argentina." *Bulletin of the European Association of Fish Pathologists* 26 (4): 186–93.

Brawn, V.M. 1982. "Behavior of Atlantic Salmon (*Salmo salar*) during Suspended Migration in an Estuary, Sheet Harbour, Nova Scotia, Observed Visually and by Ultrasonic Tracking." *Canadian Journal of Fisheries and Aquatic Sciences* 39 (2): 248–56. http://dx.doi.org/10.1139/f82-035.

Breine, J.J., J. Maes, P. Quataert, E. van den Bergh, I. Simoens, G. Van Thuyne, and C. Belpaire. 2007. "A Fish-Based Assessment Tool for the Ecological Quality of the Brackish Schelde Estuary in Flanders (Belgium)." *Hydrobiologia* 575 (1): 141–59. http://dx.doi.org/10.1007/s10750-006-0357-z.

Breine, J.J., P. Quataert, M. Stevens, F. Ollevier, F.A. Volckaert, E. van den Bergh, and J. Maes. 2010. "A Zone-Specific Fish-Based Biotic Index as a Management Tool for the Zeeschelde Estuary (Belgium)." *Marine Pollution Bulletin* 60 (7): 1099–1112. http://dx.doi.org/10.1016/j.marpolbul.2010.01.014.

Breitburg, D.L., J.G. Sanders, C.C. Gilmour, C.A. Hatfield, R.W. Osman, G.F. Riedel, Sybil P. Seitzinger, and Kevin G. Sellner. 1999. "Variability in Responses to Nutrients and Trace Elements, and Transmission of Stressor Effects through an Estuarine Food Web." *Journal of Limnology and Oceanography* 44 (3, part 2): 837–63. http://dx.doi.org/10.4319/lo.1999.44.3_part_2.0837.

Brett, J.R. 1952. "Temperature Tolerance in Young Pacific Salmon, Genus *Oncorhynchus*." *Journal of the Fisheries Research Board of Canada* 9 (6): 265–323. http://dx.doi.org/10.1139/f52-016.

–. 1979. "Environmental Factors and Growth." In *Fish Physiology: Bioenergetics and Growth*, vol. 8, ed. W.S. Hoar, D.J. Randall, and J.R. Brett, 599–675. Waltham, MA: Academic Press. http://dx.doi.org/10.1016/S1546-5098(08)60033-3.

–. 1995. "Energetics." In *Physiological Ecology of Pacific Salmon*, ed. C. Groot, L. Margolis, and W.C. Clarke, 1–68. Vancouver: UBC Press.

Brett, J.R., W.C. Clarke, and J.E. Shelbourn. 1982. "Experiments on Thermal Requirements for Growth and Food Conversion Efficiency of Juvenile Chinook Salmon, *Oncorhynchus tshawytscha*." Nanaimo, BC: Department of Fisheries and Oceans.

Brett, J.R., and T.D.D. Groves. 1979. "Physiological Energetics." In *Fish Physiology: Bioenergetics and Growth*, vol. 8, ed. W.S. Hoar, D.J. Randall, and J.R. Brett, 279–352. Waltham, MA: Academic Press. http://dx.doi.org/10.1016/S1546-5098(08)60029-1.

Bricknell, I.R., S.J. Dalesman, B. O'Shea, C.C. Pert, and A.J. Mordue Luntz. 2006. "Effect of Environmental Salinity on Sea Lice *Lepeophtheirus salmonis* Settlement Success." *Diseases of Aquatic Organisms* 71 (3): 201–12. http://dx.doi.org/10.3354/dao071201.

British Columbia Marine Conservation Analysis. 2014. "British Columbia Marine Conservation Analysis." British Columbia Conservation Foundation, http://bcmca.ca/.

British Columbia Ministry of Forests. Lands and Natural Resource Operations. 2015. "Freshwater Fishing Regulations." http://www.env.gov.bc.ca/fw/fish/regulations/docs/1517/fishing_synopsis_2015-17_region2.pdf.

Brittain, J.E., G.M. Gislason, V. Ponomarev, J. Bogen, S. Brørs, A.J. Jensen, L.G. Khokhlova, et al. 2009. "Arctic Rivers." In *Rivers of Europe*, ed. K. Tockner, U. Uehlinger, and C.T. Robinson, 337–80. Waltham, MA: Academic Press. http://dx.doi.org/10.1016/B978-0-12-369449-2.00009-6.

Brodeur, R.D., K.W. Myers, and J.H. Helle. 2003. *Research Conducted by the United States on the Early Ocean Life History of Pacific Salmon*. North Pacific Anadromous Fish Commission Bulletin No. 3, 89–131. Vancouver: North Pacific Anadromous Fish Commission.

Brooks, K.M., and S.R.M. Jones. 2008. "Perspectives on Pink Salmon and Sea Lice: Scientific Evidence Fails to Support the Extinction Hypothesis." *Reviews in Fisheries Science* 16 (4): 403–12. http://dx.doi.org/10.1080/10641260801937131.

Brown, C., and K. Laland. 2001. "Social Learning and Life Skills Training for Hatchery Reared Fish." *Journal of Fish Biology* 59 (3): 471–93. http://dx.doi.org/10.1111/j.1095-8649.2001.tb02354.x.

Brown, C., K. Laland, and J. Krause. 2011. "Fish Cognition and Behaviour." In *Fish Cognition and Behaviour*, 2nd ed., ed. C. Brown, K. Laland, and J. Krause, 1–9. Oxford: Wiley-Blackwell Publishing. http://dx.doi.org/10.1002/9781444342536.ch1.

Brown, C.A., and R.J. Ozretich. 2009. "Coupling between the Coastal Ocean and Yaquina Bay, Oregon: Importance of Oceanic Inputs Relative to Other Nitrogen Sources." *Estuaries and Coasts* 32 (2): 219–37. http://dx.doi.org/10.1007/s12237-008-9128-6.

Brown, J.A., D.M. Scott, and R.W. Wilson. 2007. "Do Estuaries Act as Saline Bridges to Allow Invasion of New Freshwater Systems by Non-Indigenous Fish Species?" In *Biological Invaders in Inland Waters: Profiles, Distribution, and Threats*, ed. F. Gherardi, 401–14. Dordrecht, Netherlands: Springer. http://dx.doi.org/10.1007/978-1-4020-6029-8_21.

Brown, J.H., P.A. Marquet, and M.L. Taper. 1993. "Evolution of Body Size: Consequences of an Energetic Definition of Fitness." *American Naturalist* 142 (4): 573–84. http://dx.doi.org/10.1086/285558.

Brown, L.R. 2003. "Will Tidal Wetland Restoration Enhance Populations of Native Fishes?" *San Francisco Estuary and Watershed Science* 1 (1). https://escholarship.org/uc/item/2cp4d8wk.

Brown, L.R., and D. Michniuk. 2007. "Littoral Fish Assemblages of the Alien-Dominated Sacramento–San Joaquin Delta, California, 1980–1983 and 2001–2003." *Estuaries and Coasts* 30 (1): 186–200. http://dx.doi.org/10.1007/BF02782979.

Brown, S.K., K.R. Buja, S.H. Jury, M.E. Monaco, and A. Banner. 2000. "Habitat Suitability Index Models for Eight Fish and Invertebrate Species in Casco and Sheepscot Bays, Maine." *North American Journal of Fisheries Management* 20 (2): 408–35. http://dx.doi.org/10.1577/1548-8675(2000)020<0408:HSIMFE>2.3.CO;2.

Bryhn, A.C., M.A. Bergenius, P.H. Dimberg, and A. Adill. 2013. "Biomass and Number of Fish Impinged at a Nuclear Power Plant by the Baltic Sea." *Environmental Monitoring and Assessment* 185 (12): 10073–84. http://dx.doi.org/10.1007/s10661-013-3313-1.

Buchanan, S., A.P. Farrell, J. Fraser, P. Gallaugher, R. Joy, and R. Routledge. 2002. "Reducing Gillnet Mortality of Incidentally Caught Coho Salmon." *North American Journal of Fisheries Management* 22 (4): 1270–75. http://dx.doi.org/10.1577/1548-8675(2002)022<1270:RGNMOI>2.0.CO;2.

Buckland, F. 1880. *Natural History of British Fishes*. London: Society for Promoting Christian Knowledge.

Budnikova, L.L. 1994. "Amphipods in the Diet of Smolt Chum Salmon and Some Other Fishes in Kalininka Bay, Southwestern Sakhalin." *Russian Journal of Marine Biology* 20 (3): 143–47. Originally published in *Biologiya Morya* 20 (3): 190–96 (1993).

Bugaev, V.F. 2004. *Results of Identification of Sockeye Salmon* (Oncorhynchus nerka) *Secondary Local Stocks and Secondary Groups of Local Stocks in the Coastal and River Catches of Kamchatka River for 1978–2001*. North Pacific Anadromous Fish

Commission Technical Report No. 5. Vancouver: North Pacific Anadromous Fish Commission.
Bugaev, V.F., and V.I. Karpenko. 1984. "Some Data on Migration and Feeding of Underyearling Sockeye Salmon, *Oncorhynchus nerka* (Salmonidae), at the Mouth of the Kamchatka River." *Journal of Ichthyology* 23 (6): 146–50.
Burdloff, D., S. Gasparini, B. Sautour, H. Etcheber, and J. Castel. 2000. "Is the Copepod Egg Production in a Highly Turbid Estuary (the Gironde, France) a Function of the Biochemical Composition of Seston?" *Aquatic Ecology* 34 (2): 165–75. http://dx.doi.org/10.1023/A:1009903702667.
Burke, J.L. 2004. "Life Histories of Juvenile Chinook Salmon in the Columbia River Estuary, 1916 to the Present." MS thesis, Oregon State University, Corvallis.
Burnham, K.P., and D.R. Anderson. 2002. *Model Selection and Multi-Model Inference: A Practical Information-Theoretic Approach*. 2nd ed. New York: Springer.
Burt, D., and T. Burns. 1995. *Assessment of Salmonid Habitat in the Lower Campbell River*. Burnaby: British Columbia Hydro Environmental Affairs.
Burt, W.H. 1943. "Territoriality and Home Range Concepts as Applied to Mammals." *Journal of Mammalogy* 24 (3): 346–52. http://dx.doi.org/10.2307/1374834.
Busby, M.S., and R.A. Barnhart. 1995. "Potential Food Sources and Feeding Ecology of Juvenile Fall Chinook Salmon in California's Mattole River Lagoon." *California Fish and Game* 81 (4): 133–46.
Butler, J.R.A., and J. Watt. 2003. "Assessing and Managing the Impacts of Marine Salmon Farms on Wild Atlantic Salmon in Western Scotland: Identifying Priority Rivers for Conservation. In *Salmon at the Edge*, ed. E.H. Mills, 93–118. Oxford: Wiley-Blackwell Publishing. http://dx.doi.org/10.1002/9780470995495.ch9.
Buysse, D., A.J. Coeck, and A.J. Maes. 2008. "Potential Re-establishment of Diadromous Fish Species in the River Scheldt (Belgium)." *Hydrobiologia* 602 (1): 155–59. http://dx.doi.org/10.1007/s10750-008-9292-5.
Byock, J.L. 2001. *Viking Age Iceland*. New York: Penguin.
Bystriansky, J.S., N.T. Frick, J.G. Richards, P.M. Schulte, and J.S. Ballantyne. 2007. "Wild Arctic Char (*Salvelinus alpinus*) Upregulate Gill Na^+, K^+, -ATPase during Freshwater Migration." *Physiological and Biochemical Zoology* 80 (3): 270–82. http://dx.doi.org/10.1086/512982.
Cabral, H.N., M.J. Costa, and J.P. Salgado. 2001. "Does the Tagus Estuary Fish Community Reflect Environmental Changes?" *Climate Research* 18 (1–2): 119–26. http://dx.doi.org/10.3354/cr018119.
Cadwallader, P.L., and A.K. Eden. 1981. "Food and Growth of Hatchery-Produced Chinook Salmon, *Oncorhynchus tshawytscha* (Walbaum), in Landlocked Lake Purrumbete, Victoria, Australia." *Journal of Fish Biology* 18 (3): 321–30. http://dx.doi.org/10.1111/j.1095-8649.1981.tb03773.x.
Calderwood, W.L. 1908. *The Life of the Salmon, with Reference More Especially to the Fish in Scotland*. London: Edward Arnold.
Callaway, J., J.B. Zedler, and D.L. Ross. 1997. "Using Tidal Salt Marsh Mesocosms to Aid Wetland Restoration." *Restoration Ecology* 5 (2): 135–46. http://dx.doi.org/10.1046/j.1526-100X.1997.09716.x.
Cameron, W.M., and D.W. Pritchard. 1963. "Estuaries." In *The Sea: Ideas and Observations on Progress in the Study of the Seas*, vol. 2, ed. M.N. Hill, 306–24. New York: Wiley-Interscience Publishers.

Canadian Council of Ministers of the Environment. 1999. *Canadian Water Quality Guidelines for the Protection of Aquatic Life: Dissolved Oxygen (Marine), Canadian Environmental Quality Guidelines, 1999*. Winnipeg: Canadian Council of Ministers of the Environment. http://www.ccme.ca/files/ceqg/en/178.pdf.

Candy, J.R., and T.P. Quinn. 1999. "Behavior of Adult Chinook Salmon (*Oncorhynchus tshawytscha*) in British Columbia Coastal Waters Determined from Ultrasonic Telemetry." *Canadian Journal of Zoology* 77 (7): 1161–69. http://dx.doi.org/10.1139/z99-043.

Cannata, S.P. 1998. *Observations of Steelhead Trout* (Oncorhynchus mykiss), *coho salmon* (O. kisutch) *and Water Quality of the Navarro River Estuary/Lagoon, May 1996 to December 1997*. Draft report. Arcata, CA: Humboldt State University Foundation.

Carey, M.P., and D.H. Wahl. 2010. "Interactions of Multiple Predators with Different Foraging Modes in an Aquatic Food Web." *Oecologia* 162 (2): 443–52. http://dx.doi.org/10.1007/s00442-009-1461-3.

Carey, M.P., B.L. Sanderson, K.A. Barnas, and J.D. Olden. 2012. "Native Invaders – Challenges for Science, Management, Policy, and Society." *Frontiers in Ecology and the Environment* 10 (7): 373–81. http://dx.doi.org/10.1890/110060.

Carl, L.M., and M.C. Healey. 1984. "Differences in Enzyme Frequency and Body Morphology among Three Juvenile Life History Types of Chinook Salmon (*Oncorhynchus tshawytscha*) in the Nanaimo River, British Columbia." *Canadian Journal of Fisheries and Aquatic Sciences* 41 (7): 1070–77. http://dx.doi.org/10.1139/f84-125.

Carls, M.G., S.D. Rice, G.D. Marty, and D.K. Naydan. 2004. "Pink Salmon Spawning Habitat Is Recovering a Decade after the *Exxon Valdez* Oil Spill." *Transactions of the American Fisheries Society* 133 (4): 834–44. http://dx.doi.org/10.1577/T03-125.1.

Carlson, S.M., and T.R. Seamons. 2008. "A Review of Quantitative Genetic Components of Fitness in Salmonids: Implications for Adaptation to Future Change." *Evolutionary Applications* 1 (2): 222–38. http://dx.doi.org/10.1111/j.1752-4571.2008.00025.x.

Carmack, E.C., and R.W. Macdonald. 2002. "Oceanography of the Canadian Shelf of the Beaufort Sea: A Setting for Marine Life." *Arctic* 55 (1): 29–45.

Carr, J., F. Whoriskey, and D. Courtemanche. 2004. "Landlocked Atlantic Salmon: Movements to Sea by a Putative Freshwater Life History Form." In *Aquatic Telemetry: Advances and Applications: Proceedings of the Fifth Conference on Fish Telemetry Held in Europe, Ustica, Italy, 9–13 June 2003*, ed. M.T. Spedicato, G. Lembo, and G. Marmulla, 141–50. Rome: Food and Agriculture Organization of the United Nations.

Cattrijsse, A., I. Codling, A. Conides, S. Duhamel, R. Gibson, K. Hostens, and D. McLusky. 2002. "Estuarine Development/Habitat Restoration and Re-creation and Their Role in Estuarine Management for the Benefit of Aquatic Resources." In *Fishes in Estuaries*, ed. K. Hemingway and M. Elliott, 266–321. Oxford: Blackwell Science. http://dx.doi.org/10.1002/9780470995228.ch6.

Cederholm, C.J., D.H. Johnson, R.E. Bilby, L.G. Dominguez, A.M. Garrett, W.H. Graeber, et al. 2000. *Pacific Salmon and Wildlife – Ecological Contexts, Relationships, and Implications for Management*. Special Edition Technical Report, prepared for D.H. Johnson and T.A. O'Neil, Managing Directors of Wildlife-Habitat

Relationships in Oregon and Washington. Olympia: Washington Department of Fish and Wildlife.

Celtic Sea Trout Project. 2014. "A Short Summary of the Celtic Sea Trout Project at August 2013. Project Partner Lead." http://celticseatrout.com.

Champ, W., F. Kelly, and J. King. 2009. "The Water Framework Directive: Using Fish as a Management Tool." *Biology and Environment* (Dublin) 109 (3): 191–206.

Chan, K., T. Satterfield, and J. Goldstein. 2012. "Rethinking Ecosystem Services to Better Address and Navigate Cultural Values." *Ecological Economics* 74: 8–18. http://dx.doi.org/10.1016/j.ecolecon.2011.11.011.

Chapman, D.W. 1986. "Salmon and Steelhead Abundance in the Columbia River in the Nineteenth Century." *Transactions of the American Fisheries Society* 115 (5): 662–70. http://dx.doi.org/10.1577/1548-8659(1986)115<662:SASAIT>2.0.CO;2.

Chernitsky, A.G., S.P. Gambaryan, L.A. Karpenko, E.A. Lavrova, and D.S. Shkurko. 1993. "The Effect of Abrupt Salinity Changes on Blood and Muscle Electrolyte Content in the Smolts of the Atlantic Salmon, *Salmo salar*." *Comparative Biochemistry and Physiology. Part A, Physiology* 104 (3): 551–54. http://dx.doi.org/10.1016/0300-9629(93)90462-D.

Chernitsky, A.G., G.V. Zabruskov, V.V. Ermolaev, and D.S. Shkurko. 1995. "Life History of Trout, *Salmo trutta* L., in the Varsina River Estuary (the Barents Sea)." *Nordic Journal of Freshwater Research* 71: 183–89.

Cheung, W.W.L., J.L. Sarmiento, J. Dunne, T.L. Frölicher, V.W.Y. Lam, M.L.D. Palomares, et al. 2013. "Shrinking of Fishes Exacerbates Impacts of Global Ocean Changes on Marine Ecosystems." *Nature Climate Change* 3 (3): 254–58. http://dx.doi.org/10.1038/nclimate1691.

Chistiakov, D.A., B. Hellemans, and F.A. Volckaert. 2006. "Microsatellites and Their Genomic Distribution, Evolution, Function and Applications: A Review with Special Reference to Fish Genetics." *Aquaculture* (Amsterdam) 255 (1–4): 1–29. http://dx.doi.org/10.1016/j.aquaculture.2005.11.031.

Chittenden, C.M., S. Sura, K.G. Butterworth, K.F. Cubitt, N. Plantalech Manel-la, S. Balfry, F. Økland, and R.S. McKinley. 2008. "Riverine, Estuarine and Marine Migratory Behaviour and Physiology of Wild and Hatchery-Reared Coho Salmon *Oncorhynchus kisutch* (Walbaum) Smolts Descending the Campbell River, BC, Canada." *Journal of Fish Biology* 72 (3): 614–28. http://dx.doi.org/10.1111/j.1095-8649.2007.01729.x.

Christensen, V., and C.J. Walters. 2004. "Ecopath with Ecosim: Methods, Capabilities, and Limitations." *Ecological Modelling* 172 (2–4): 109–39. http://dx.doi.org/10.1016/j.ecolmodel.2003.09.003.

City of Sapporo. 2015. "Ashiri Cheppu Nomi" [Ceremony of the receiving of the first salmon]. http://www.city.sapporo.jp/keizai/kanko/event/event_calendar_english2009-2010.html.

Claridge, P.N., and I.C. Potter. 1994. "Abundance, Seasonality and Size of Atlantic Salmon Smolts Entrained on Power Station Intake Screens in the Severn Estuary." *Journal of the Marine Biological Association of the United Kingdom* 74 (03): 527–34. http://dx.doi.org/10.1017/S0025315400047640.

Clarke, K.R., and R.N. Gorley. 2006. *Plymouth Routine in Multivariate Ecological Research, v6: User Manual/Tutorial*. Ivybridge, UK: PRIMER-E, Plymouth. http://www.primer-e.com/.

Clarke, W.C., J.E. Shelbourn, and J.R. Brett. 1981. "Effect of Artificial Photoperiod Cycles, Temperature, and Salinity on Growth and Smolting in Underyearling Coho (*Oncorhynchus kisutch*), Chinook (*O. tshawytscha*), and Sockeye (*O. nerka*) Salmon." *Aquaculture* (Amsterdam) 22 (1): 105–16. http://dx.doi.org/10.1016/0044-8486(81)90137-X.

Claxton, A., K.C. Jacobson, M. Bhuthimethee, D. Teel, and D. Bottom. 2013. "Parasites in Subyearling Chinook Salmon (*Oncorhynchus tshawytscha*) Suggest Increased Habitat Use in Wetlands Compared to Sandy Beach Habitats in the Columbia River Estuary." *Hydrobiologia* 717 (1): 27–39. http://dx.doi.org/10.1007/s10750-013-1564-z.

Cohen Commission. 2012. *Commission of Inquiry into the Decline of Sockeye Salmon in the Fraser River (Canada). The Uncertain Future of Fraser River Sockeye. Vol. 1: The Sockeye Fishery. Final Report – October 2012.* Catalogue No. CP32-93/2012E-1. Ottawa: Public Works and Government Services Canada. http://epe.lac-bac.gc.ca/100/206/301/pco-bcp/commissions/cohen/cohen_commission/LOCALHOS/EN/INDEX.HTM.

Colautti, R.I., and H.J. MacIsaac. 2004. "A Neutral Terminology to Define 'Invasive' Species." *Diversity and Distributions* 10 (2): 135–41. http://dx.doi.org/10.1111/j.1366-9516.2004.00061.x.

Colclough, S.R., G. Gray, A. Bark, and B. Knights. 2002. "Fish and Fisheries of the Tidal Thames: Management of the Modern Resource, Research Aims and Future Pressures." *Journal of Fish Biology* 61 (Suppl. sA): 64–73. http://dx.doi.org/10.1111/j.1095-8649.2002.tb01762.x.

Colclough, S., L. Fonseca, T. Astley, K. Thomas, and W. Watts. 2005. "Fish Utilisation of Managed Realignments." *Fisheries Management and Ecology* 12 (6): 351–60. http://dx.doi.org/10.1111/j.1365-2400.2005.00467.x.

Congleton, J.L., S.K. Davis, and S.R. Foley. 1982. "Distribution, Abundance and Outmigration Timing of Chum and Chinook Salmon Fry in the Skagit Salt Marsh." In *Proceedings of the Salmon and Trout Migratory Behavior Symposium*, ed. E.L. Brannon and E.O. Salo, 153–63. Seattle: School of Fisheries, University of Washington.

Conover, D.O., L.M. Clarke, S.B. Munch, and G.N. Wagner. 2006. "Spatial and Temporal Scales of Adaptive Divergence in Marine Fishes and the Implications for Conservation." *Journal of Fish Biology* 69 (sc): 21–47. http://dx.doi.org/10.1111/j.1095-8649.2006.01274.x.

Consuegra, S., N. Phillips, G. Gajardo, and C. Garcia de Leaniz. 2011. "Winning the Invasion Roulette: Escapes from Fish Farms Increase Admixture and Facilitate Establishment of Non Native Rainbow Trout." *Evolutionary Applications* 4 (5): 660–71. http://dx.doi.org/10.1111/j.1752-4571.2011.00189.x.

Convention on Biological Diversity. 2012. *Biodiversity and Climate Change: Examples of Bioclimatic Models. Sixteenth Meeting of Subsidiary Body of Scientific, Technical and Technological Advice, Montreal, Canada.* UNEP/CBD/SBSTTA/16/INF/26. Montreal: Convention on Biological Diversity. https://www.cbd.int/doc/meetings/sbstta/sbstta-16/information/sbstta-16-inf-26-en.pdf.

Cooke, S.J., S.G. Hinch, A.P. Farrell, M. Lapointe, S.R.M. Jones, J.S. Macdonald, David A. Patterson, Michael C. Healey, and Glen Van Der Kraak. 2004. "Abnormal Migration Timing and High En Route Mortality of Sockeye Salmon in the Fraser River, British Columbia." *Fisheries* (Bethesda, MD) 29 (2): 22–33. http://dx.doi.org/10.1577/1548-8446(2004)29[22:AMTAHE]2.0.CO;2.

Cooksey, J.M. 2006. "Community Use of Created Intertidal Habitats in an Urban Estuary: Abundance Patterns and Diet Composition of Common Estuarine Fishes in the Lower Duwamish Waterway, Seattle, Washington." MS thesis, University of Washington.

Cooney, R.T. 1993. "A Theoretical Evaluation of the Carrying Capacity of Prince William Sound, Alaska, for Juvenile Pacific Salmon." *Fisheries Research* 18 (1–2): 77–87. http://dx.doi.org/10.1016/0165-7836(93)90041-5.

Copeland, T., and D.A. Venditti. 2009. "Contribution of Three Life History Types to Smolt Production in a Chinook Salmon (*Oncorhynchus tshawytscha*) Population." *Canadian Journal of Fisheries and Aquatic Sciences* 66 (10): 1658–65. http://dx.doi.org/10.1139/F09-110.

Copp, G.H., L. Vilizzi, J.D. Mumford, G. Fenwick, M.J. Godard, and R.E. Gozlan. 2009. "Calibration of FISK, an Invasiveness Screening Tool for Non-Native Freshwater Fishes." *Risk Analysis* 29 (3): 457–67. http://dx.doi.org/10.1111/j.1539-6924.2008.01159.x.

Correa, C., and M. Gross. 2008. "Chinook Salmon Invade Southern South America." *Biological Invasions* 10 (5): 615–39. http://dx.doi.org/10.1007/s10530-007-9157-2.

Costello, M.J. 2006. "Ecology of Sea Lice Parasitic on Farmed and Wild Fish." *Trends in Parasitology* 22 (10): 475–83. http://dx.doi.org/10.1016/j.pt.2006.08.006.

Coull, B.C. 1999. "Role of Meiofauna in Estuarine Soft Bottom Habitats." *Australian Journal of Ecology* 24 (4): 327–43. http://dx.doi.org/10.1046/j.1442-9993.1999.00979.x.

Cowardin, L.M., V. Carter, F.C. Golet, and E.T. LaRoe. 1979. *Classification of Wetlands and Deepwater Habitats of the United States.* Washington, DC: US Department of the Interior, Fish and Wildlife Service; Jamestown, ND: Northern Prairie Wildlife Research Center. http://www.fws.gov/wetlands/Documents/Classification-of-Wetlands-and-Deepwater-Habitats-of-the-United-States.pdf.

Cowx, I.G., ed. 2003. *Interactions between Fish and Birds: Implications for Management.* Oxford: Blackwell Publishing. http://dx.doi.org/10.1002/9780470995372.

Cowx, I.G., J.P. Harvey, R.A. Noble, and A.D. Nunn. 2009. "Establishing Survey and Monitoring Protocols for the Assessment of Conservation Status of Fish Populations in River Special Areas of Conservation in the UK." *Aquatic Conservation: Marine and Freshwater Ecosystems* 19 (1): 96–103. http://dx.doi.org/10.1002/aqc.968.

Craig, B.E., C.A. Simenstad, and D.L. Bottom. 2014. "Rearing in Natural and Recovering Tidal Wetlands Enhances Growth and Life history Diversity of Columbia Estuary Tributary Coho Salmon *Oncorhynchus kisutch* Population." *Journal of Fish Biology* 85 (1): 31–51. http://dx.doi.org/10.1111/jfb.12433.

Craig, P.C., W.B. Griffiths, S.R. Johnson, and D.M. Schell. 1984. "Trophic Dynamics in an Arctic Lagoon." In *The Alaskan Beaufort Sea and Ecosystems,* ed. P.W. Barnes, D.M. Schell, and E. Reimnitz, 347–80. Orlando, FL: Academic Press. http://dx.doi.org/10.1016/B978-0-12-079030-2.50023-X.

Crain, C.M., K. Kroeker, and B.S. Halpern. 2008. "Interactive and Cumulative Effects of Multiple Human Stressors in Marine Systems." *Ecology Letters* 11 (12): 1304–15. http://dx.doi.org/10.1111/j.1461-0248.2008.01253.x.

Crane, M., and A. Hyatt. 2011. "Viruses of Fish: An Overview of Significant Pathogens." *Viruses* 3 (11): 2025–46. http://dx.doi.org/10.3390/v3112025.

Crawford, S.S., and A.M. Muir. 2008. "Global Introductions of Salmon and Trout in the Genus *Oncorhynchus*: 1870–2007." *Reviews in Fish Biology and Fisheries* 18 (3): 313–44. http://dx.doi.org/10.1007/s11160-007-9079-1.

Crooks, J.A. 2002. "Characterizing Ecosystem-Level Consequences of Biological Invasions: The Role of Ecosystem Engineers." *Oikos* 97 (2): 153–66. http://dx.doi.org/10.1034/j.1600-0706.2002.970201.x.

Crossin, G.T., S.G. Hinch, S.J. Cooke, D.W. Welch, et al. 2008. "Exposure to High Temperature Influences the Behaviour, Physiology, and Survival of Sockeye Salmon during Spawning Migration." *Canadian Journal of Zoology* 86 (2): 127–40. http://dx.doi.org/10.1139/Z07-122.

Crozier, L.G., A.P. Hendry, P.W. Lawson, T.P. Quinn, N.J. Mantua, J. Battin, R.G. Shaw, and R.B. Huey. 2008. "Potential Responses to Climate Change in Organisms with Complex Life Histories: Evolution and Plasticity in Pacific Salmon." *Evolutionary Applications* 1 (2): 252–70. http://dx.doi.org/10.1111/j.1752-4571.2008.00033.x.

Crozier, W.W., P.J. Schön, G. Chaput, E.C.E. Potter, N.Ó. Maoiléidigh, and J.C. MacLean. 2004. "Managing Atlantic Salmon (*Salmo salar* L.) in the Mixed Stock Environment: Challenges and Considerations." *ICES Journal of Marine Science* 61 (8): 1344–58. http://dx.doi.org/10.1016/j.icesjms.2004.08.013.

Cunjak, R.A. 1992. "Comparative Feeding, Growth and Movements of Atlantic Salmon (*Salmo salar*) Parr from Riverine and Estuarine Environments." *Ecology of Freshwater Fish* 1 (1): 26–34. http://dx.doi.org/10.1111/j.1600-0633.1992.tb00004.x.

Cunjak, R.A., E.M.P. Chadwick, and M. Shears. 1989. "Downstream Movements and Estuarine Residence by Atlantic Salmon Parr (*Salmo salar*)." *Canadian Journal of Fisheries and Aquatic Sciences* 46 (9): 1466–71. http://dx.doi.org/10.1139/f89-187.

Cuo, L., D.P. Lettenmaier, M. Alberti, and J.E. Richey. 2009. "Effects of a Century of Land Cover and Climate Change on the Hydrology of the Puget Sound Basin." *Hydrological Processes* 23 (6): 907–33. http://dx.doi.org/10.1002/hyp.7228.

Curran, J.C., and A.R. Henderson. 1988. "The Oxygen Requirements of a Polluted Estuary for the Establishment of a Migratory Salmon, *Salmo salar* L., population." *Journal of Fish Biology* 33 (Suppl. sA): 63–69. http://dx.doi.org/10.1111/j.1095-8649.1988.tb05559.x.

Curran, J.W., and S.P. Keeney. 2006. "Replacement of Fluorescent Lamps with High-Brightness LEDs in a Bridge Lighting Application." In *Proceeding of Sixth International Conference on Solid State Lighting*, vol. 633719, ed. I.T. Ferguson, N. Narendran, T. Taguchi, and I. E. Ashdown. Bellingham, WA: Society for Photo-Optical Instrumentation Engineers. http://dx.doi.org/10.1117/12.682727.

Curry, R.A., J. van de Sande, and F.G. Whoriskey. 2006. "Temporal and Spatial Habitats of Anadromous Brook Charr in the Laval River and Its Estuary." *Environmental Biology of Fishes* 76 (2–4): 361–70. http://dx.doi.org/10.1007/s10641-006-9041-4.

Curry, R.A., L. Bernatchez, F. Whoriskey Jr., and C. Audet. 2010. "The Origins and Persistence of Anadromy in Brook Charr." *Reviews in Fish Biology and Fisheries* 20 (4): 557–70. http://dx.doi.org/10.1007/s11160-010-9160-z.
Cushing, C.E., K.W. Cummins, and G.W. Minshall, eds. 2006. *River and Stream Ecosystems of the World*. Los Angeles: University of California Press.
D'Or, R.K., and M.J. Stokesbury. 2009. "The Ocean Tracking Network – Adding Marine Animal Movements to the Global Ocean Observing System." In *Tagging and Tracking of Marine Animals with Electronic Devices*, ed. J.L. Nielsen, H. Arrizabalaga, N. Fragoso, A. Hobday, M. Lutcavage, and J. Sibert, 91–100. Amsterdam: Springer.
Dalrymple, R.W., and R.N. Rhodes. 1995. "Estuarine Dunes and Bars." In *Geomorphology and Sedimentology of Estuaries: Developments in Sedimentology*, vol. 53, ed. G.M.E. Perillo, 359–422. Amsterdam: Elsevier. http://dx.doi.org/10.1016/S0070-4571(05)80033-0.
Darwin, C. 1871. *The Descent of Man, and Selection in Relation to Sex*, 2: 3–6. London: John Murray. http://dx.doi.org/10.1037/12293-000.
Davis, J.C. 1975. "Minimal Dissolved Oxygen Requirements of Aquatic Life with Emphasis on Canadian Species: A Review." *Journal of the Fisheries Research Board of Canada* 32 (12): 2295–2332. http://dx.doi.org/10.1139/f75-268.
Dawe, N.K., G.E. Bradfield, W.S. Boyd, D.E.C. Trethewey, and A.N. Zolbrod. 2000. "Marsh Creation in a Northern Pacific Estuary: Is Thirteen Years of Monitoring Vegetation Dynamics Enough?" *Conservation Ecology* 4 (2): 1–12.
De Leeuw, J.J., R. ter Hofstede, and H.V. Winter. 2007. "Sea Growth of Anadromous Brown Trout (*Salmo trutta*)." *Journal of Sea Research* 58 (2): 163–65. http://dx.doi.org/10.1016/j.seares.2006.12.001.
Degerman, E., K. Leonardsson, and H. Lundqvist. 2012. "Coastal Migrations, Temporary Use of Neighbouring Rivers, and Growth of Sea Trout (*Salmo trutta*) from Nine Northern Baltic Sea Rivers." *ICES Journal of Marine Science* 69 (6): 971–80. http://dx.doi.org/10.1093/icesjms/fss073.
Dempson, J.B. 1993. "Salinity Tolerance of Freshwater Acclimated, Small Sized Arctic Charr, *Salvelinus alpinus* from Northern Labrador." *Journal of Fish Biology* 43 (3): 451–62. http://dx.doi.org/10.1111/j.1095-8649.1993.tb00580.x.
Department of Environment. 1972. "The Estuarine Food Web." In *Effects of Existing and Proposed Industrial Development on the Aquatic Ecosystem of the Squamish Estuary: Prepared for the Federal-Provincial Task Force on the Squamish Estuary Harbour Development*, vol. 21. Vancouver: Department of Environment, Canada Fisheries Service, Fisheries Research Board Pacific Region.
Department of Fisheries and Oceans. 1995. *Fraser River Chinook: Fraser River Action Plan, Fishery Management Group*. Vancouver: Department of Fisheries and Oceans.
–. 2001. *Selective (Salmon) Fisheries Program*. Vancouver: Fisheries and Oceans Canada, Pacific Region.
Depetris, P.J., D.M. Gaiero, J.L. Probst, J. Hartmann, and S. Kempe. 2005. "Biogeochemical Output and Typology of Rivers Draining Patagonia's Atlantic Seaboard." *Journal of Coastal Research* 21 (4): 835–44. http://dx.doi.org/10.2112/015-NIS.1.
Devlin, R.H., L.F. Sundström, and W.M. Muir. 2006. "Interface of Biotechnology and Ecology for Environmental Risk Assessments of Transgenic Fish." *Trends in Biotechnology* 24 (2): 89–97. http://dx.doi.org/10.1016/j.tibtech.2005.12.008.

Dexter, E., S.M. Bollens, G. Rollwagen-Bollens, J. Emerson, and J. Zimmerman. 2015. "Persistent vs. Ephemeral Invasions: 8.5 Years of Zooplankton Community Dynamics in the Columbia River." *Limnology and Oceanography* 60 (2): 527–39. http://dx.doi.org/10.1002/lno.10034.

Di Bacco, C., D.B. Humphrey, L.E. Nasmith, and C.D. Levings. 2012. "Ballast Water Transport of Non-Indigenous Zooplankton to Canadian Ports." *ICES Journal of Marine Science* 69 (3): 483–91. http://dx.doi.org/10.1093/icesjms/fsr133.

Diefenderfer, H.L., and D.R. Montgomery. 2009. "Pool Spacing, Channel Morphology, and the Restoration of Tidal Forested Wetlands of the Columbia River, USA." *Restoration Ecology* 17 (1): 158–68. http://dx.doi.org/10.1111/j.1526-100X.2008.00449.x.

Diekmann, M. 2003. "Species Indicator Values as an Important Tool in Applied Plant Ecology – A Review." *Basic and Applied Ecology* 4 (6): 493–506. http://dx.doi.org/10.1078/1439-1791-00185.

Dijkema, K.S. 1990. "Salt and Brackish Marshes around the Baltic Sea and Adjacent Parts of the North Sea: Their Vegetation and Management." *Biological Conservation* 51 (3): 191–209. http://dx.doi.org/10.1016/0006-3207(90)90151-E.

Dingle, H. 1980. "Ecology and Evolution of Migration." In *Animal Migration, Orientation and Navigation*, ed. S.A. Gauthreaux Jr., 1–101. New York: Academic Press.

Dionne, Michele, E. Bonebakker, and K. Whiting-Grant. 2003. "Maine's Salt Marshes: Their Functions, Values, and Restoration – A Resource Guide." *Maine Sea Grant Publication* 27: 1–17.

Dionne, Mélanie, F. Caron, J.J. Dodson, and L. Bernatchez. 2008. "Landscape Genetics and Hierarchical Genetic Structure in Atlantic Salmon: The Interaction of Gene Flow and Local Adaptation." *Molecular Ecology* 17 (10): 2382–96. http://dx.doi.org/10.1111/j.1365-294X.2008.03771.x.

Dobrynina, M.V., S.A. Gorshkov, and N.M. Kinas. 1989. "Effect of Density of Juvenile Pink Salmon, *Oncorhynchus gorbuscha*, on Their Vulnerability to Predators in the Utka River (Kamchatka)." *Journal of Ichthyology* 29 (1): 148–55. Translated from Russian; originally published in *Voprosy Ikthiologii* 28 (6): 971–77 (1988).

Dobrzycka, A., and A. Szaniawska. 1995. "Energy Strategy of *Corophium volutator* (Pallas, 1766) (Amphipoda) Population from the Gulf of Gdańsk." *Thermochimica Acta* 251 (1): 11–20. http://dx.doi.org/10.1016/0040-6031(94)02073-W.

Dodson, J.J. 1997. "Fish Migration: An Evolutionary Perspective." In *Behavioural Ecology of Teleost Fishes*, ed. J.G. Godin, 10–36. Oxford: Oxford University Press.

Dodson, J.J., J. Laroche, and F. Lecomte. 2009. "Contrasting Evolutionary Pathways of Anadromy in Euteleostean Fishes." In *Challenges for Diadromous Fishes in a Dynamic Global Environment, Symposium 69*, ed. A.J. Haro, K.L. Smith, R.A. Rulifson, C.M. Moffitt, R.J. Klauda, M.J. Dadswell, et al., 63–77. Bethesda, MD: American Fisheries Society.

Dolgopolova, E.N., and M.V. Isupova. 2010. "Classification of Estuaries by Hydrodynamic Processes." *Water Resources* 37 (3): 268–84. http://dx.doi.org/10.1134/S0097807810030024.

Dolloff, C.A., P.A. Flebbe, and J.E. Thorpe. 1994. "Salmonid Flexibility: Responses to Environmental Extremes." *Transactions of the American Fisheries Society* 123

(4): 606–12. http://dx.doi.org/10.1577/1548-8659(1994)123<0606:SFRTEE>2.3.CO;2.

Dorcey, A.H.J. 2010. "Sustainability Governance: Surfing the Waves of Transformation." In *Resource and Environmental Management in Canada: Addressing Conflict and Uncertainty*, 4th ed., ed. B. Mitchell, 528–54. Oxford: Oxford University Press.

Doucett, R.R., M. Power, G. Power, F. Caron, and J.D. Reist. 1999. "Evidence for Anadromy in a Southern Relict Population of Arctic Charr from North America." *Journal of Fish Biology* 55 (1): 84–93.

Drenner, S.M., T.D. Clark, C.K. Whitney, E.G. Martins, S.J. Cooke, and S.G. Hinch. 2012. "A Synthesis of Tagging Studies Examining the Behaviour and Survival of Anadromous Salmonids in Marine Environments." *PLoS One* 7 (3): e31311. http://dx.doi.org/10.1371/journal.pone.0031311.

Ducrotoy, J.P. 2010. "Ecological Restoration of Tidal Estuaries in North Western Europe: An Adaptive Strategy to Multi-Scale Changes." *Plankton and Benthos Research* 5 (Suppl. 1): 174–84. http://dx.doi.org/10.3800/pbr.5.174.

Ducrotoy, J.P., and J.C. Dauvin. 2008. "Estuarine Conservation and Restoration: The Somme and the Seine Case Studies (English Channel, France)." *Marine Pollution Bulletin* 57 (1–5): 208–18. http://dx.doi.org/10.1016/j.marpolbul.2008.04.031.

Duffy, E.J., and D.A. Beauchamp. 2008. "Seasonal Patterns of Predation on Juvenile Pacific Salmon by Anadromous Cutthroat Trout in Puget Sound." *Transactions of the American Fisheries Society* 137 (1): 165–81. http://dx.doi.org/10.1577/T07-049.1.

–. 2011. "Rapid Growth in the Early Marine Period Improves the Marine Survival of Chinook Salmon (*Oncorhynchus tshawytscha*) in Puget Sound, Washington." *Canadian Journal of Fisheries and Aquatic Sciences* 68 (2): 232–40. http://dx.doi.org/10.1139/F10-144.

Dumont, P., and J.R. Mongeau. 1990. "Bilan des efforts d'introduction de la truite brune (*Salmo trutta*) dans les eaux de la plaine de Montréal, Québec." *Bulletin Français de la Pêche et de la Pisciculture* 319 (319): 153–66. http://dx.doi.org/10.1051/kmae:1990001.

Dumont, P., J.F. Bergeron, P. Dulude, Y. Mailhot, A. Rouleau, G. Ouellet, and J.P. Lebel. 1988. "Introduced Salmonids: Where Are They Going in Québec Watersheds of the Saint-Laurent River?" *Fisheries* (Bethesda, MD) 13 (3): 9–17. http://dx.doi.org/10.1577/1548-8446(1988)013<0009:ISWATG>2.0.CO;2.

Dunton, K.H., T. Weingartner, and E.C. Carmack. 2006. "The Nearshore Western Beaufort Sea Ecosystem: Circulation and Importance of Terrestrial Carbon in Arctic Coastal Food Webs." *Progress in Oceanography* 71 (2–4): 362–78. http://dx.doi.org/10.1016/j.pocean.2006.09.011.

Dürr, H.H., G.G. Laruelle, C.M. van Kempen, C.P. Slomp, M. Meybeck, and H. Middelkoop. 2011. "Worldwide Typology of Nearshore Coastal Systems: Defining the Estuarine Filter of River Inputs to the Oceans." *Estuaries and Coasts* 34 (3): 441–58. http://dx.doi.org/10.1007/s12237-011-9381-y.

Dyer, B.S. 2000. "Systematic Review and Biogeography of the Freshwater Fishes of Chile" [Revisión Sistemática y Biogeografica de los Peces Dulceacuicloas de Chile]. *Estudios Oceanológicos* 19: 77–98.

Dyer, K.R. 1998. *Estuaries: A Physical Introduction*. Oxford: Wiley Publishing.

Eastern Brook Trout Joint Venture. 2014. *Eastern Brook Trout: Roadmap to Restoration*. Washington, DC: United States Fish and Wildlife Service. http://easternbrooktrout.org/reports/ebtjv-roadmap-to-restoration/view.

Ebert, D. 2005. *Ecology, Epidemiology, and Evolution of Parasitism in* Daphnia. National Library of Medicine (US), National Center for Biotechnology Information, http://www.ncbi.nlm.nih.gov/books/NBK2036/.

Eggers, D.M., and J.H. Clark. 2006. *Assessment of Historical Runs and Escapement Goals for Kotzebue Area Chum Salmon*. Game, Fishery Manuscript No. 06–01. Anchorage: Department of Fish and Game.

Einum, S., and I.A. Fleming. 2004. "Environmental Unpredictability and Offspring Size: Conservative versus Diversified Bet-Hedging." *Evolutionary Ecology Research* 6 (3): 443–55.

Eionet. 2014. "ETC/ICM European Topic Centre on Inland, Coastal and Marine Waters." http://icm.eionet.europa.eu/.

Eldon, G.A., and A.J. Greager. 1983. *Fishes of the Rakaia Lagoon*. Fisheries Environmental Report No. 30. New Zealand: Ministry of Agriculture and Fisheries.

Eldon, G.A., and G.R. Kelly. 1985. *Fishes of the Waimakariri River Estuary*. Fisheries Environmental Report No. 56. New Zealand: Ministry of Agriculture and Fisheries.

Elliott, J.M. 1991. "Tolerance and Resistance to Thermal Stress in Juvenile Atlantic Salmon, *Salmo salar*." *Freshwater Biology* 25 (1): 61–70. http://dx.doi.org/10.1111/j.1365-2427.1991.tb00473.x.

–. 1997. "Stomach Contents of Adult Sea Trout Caught in Six English Rivers." *Journal of Fish Biology* 50 (5): 1129–32.

Elliott, M., and F. Dewailly. 1995. "The Structure and Components of European Estuarine Fish Assemblages." *Netherlands Journal of Aquatic Ecology* 29 (3–4): 397–417. http://dx.doi.org/10.1007/BF02084239.

Elliott, M., and K.L. Hemingway, eds. 2002. *Fishes in Estuaries*. Oxford: Blackwell Science. http://dx.doi.org/10.1002/9780470995228.

Elliott, M., K.L. Hemingway, M.J. Costello, S. Duhamel, K. Hostens, M. Labropoulou, et al. 2002. "Links between Fishes and Other Trophic Levels." In *Fishes in Estuaries*, ed. M. Elliott and K. Hemingway, 124–216. Oxford: Blackwell Science. http://dx.doi.org/10.1002/9780470995228.ch4.

Emmerton, C.A., L.F. Lesack, and W.F. Vincent. 2008. "Nutrient and Organic Matter Patterns across the Mackenzie River, Estuary and Shelf during the Seasonal Recession of Sea-Ice." *Journal of Marine Systems* 74 (3–4): 741–55. http://dx.doi.org/10.1016/j.jmarsys.2007.10.001.

Emmett, R.L. 1997. "Estuarine Survival of Salmonids: The Importance of Interspecific and Intraspecific Predation and Competition." In *Estuarine and Ocean Survival of Northeastern Pacific Salmon: Proceedings of the Workshop* (NOAA Technical Memorandum NMFS-NWFSC-29), ed. R.L. Emmett and M.H. Schiewe, 147–58. Seattle: US Department of Commerce.

Emmett, R.L., R.D. Brodeur, and P.M. Orton. 2004. "The Vertical Distribution of Juvenile Salmon (*Oncorhynchus* spp.) and Associated Fishes in the Columbia River Plume." *Fisheries Oceanography* 13 (6): 392–402. http://dx.doi.org/10.1111/j.1365-2419.2004.00294.x.

Emmett, R.L., G.K. Krutzikowsky, and P. Bentley. 2006. "Abundance and Distribution of Pelagic Piscivorous Fishes in the Columbia River Plume during Spring/Early Summer 1998–2003: Relationship to Oceanographic Conditions, Forage Fishes, and Juvenile Salmonids." *Progress in Oceanography* 68 (1): 1–26. http://dx.doi.org/10.1016/j.pocean.2005.08.001.

Emmett, Robert, Roberto Llanso, Jan Newton, Ron Thom, Michelle Hornberger, Cheryl Morgan, Colin Levings, Andrea Copping, and Paul Fishman. 2000. "Geographic Signatures of North American West Coast Estuaries." *Estuaries* 23 (6): 765–92. http://dx.doi.org/10.2307/1352998.

Engle, N.L. 2011. "Adaptive Capacity and Its Assessment." *Global Environmental Change* 21 (2): 647–56. http://dx.doi.org/10.1016/j.gloenvcha.2011.01.019.

Engle, V.D., and J.K. Summers. 1999. "Latitudinal Gradients in Benthic Community Composition in Western Atlantic Estuaries." *Journal of Biogeography* 26 (5): 1007–23. http://dx.doi.org/10.1046/j.1365-2699.1999.00341.x.

Engle, V.D., J.C. Kurtz, L.M. Smith, C. Chancy, and P. Bourgeois. 2007. "A Classification of US Estuaries Based on Physical and Hydrologic Attributes." *Environmental Monitoring and Assessment* 129 (1–3): 397–412. http://dx.doi.org/10.1007/s10661-006-9372-9.

Erkinaro, J., E. Niemelä, J.P. Vähä, C.R. Primmer, S. Brørs, and E. Hassinen. 2010. "Distribution and Biological Characteristics of Escaped Farmed Salmon in a Major Subarctic Wild Salmon River: Implications for Monitoring." *Canadian Journal of Fisheries and Aquatic Sciences* 67 (1): 130–42. http://dx.doi.org/10.1139/F09-173.

Esteban, E.M., and M.P. Marchetti. 2004. "What's on the Menu? Evaluating a Food Availability Model with Young-of-the-Year Chinook Salmon in the Feather River, California." *Transactions of the American Fisheries Society* 133 (3): 777–88. http://dx.doi.org/10.1577/T03-115.1.

European Commission DG Environment. 2013. *Interpretation Manual of European Union Habitats, version EUR 28, April, 2013*. European Environment Agency, http://ec.europa.eu/environment/nature/legislation/habitatsdirective/docs/Int_Manual_EU28.pdf.

Evans, D. 2012. "Building the European Union's Natura 2000 network." *Natureza & Conservação* 1: 11–26. http://dx.doi.org/10.3897/natureconservation.1.1808.

Evans, T.G., E. Hammill, K. Kaukinen, A.D. Schulze, D.A. Patterson, K.K. English, Janelle M.R. Curtis, and Kristina M. Miller. 2011. "Transcriptomics of Environmental Acclimatization and Survival in Wild Adult Pacific Sockeye Salmon (*Oncorhynchus nerka*) during Spawning Migration." *Molecular Ecology* 20 (21): 4472–89. http://dx.doi.org/10.1111/j.1365-294X.2011.05276.x.

Fabry, V.J., B.A. Seibel, R.A. Feely, and J.C. Orr. 2008. "Impacts of Ocean Acidification on Marine Fauna and Ecosystems Processes." *ICES Journal of Marine Science* 65 (3): 414–32. http://dx.doi.org/10.1093/icesjms/fsn048.

Fagerlund, U.H.M., J.R. McBride, and I.V. Williams. 1995. "Stress and Tolerance." In *Physiological Ecology of Pacific Salmon*, ed. C. Groot, L. Margolis, and W.C. Clarke, 461–503. Vancouver: UBC Press.

Fahy, E. 1981. *The Beltra Fishery, Co. Mayo and Its Sea Trout* Salmo trutta *Stocks*. Irish Fisheries Bulletin No. 4. Dublin: Stationery Office.

Farrell, A.P., S.G. Hinch, S.J. Cooke, D.A. Patterson, et al. 2008. "Pacific Salmon in Hot Water: Applying Aerobic Scope Models and Biotelemetry to Predict the Success of Spawning Migrations." *Physiological and Biochemical Zoology* 81 (6): 697–708. http://dx.doi.org/10.1086/592057.

Fausch, K.D. 2007. "Introduction, Establishment and Effects of Non Native Salmonids: Considering the Risk of Rainbow Trout Invasion in the United Kingdom." *Journal of Fish Biology* 71 (Supp. sd): 1–32. http://dx.doi.org/10.1111/j.1095-8649.2007.01682.x.

Fausch, K.D., Y. Taniguchi, S. Nakano, G.D. Grossman, and C.R. Townsend. 2001. "Flood Disturbance Regimes Influence Rainbow Trout Invasion Success among Five Holarctic Regions." *Ecological Applications* 11 (5): 1438–55. http://dx.doi.org/10.1890/1051-0761(2001)011[1438:FDRIRT]2.0.CO;2.

Fechhelm, R.G., J.D. Bryan, W.B. Griffiths, and L.R. Martin. 1997. "Summer Growth Patterns of Northern Dolly Varden (*Salvelinus malma*) Smolts from the Prudhoe Bay Region of Alaska." *Canadian Journal of Fisheries and Aquatic Sciences* 54 (5): 1103–10. http://dx.doi.org/10.1139/f97-022.

Feely, R.A., S.R. Alin, J. Newton, C.L. Sabine, M. Warner, A. Devol, Christopher Krembs, and Carol Maloy. 2010. "The Combined Effects of Ocean Acidification, Mixing, and Respiration on pH and Carbonate Saturation in an Urbanized Estuary." *Estuarine, Coastal and Shelf Science* 88 (4): 442–49. http://dx.doi.org/10.1016/j.ecss.2010.05.004.

Ferguson, A. 2006. "Genetics of Sea Trout, with Particular Reference to Britain and Ireland." In *Sea Trout: Biology, Conservation and Management: Proceedings of the First International Sea Trout Symposium, Cardiff, UK, July 2004*, ed. G. Harris and N. Milner, 157–82. Oxford: Blackwell Publishing.

Feyrer, F., B. Herbold, S.A. Matern, and P.B. Moyle. 2003. "Dietary Shifts in a Stressed Fish Assemblage: Consequences of a Bivalve Invasion in the San Francisco Estuary." *Environmental Biology of Fishes* 67 (3): 277–88. http://dx.doi.org/10.1023/A:1025839132274.

Finstad, A.G., and C.L. Hein. 2012. "Migrate or Stay: Terrestrial Primary Productivity and Climate Drive Anadromy in Arctic Char." *Global Change Biology* 18 (8): 2487–97. http://dx.doi.org/10.1111/j.1365-2486.2012.02717.x.

Finstad, B., P.A. Bjørn, C.D. Todd, F. Whoriskey, P.G. Gargan, G. Forde, and C.W. Revie. 2011. "The Effect of Sea Lice on Atlantic Salmon and Other Salmonid Species." In *Atlantic Salmon Ecology*, ed. Ø. Aas, S. Einum, A. Klemetsen, and J. Skurdal, 253–76. Oxford: Wiley-Blackwell Publishing.

Fisher, B., R.K. Turner, and P. Morling. 2009. "Defining and Classifying Ecosystem Services for Decision Making." *Ecological Economics* 68 (3): 643–53. http://dx.doi.org/10.1016/j.ecolecon.2008.09.014.

Fleming, I.A. 1998. "Pattern and Variability in the Breeding System of Atlantic Salmon (*Salmo salar*), with Comparisons to Other Salmonids." *Canadian Journal of Fisheries and Aquatic Sciences* 55 (S1): 59–76. http://dx.doi.org/10.1139/d98-009.

Food and Agriculture Organization of the United Nations. 2014. "World Aquaculture Production of Fish, Crustaceans, Molluscs, etc., by Principal Species." In *Fishery and Agriculture Statistics Organization of the United Nations Yearbook of Fishery Statistics 2011*, 46. ftp://ftp.fao.org/FI/STAT/summary/a-6.pdf.

Ford, M.J., H. Fuss, B. Boelts, E. LaHood, J. Hard, and J. Miller. 2006. "Changes in Run Timing and Natural Smolt Production in a Naturally Spawning Coho Salmon (*Oncorhynchus kisutch*) Population after 60 Years of Intensive Hatchery Supplementation." *Canadian Journal of Fisheries and Aquatic Sciences* 63 (10): 2343–55. http://dx.doi.org/10.1139/f06-119.

Forseth, T., O. Ugedal, B. Jonsson, A. Langeland, and O. Njastad. 1991. "Radiocaesium Turnover in Arctic Charr (*Salvelinus alpinus*) and Brown Trout (*Salmo trutta*) in a Norwegian Lake." *Journal of Applied Ecology* 28 (3): 1053–67. http://dx.doi.org/10.2307/2404225.

Foundation for Water Research. 2014. "Information Centre for Water, Wastewater, and Related Environmental Issues." http://www.fwr.org/about.html.

Fowey Estuary Partnership. 2014. *Fowey Estuary Management Plan.* https://www.foweyharbour.co.uk/assets/file/pdfs/Environmental/FEMP2012.pdf.

Francis, R.A., and S.P.G. Hoggart. 2009. "Urban River Wall Habitat and Vegetation: Observations from the River Thames through Central London." *Urban Ecosystems* 12 (4): 465–85. http://dx.doi.org/10.1007/s11252-009-0096-9.

Fraser, D.J., L.K. Weir, L. Bernatchez, M.M. Hansen, and E.B. Taylor. 2011. "Extent and Scale of Local Adaptation in Salmonid Fishes: Review and Meta-Analysis." *Heredity* 106 (3): 404–20. http://dx.doi.org/10.1038/hdy.2010.167.

Fraser, F.J., P.J. Starr, and A.Y. Fedorenko. 1982. *A Review of the Chinook and Coho Salmon of the Fraser River.* Canadian Technical Report of Fisheries and Aquatic Sciences, No. 1126. New Westminster, BC: Fisheries and Oceans Canada.

Fraser River Estuary Management Program. 2014. "Estuary Management Plan." http://www.bieapfremp.org/fremp/managementplan/colourcoding.html.

Frenkel, R.E., and J.C. Morlan. 1991. "Can We Restore Our Salt Marshes? Lessons from the Salmon River, Oregon." *Northwest Environmental Journal* 7 (1): 119–35.

Fresh, K.L. 1997. "The Role of Competition and Predation in the Decline of Pacific Salmon and Steelhead." In *Pacific Salmon and Their Ecosystems: Status and Future Options,* ed. D.J. Stouder, P.A. Bisson, and R.J. Naiman, 245–75. New York: Chapman and Hall. http://dx.doi.org/10.1007/978-1-4615-6375-4_16.

–. 2006. *Juvenile Pacific Salmon in Puget Sound: Puget Sound Nearshore Partnership/ Puget Sound Nearshore Ecosystem Restoration Project.* Technical Report No. 2006–06. Seattle: US Army Corps of Engineers.

Friesen, T.A., J.S. Vile, and A.L. Pribyl. 2007. "Outmigration of Juvenile Chinook Salmon in the Lower Willamette River, Oregon." *Northwest Science* 81 (3): 173–90. http://dx.doi.org/10.3955/0029-344X-81.3.173.

Froese, R., and D. Pauly, eds. 2014. *FishBase.* Vancouver: University of British Columbia. http://www.fishbase.org.

Fukushima, M., H. Shimazaki, P.S. Rand, and M. Kaeriyama. 2011. "Reconstructing Sakhalin Taimen (*Parahucho parryi*) Historical Distribution and Identifying Causes for Local Extinctions." *Transactions of the American Fisheries Society* 140 (1): 1–13.

Fukuwaka, M., and T. Suzuki. 2000. *Density-Dependence of Chum Salmon in Coastal Waters of the Japan Sea.* North Pacific Anadromous Fish Commission Bulletin No. 2, 75–81. Vancouver: North Pacific Anadromous Fish Commission.

Funakoshi, G., and S. Kasuya. 2009. "Influence of an Estuary Dam on the Dynamics of Bisphenol A and Alkylphenols." *Chemosphere* 75 (4): 491–97. http://dx.doi.org/10.1016/j.chemosphere.2008.12.050.

Gallagher, C.P., and T.A. Dick. 2010. "Historical and Current Population Characteristics and Subsistence Harvest of Arctic Char from the Sylvia Grinnell River, Nunavut, Canada." *North American Journal of Fisheries Management* 30 (1): 126–41. http://dx.doi.org/10.1577/M09-027.1.

Galland, D., and T. McDaniels. 2008. "Are New Industry Policies Precautionary? The Case of Salmon Aquaculture Siting Policy in British Columbia." *Environmental Science and Policy* 11 (6): 517–32. http://dx.doi.org/10.1016/j.envsci.2008.05.002.

Garcia de Leaniz, C., G. Gajardo, and S. Consuegra. 2010. "From Best to Pest: Changing Perspectives on the Impact of Exotic Salmonids in the Southern Hemisphere." *Systematics and Biodiversity* 8 (4): 447–59. http://dx.doi.org/10.1080/14772000.2010.537706.

Garcia de Leaniz, C., I.A. Fleming, S. Einum, E. Verspoor, W.C. Jordan, S. Consuegra, et al. 2007. "A Critical Review of Adaptive Genetic Variation in Atlantic Salmon: Implications for Conservation." *Biological Reviews of the Cambridge Philosophical Society* 82 (2): 173–211. http://dx.doi.org/10.1111/j.1469-185X.2006.00004.x.

Garono, R.J., and R. Robinson. 2003. *Estuarine and Tidal Freshwater Habitat Cover Types along the Lower Columbia River Estuary Determined from Landsat 7 ETM+ Imagery*. Corvallis, OR: Wetlands and Watershed Assessment Group. http://dx.doi.org/10.2172/962829.

Gausland, I. 2000. "The Impact of Seismic Surveys on Marine Life." *Leading Edge* 19 (8): 903–5. http://dx.doi.org/10.1190/1.1438746.

Gee, A.K., K. Wasson, S.L. Shaw, and J. Haskins. 2010. "Signatures of Restoration and Management Changes in the Water Quality of a Central California Estuary." *Estuaries and Coasts* 33 (4): 1004–24. http://dx.doi.org/10.1007/s12237-010-9276-3.

Gende, S.M., T.P. Quinn, M.F. Willson, R. Heintz, and T.M. Scott. 2004. "Magnitude and Fate of Salmon-Derived Nutrients and Energy in a Coastal Stream Ecosystem." *Journal of Freshwater Ecology* 19 (1): 149–60. http://dx.doi.org/10.1080/02705060.2004.9664522.

Genovart, M., N. Negre, G. Tavecchia, A. Bistuer, L. Parpal, and D. Oro. 2010. "The Young, the Weak and the Sick: Evidence of Natural Selection by Predation." *PLoS One* 5 (3): e9774. http://dx.doi.org/10.1371/journal.pone.0009774.

Gerasimov, N.N., and Y.N. Gerasimov. 1998. "The International Significance of Wetland Habitats in the Lower Moroshechnaya River (West Kamchatka, Russia) for Waders." *International Wader Studies* 10: 237–42.

Gewant, D., and S.M. Bollens. 2012. "Fish Assemblages of Interior Tidal Marsh Channels in Relation to Environmental Variables in the Upper San Francisco Estuary." *Environmental Biology of Fishes* 94 (2): 483–99. http://dx.doi.org/10.1007/s10641-011-9963-3.

Gilbert, C.H. 1913. *Age of Maturity of the Pacific Salmon* Oncorhynchus. Bulletin of the US Bureau of Fish No. 32, 1–22. Washington, DC: Government Printing Office.

Gilbert, S., and R. Horner. 1986. *The Thames Barrier*. London: Thomas Telford.

Gill, A.B. 2003. "The Dynamics of Prey Choice in Fish: The Importance of Prey Size and Satiation." *Journal of Fish Biology* 63 (s1): 105–16. http://dx.doi.org/10.1111/j.1095-8649.2003.00214.x.

Gjelland, K.Ø., R.M. Serra-Llinares, R.D. Hedger, P. Arechavala-Lopez, R. Nilsen, B. Finstad, I. Uglem, O.T. Skilbrei, and P.A. Bjørn. 2014. "Effects of Salmon Lice Infection on the Behaviour of Sea Trout in the Marine Phase." *Aquaculture Environment Interactions* 5 (3): 221–33. http://dx.doi.org/10.3354/aei00105.

Glasgow, H.B., J.M. Burkholder, R.E. Reed, A.J. Lewitus, and J.E. Kleinman. 2004. "Real-Time Remote Monitoring of Water Quality: A Review of Current Applications, and Advancements in Sensor, Telemetry, and Computing Technologies." *Journal of Experimental Marine Biology and Ecology* 300 (1–2): 409–48. http://dx.doi.org/10.1016/j.jembe.2004.02.022.

Godbout, L., C.C. Wood, R.E. Withler, S. Latham, R.J. Nelson, L. Wetzel, R. Barnett-Johnson, M.J. Grove, A.K. Schmitt, K.D. McKeegan, et al. 2011. "Sockeye Salmon (*Oncorhynchus nerka*) Return after an Absence of Nearly 90 Years: A Case of Reversion to Anadromy." *Canadian Journal of Fisheries and Aquatic Sciences* 68 (9): 1590–1602. http://dx.doi.org/10.1139/f2011-089.

Godin, J.G.J. 1981. "Daily Patterns of Feeding Behavior, Daily Rations, and Diets of Juvenile Pink Salmon (*Oncorhynchus gorbuscha*) in Two Marine Bays of British Columbia." *Canadian Journal of Fisheries and Aquatic Sciences* 38 (1): 10–15. http://dx.doi.org/10.1139/f81-002.

Good, T.P., M.M. McClure, B.P. Sandford, K.A. Barnas, D.M. Marsh, B.A. Ryan, and E. Casillas. 2007. "Quantifying the Effect of Caspian Tern Predation on Threatened and Endangered Pacific Salmon in the Columbia River Estuary." *Endangered Species Research* 3 (1): 11–21. http://dx.doi.org/10.3354/esr003011.

Gordeeva, N.V., E.A. Salmenkova, Y.P. Altukhov, A.A. Makhrov, and S.P. Pustovoit. 2003. "Genetic Changes in Pink Salmon *Oncorhynchus gorbuscha* Walbaum during Acclimatization in the White Sea Basin." *Russian Journal of Genetics* 39 (3): 322–32. Translated from Russian; originally published in *Genetika* 39 (3): 402–12. http://dx.doi.org/10.1023/A:1023283919776.

Gordon, D.K., and C.D. Levings. 1984. *Seasonal Changes of Inshore Fish Populations on Sturgeon and Roberts Bank, Fraser River Estuary, British Columbia*. Canadian Technical Report of Fisheries and Aquatic Sciences, No. 1240. West Vancouver: Fisheries and Oceans Canada.

Grant, J.W.A., and I. Imre. 2005. "Patterns of Density Dependent Growth in Juvenile Stream Dwelling Salmonids." *Journal of Fish Biology* 67 (sB): 100–10. http://dx.doi.org/10.1111/j.0022-1112.2005.00916.x.

Gray, A. 2005. "The Salmon River Estuary: Restoring Tidal Inundation and Tracking Ecosystem Response." PhD dissertation, University of Washington.

Gray, A., C.A. Simenstad, D.L. Bottom, and T.J. Cornwell. 2002. "Contrasting Functional Performance of Juvenile Salmon in Recovering Wetlands of the Salmon River Estuary, Oregon, USA." *Restoration Ecology* 10 (3): 514–26. http://dx.doi.org/10.1046/j.1526-100X.2002.01039.x.

Green, D.R., and S.D. King. 2005. "Applying the Geospatial Technologies to Estuary Environments." In *GIS for Coastal Zone Management*, ed. J. Smith and D. Barlett, 239–56. Boca Raton, FL: CRC Press.

Green, R.H. 1979. *Sampling Design and Statistical Methods for Environmental Biologists*. Oxford: John Wiley and Sons.

Greene, C.M., and T. Beechie. 2004. "Consequences of Potential Density-Dependent Mechanisms on Recovery of Ocean-Type Chinook Salmon

(*Oncorhynchus tshawytscha*)." *Canadian Journal of Fisheries and Aquatic Sciences* 61 (4): 590–602. http://dx.doi.org/10.1139/f04-024.

Greene, C.M., D.W. Jensen, G.R. Pess, E.A. Steel, and E. Beamer. 2005. "Effects of Environmental Conditions during Stream, Estuary, and Ocean Residency on Chinook Salmon Return Rates in the Skagit River, Washington." *Transactions of the American Fisheries Society* 134 (6): 1562–81. http://dx.doi.org/10.1577/T05-037.1.

Gregor, J.W. 1944. "The Ecotype." *Biological Reviews of the Cambridge Philosophical Society* 19 (1): 20–30. http://dx.doi.org/10.1111/j.1469-185X.1944.tb00299.x.

Gregory, R., L. Failing, M. Harstone, G. Long, T. McDaniels, and D. Ohlson. 2012. *Structured Decision Making: A Practical Guide to Environmental Management Choices*. Oxford: Wiley-Blackwell Publishing. http://dx.doi.org/10.1002/9781444398557.

Gregory, R.S. 1993. "Effect of Turbidity on the Predator Avoidance Behaviour of Juvenile Chinook Salmon (*Oncorhynchus tshawytscha*)." *Canadian Journal of Fisheries and Aquatic Sciences* 50 (2): 241–46. http://dx.doi.org/10.1139/f93-027.

Gregory, R.S., and C.D. Levings. 1996. "The Effects of Turbidity and Vegetation on the Risk of Juvenile Salmonids, *Oncoryhnchus* spp., to Predation by Adult Cutthroat Trout, *O. clarkii*." *Environmental Biology of Fishes* 47 (3): 279–88. http://dx.doi.org/10.1007/BF00000500.

–. 1998. "Turbidity Reduces Predation on Migrating Juvenile Pacific Salmon." *Transactions of the American Fisheries Society* 127 (2): 275–85. http://dx.doi.org/10.1577/1548-8659(1998)127<0275:TRPOMJ>2.0.CO;2.

Grimaldo, L.F., A.R. Stewart, and W. Kimmerer. 2009. "Dietary Segregation of Pelagic and Littoral Fish Assemblages in a Highly Modified Tidal Freshwater Estuary." *Marine and Coastal Fisheries* 1 (1): 200–17. http://dx.doi.org/10.1577/C08-013.1.

Grimes, C.B., R.D. Brodeur, S.M. McKinnell, and L.J. Haldorson. 2007. "An Introduction to the Ecology of Juvenile Salmon in the Northeast Pacific Ocean: Regional Comparisons." In *Ecology of Juvenile Salmon in the Northeast Pacific Ocean: Regional Comparisons* (American Fisheries Society Symposium 57), ed. C.B. Grimes, R.D. Brodeur, L.J. Haldorson, and S.M. McKinnell, 1–6. Bethesda, MD: American Fisheries Society.

Gritsenko, O.F., and A.A. Churikov. 1977. "Biology of the Trout *Salvelinus* and Their Place in the Ichthyocoenosis of Bays in Northeastern Sakhalin, 2: Feeding." *Journal of Ichthyology* 17 (4): 591–99. Translated from Russian; originally published in *Voprosy Ikhtiologii* 17 (4): 668–76 (1976).

Gritsenko, O.F., E.M. Malkin, and A.A. Churikov. 1974. "Sakhalin Taimen *Hucho perryi* (Brevoort) from Bogataya River (the Eastern Coast of Sakhalin)." *Izvestiya Tikho Okeansovo Nauchnovo Instituta Ribalovochnih I Okeanographov* [*Proceedings of the Pacific Scientific Research Institute of Fisheries and Oceanography*] 93: 91–101 [in Russian].

Grønvik, S., and A. Klemetsen. 1987. "Marine Food and Diet Overlap of Co-occurring Arctic Charr *Salvelinus alpinus* (L.), Brown Trout *Salmo trutta* L. and Atlantic Salmon *S. salar* L. off Senja, N. Norway." *Polar Biology* 7 (3): 173–77. http://dx.doi.org/10.1007/BF00259205.

Groot, C., W.C. Clarke, and L. Margolis, eds. 1995. *Physiological Ecology of Pacific Salmon*. Vancouver: UBC Press.

Groot, C., and L. Margolis, eds. 1991. *Pacific Salmon Life Histories*. Vancouver: UBC Press.

Groot, E.P. 1989. "Intertidal Spawning of Chum Salmon: Saltwater Tolerance of the Early Life Stages to Actual and Simulated Intertidal Conditions." MSc thesis, University of British Columbia.

Gruber, J.S. 2010. "Key Principles of Community-Based Natural Resource Management: A Synthesis and Interpretation of Identified Effective Approaches for Managing the Commons." *Environmental Management* 45 (1): 52–66. http://dx.doi.org/10.1007/s00267-008-9235-y.

Guarinello, M.L., E.J. Shumchenia, and J.W. King. 2010. "Marine Habitat Classification for Ecosystem-Based Management: A Proposed Hierarchical Framework." *Environmental Management* 45 (4): 793–806. http://dx.doi.org/10.1007/s00267-010-9430-5.

Gulseth, O.A., and K.J. Nilssen. 2000. "The Brief Period of Spring Migration, Short Marine Residence, and High Return Rate of a Northern Svalbard Population of Arctic Char." *Transactions of the American Fisheries Society* 129 (3): 782–96. http://dx.doi.org/10.1577/1548-8659(2000)129<0782:TBPOSM>2.3.CO;2.

Haedrich, R.L. 1983. "Estuarine Fishes." In *Estuaries and Enclosed Seas: Ecosystems of the World*, ed. G.H. Ketchum, 26: 183–207. Amsterdam: Elsevier Scientific.

Hamburger, K., C. Lindegaard, and P.C. Dall. 1996. "The Role of Glycogen during the Ontogenesis of *Chironomus anthracinus* (Chironomidae, Diptera)." *Hydrobiologia* 318 (1–3): 51–59. http://dx.doi.org/10.1007/BF00014131.

Hamon, T.R., C.J. Foote, R. Hilborn, and D.E. Rogers. 2000. "Selection on Morphology of Spawning Wild Sockeye Salmon by a Gill-Net Fishery." *Transactions of the American Fisheries Society* 129 (6): 1300–15. http://dx.doi.org/10.1577/1548-8659(2000)129<1300:SOMOSW>2.0.CO;2.

Handeland, S.O., A.M. Arnesen, and S.O. Stefansson. 2003. "Seawater Adaptation and Growth of Post-Smolt Atlantic Salmon (*Salmo salar*) of Wild and Farmed Strains." *Aquaculture* (Amsterdam) 220 (1–4): 367–84. http://dx.doi.org/10.1016/S0044-8486(02)00508-2.

Handeland, S.O., T. Jarvi, A. Ferno, and S.O. Stefansson. 1996. "Osmotic Stress, Antipredator Behaviour, and Mortality of Atlantic Salmon (*Salmo salar*) Smolts." *Canadian Journal of Fisheries and Aquatic Sciences* 53 (12): 2673–80. http://dx.doi.org/10.1139/f96-227.

Hanna, S.S. 2008. "Institutions for Managing Resilient Salmon (*Oncorhynchus* spp.) Ecosystems: The Role of Incentives and Transaction Costs." *Ecology and Society* 13 (2): 35.

Hansen, L.P., and T.P. Quinn. 1998. "The Marine Phase of the Atlantic Salmon (*Salmo salar*) Life Cycle, with Comparisons to Pacific Salmon." *Canadian Journal of Fisheries and Aquatic Sciences* 55 (S1): 104–18. http://dx.doi.org/10.1139/d98-010.

Hanson, J.M., and S.C. Courtenay. 1995. "Seasonal Abundance and Distributions of Fishes in the Miramichi Estuary." In *Water, Science, and the Public: The Miramichi Ecosystem* (Canadian Special Publication of Fisheries and Aquatic Sciences No. 123), ed. E.M.P. Chadwick, 141–60. Ottawa: National Research Council of Canada.

Hanson, K.C., K.G. Ostrand, and R.A. Glenn. 2012. "Physiological Characterization of Juvenile Chinook Salmon Utilizing Different Habitats during Migration through the Columbia River Estuary." *Comparative Biochemistry and Physiology*.

Part A, Molecular and Integrative Physiology 163 (3–4): 343–49. http://dx.doi.org/10.1016/j.cbpa.2012.07.008.

Hanson, P.C., T.B. Johnson, D.E. Schindler, and J.F. Kitchell. 1997. *Fish Bioenergetics 3.0*. Technical Report WISCU-T-97-001; C6, Appendix A. Madison: University of Wisconsin Sea Grant Institute.

Harache, Y. 1992. "Pacific Salmon in Atlantic Waters." In *Proceedings of Introductions and Transfers of Aquatic Species: Selected Papers from a Symposium Held in Halifax, Nova Scotia, June 12–13, 1990*. Marine Science Symposia Series, 194: 31–55. Copenhagen: International Council for the Exploration of the Sea.

Hargreaves, N.B., and R.J. LeBrasseur. 1986. "Size Selectivity of Coho (*Oncorhynchus kisutch*) Preying on Juvenile Chum Salmon (*O. keta*)." *Canadian Journal of Fisheries and Aquatic Sciences* 43 (3): 581–86. http://dx.doi.org/10.1139/f86-069.

Harnish, R.A., G.E. Johnson, G.A. McMichael, M.S. Hughes, and B.D. Ebberts. 2012. "Effect of Migration Pathway on Travel Time and Survival of Acoustic-Tagged Juvenile Salmonids in the Columbia River Estuary." *Transactions of the American Fisheries Society* 141 (2): 507–19. http://dx.doi.org/10.1080/00028487.2012.670576.

Harris, G., and N. Milner, eds. 2006. *Sea Trout: Biology, Conservation and Management: Proceedings of the First International Sea Trout Symposium, Cardiff, UK, July 2004*. Oxford: Blackwell Publishing.

Harris, P.D., L. Bachmann, and T.A. Bakke. 2011. "The Parasites and Pathogens of the Atlantic Salmon: Lessons from *Gyrodactylus salaris*." In *Atlantic Salmon Ecology*, ed. Ø. Aas, S. Einum, A. Klemetsen, and J. Skurdal, 221–52. Oxford: Wiley-Blackwell Publishing.

Hart, D.E. 2007. "River Mouth Lagoon Dynamics on Mixed Sand and Gravel Barrier Coasts." *Journal of Coastal Research* SI50: 927–31.

Harvell, D., R. Aronson, N. Baron, J. Connell, A. Dobson, S. Ellner, Leah Gerber, Kiho Kim, Armand Kuris, Hamish McCallum, et al. 2004. "The Rising Tide of Ocean Diseases: Unsolved Problems and Research Priorities." *Frontiers in Ecology and the Environment* 2 (7): 375–82. http://dx.doi.org/10.1890/1540-9295(2004)002[0375:TRTOOD]2.0.CO;2.

Harvey, B.C. 1991. "Interactions among Stream Fishes: Predator-Induced Habitat Shifts and Larval Survival." *Oecologia* 87 (1): 29–36. http://dx.doi.org/10.1007/BF00323776.

Harvey, C.J., G.D. Williams, and P.S. Levin. 2012. "Food Web Structure and Trophic Control in Central Puget Sound." *Estuaries and Coasts* 35 (3): 821–38. http://dx.doi.org/10.1007/s12237-012-9483-1.

Håstein, T., and T. Lindstad. 1991. "Diseases in Wild and Cultured Salmon: Possible Interaction." *Aquaculture* (Amsterdam) 98 (1–3): 277–88. http://dx.doi.org/10.1016/0044-8486(91)90392-K.

Hatton-Ellis, T. 2008. "The Hitchhiker's Guide to the Water Framework Directive." *Aquatic Conservation: Marine and Freshwater Ecosystems* 18 (2): 111–16. http://dx.doi.org/10.1002/aqc.947.

Hawlena, D., and O.J. Schmitz. 2010. "Physiological Stress as a Fundamental Mechanism Linking Predation to Ecosystem Functioning." *American Naturalist* 176 (5): 537–56. http://dx.doi.org/10.1086/656495.

Hayes, S.A., M.H. Bond, C.V. Hanson, E.V. Freund, J.J. Smith, E.C. Anderson, Arnold J. Ammann, and R. Bruce MacFarlane. 2008. "Steelhead Growth in a Small Central California Watershed: Upstream and Estuarine Rearing Patterns." *Transactions of the American Fisheries Society* 137 (1): 114–28. http://dx.doi.org/10.1577/T07-043.1.

Hayes, S.A., M.H. Bond, C.V. Hanson, A.W. Jones, A.J. Ammann, J.A. Harding, Alison L. Collins, Jeffrey Perez, and R. Bruce MacFarlane. 2011. "Down, Up, Down and 'Smolting' Twice? Seasonal Movement Patterns by Juvenile Steelhead (*Oncorhynchus mykiss*) in a Coastal Watershed with a Bar Closing Estuary." *Canadian Journal of Fisheries and Aquatic Sciences* 68 (8): 1341–50. http://dx.doi.org/10.1139/f2011-062.

Hayes, M.C., S.P. Rubin, R.R. Reisenbichler, F.A. Goetz, E. Jeanes, and A. McBride. 2011. "Marine Habitat Use by Anadromous Bull Trout from the Skagit River, Washington." *Marine and Coastal Fisheries* 3 (1): 394–410. http://dx.doi.org/10.1080/19425120.2011.640893.

Healey, M.C. 1979. "Detritus and Juvenile Salmon Production in the Nanaimo Estuary: I. Production and Feeding Rates of Juvenile Chum Salmon (*Oncorhynchus keta*)." *Journal of the Fisheries Research Board of Canada* 36 (5): 488–96. http://dx.doi.org/10.1139/f79-072.

–. 1980a. "Utilization of the Nanaimo River Estuary by Juvenile Chinook Salmon, *Oncorhynchus tshawytscha*." *Fish Bulletin* 77 (3): 653–68.

–. 1980b. "The Ecology of Juvenile Salmon in Georgia Strait, British Columbia." In *Salmonid Ecosystems of the North Pacific*, ed. W.J. McNeil and D.C. Himsworth, 203–9. Corvallis: Oregon State University Press.

–. 1982. "Juvenile Pacific Salmon in Estuaries: The Life Support System." In *Estuarine Comparisons*, ed. V.S. Kennedy, 315–41. New York: Academic Press. http://dx.doi.org/10.1016/B978-0-12-404070-0.50025-9.

–. 1991. "Life History of Chinook Salmon (*Oncorhynchus tshawytscha*)." In *Pacific Salmon Life Histories*, ed. C. Groot and L. Margolis, 311–93. Vancouver: UBC Press.

–. 2009. "Resilient Salmon, Resilient Fisheries for British Columbia, Canada." *Ecology and Society* 14 (1): 2.

Healey, M.C., and A. Prince. 1995. "Scales of Variation in Life History Tactics of Pacific Salmon and the Conservation of Phenotype and Genotype." In *Evolution and the Aquatic Ecosystem: Defining Unique Units in Population Conservation*, ed. J.L. Nielsen and D.A. Powers, 176–84. Bethesda, MD: American Fisheries Society.

Healy, M.G., and K.R. Hickey. 2002. "Historic Land Reclamation in the Intertidal Wetlands of the Shannon Estuary, Western Ireland." *Journal of Coastal Research* SI36: 365–73.

Heard, W.R. 1991. "Life History of Pink Salmon (*Oncorhynchus gorbuscha*)." In *Pacific Salmon Life Histories*, ed. C. Groot and L. Margolis, 121–230. Vancouver: UBC Press.

Heck, K.L. Jr., and J.F. Valentine. 2006. "Plant-Herbivore Interactions in Seagrass Meadows." *Journal of Experimental Marine Biology and Ecology* 330 (1): 420–36. http://dx.doi.org/10.1016/j.jembe.2005.12.044.

Hedger, R.D., F. Martin, D. Hatin, F. Caron, F.G. Whoriskey, and J.J. Dodson. 2008. "Active Migration of Wild Atlantic Salmon *Salmo salar* Smolt through a

Coastal Embayment." *Marine Ecology Progress Series* 355: 235–46. http://dx.doi.org/10.3354/meps07239.

Hedger, R.D., I. Uglem, E.B. Thorstad, B. Finstad, C.M. Chittenden, P. Arechavala-Lopez, A.J. Jensen, R. Nilsen, and F. Okland. 2011. "Behaviour of Atlantic Cod, a Marine Fish Predator, during Atlantic Salmon Post-Smolt Migration." *ICES Journal of Marine Science* 68 (10): 2152–62. http://dx.doi.org/10.1093/icesjms/fsr143.

Heggberget, T.G., N.A. Hvidsten, T.B. Gunnerød, and P.I. Møkkelgjerd. 1991. "Distribution of Adult Recaptures from Hatchery-Reared Atlantic Salmon (*Salmo salar*) Smolts Released In and Off-Shore of the River Surna, Western Norway." *Aquaculture* (Amsterdam) 98 (1–3): 89–96. http://dx.doi.org/10.1016/0044-8486(91)90374-G.

Heggenes, J., J.L. Bagliniere, and R.A. Cunjak. 1999. "Spatial Niche Variability for Young Atlantic Salmon (*Salmo salar*) and Brown Trout (*S. trutta*) in Heterogeneous Streams." *Ecology of Freshwater Fish* 8 (1): 1–21. http://dx.doi.org/10.1111/j.1600-0633.1999.tb00048.x.

Heggenes, J., T.G. Northcote, and A. Peter. 1991. "Seasonal Habitat Selection and Preferences by Cutthroat Trout (*Oncorhynchus clarkii*) in a Small, Coastal Stream." *Canadian Journal of Fisheries and Aquatic Sciences* 48 (8): 1364–70. http://dx.doi.org/10.1139/f91-163.

Helfman, G.S. 1986. "Fish Behaviour by Day, Night and Twilight." In *The Behaviour of Teleost Fishes*, ed. T.J. Pitcher, 366–87. London: Croom Helm. http://dx.doi.org/10.1007/978-1-4684-8261-4_14.

Helle, J.H., R.S. Williamson, and J.E. Bailey. 1964. *Intertidal Ecology and Life History of Pink Salmon at Olsen Creek, Prince William Sound, Alaska*. US Fish and Wildlife Service, Bureau of Commercial Fisheries, Special Scientific Report – Fisheries No. 483. Washington, DC: US Department of the Interior, Bureau of Commercial Fisheries.

Hem, J.D. 1986. *Study and Interpretation of the Chemical Characteristics of Natural Water*. 3rd ed. New Cumberland, PA: US Geological Survey Water-Supply.

Henderson, P.A., and D.J. Bird. 2010. "Fish and Macro-Crustacean Communities and Their Dynamics in the Severn Estuary." *Marine Pollution Bulletin* 61 (1–3): 100–14. http://dx.doi.org/10.1016/j.marpolbul.2009.12.017.

Hendry, A.P., B.H. Letcher, and G. Gries. 2003. "Estimating Natural Selection Acting on Stream Dwelling Atlantic Salmon: Implications for the Restoration of Extirpated Populations." *Conservation Biology* 17 (3): 795–805. http://dx.doi.org/10.1046/j.1523-1739.2003.02075.x.

Hendry, A.P., and S.C. Stearns, eds. 2004. *Evolution Illuminated: Salmon and Their Relatives*. Oxford: Oxford University Press.

Hering, D.K., D.L. Bottom, E.F. Prentice, K.K. Jones, and I.A. Fleming. 2010. "Tidal Movements and Residency of Subyearling Chinook Salmon (*Oncorhynchus tshawytscha*) in an Oregon Salt Marsh Channel." *Canadian Journal of Fisheries and Aquatic Sciences* 67 (3): 524–33. http://dx.doi.org/10.1139/F10-003.

Hess, J.E., and S.R. Narum. 2011. "Single-Nucleotide Polymorphism (snp) Loci Correlated with Run Timing in Adult Chinook Salmon from the Columbia River Basin." *Transactions of the American Fisheries Society* 140 (3): 855–64. http://dx.doi.org/10.1080/00028487.2011.588138.

Higgins, P.J., and C. Talbot. 1985. "Growth and Feeding in Juvenile Atlantic Salmon (*Salmo salar* L.)." In *Nutrition and Feeding in Fish*, ed. C.B. Cowey and A.M. Mackie, 243–63. London: Academic Press.

Higgs, D.A., J.S. Macdonald, C.D. Levings, and B.S. Dosanjh. 1995. "Nutrition and Feeding Habits in Relation to Life History Stage." In *Physiological Ecology of Pacific Salmon*, ed. C. Groot, L. Margolis, and W.C. Clarke, 161–315. Vancouver: UBC Press.

Hildebrand, R.H., A.C. Watts, and A.M. Randle. 2005. "The Myths of Restoration Ecology." *Ecology and Society* 10 (1): 19.

Hildén, M., and D. Rapport. 1993. "Four Centuries of Cumulative Impacts on a Finnish River and Its Estuary: An Ecosystem Health-Approach." *Journal of Aquatic Ecosystem Health* 2 (4): 261–75. http://dx.doi.org/10.1007/BF00044030.

Hinch, S.G., S.J. Cooke, M.C. Healey, and A.P. Farrell. 2006. "Behavioural Physiology of Fish Migrations: Salmon as a Model Approach." In *Fish Physiology, Vol. 24: Behaviour and Physiology of Fish*, ed. K.A. Sloman, R.W. Wilson, and S. Balshine, 239–95. Amsterdam: Elsevier Press.

Hindar, K. 2003. "Wild Atlantic Salmon in Europe: Status and Perspectives." In *Proceedings of Speaking for the Salmon: The World Summit on Salmon, June 10–13, 2003*, ed. P. Gallaugher and L. Wood, 47–52. Burnaby, BC: Simon Fraser University.

Hindar, K., C. García de Leániz, M. Koljonen, J. Tufto, and A.F. Youngson. 2007. "Fisheries Exploitation." In *The Atlantic Salmon: Genetics, Conservation and Management*, ed. E. Verspoor, L. Stradmeyer, and J.L. Nielsen, 299–324. Oxford: Blackwell Publishing. http://dx.doi.org/10.1002/9780470995846.ch10.

Hirakawa, K. 1990. "Feeding Habit of Juvenile Chum Salmon, *Oncorhynchus keta*, Released into a Chilean Fjord, South America." *Aquaculture Science* 38: 157–63 (in Japanese with English abstract).

Hiwatari, T., H. Koshikawa, R. Nagata, Y. Suda, S. Hamaoka, and K. Kohata. 2011. "Trophic Relationships in Early Spring along the Okhotsk Coast of Hokkaido, Japan, as Traced by Stable Carbon and Nitrogen Isotopes." *Plankton and Benthos Research* 6 (1): 56–67. http://dx.doi.org/10.3800/pbr.6.56.

Hoar, W.S. 1976. "Smolt Transformation: Evolution, Behavior, and Physiology." *Journal of the Fisheries Board of Canada* 33 (5): 1233–52. http://dx.doi.org/10.1139/f76-158.

Hodge, B.W., M.A. Wilzbach, and W.G. Duffy. 2014. "Potential Fitness Benefits of the Half-Pounder Life History in Klamath River Steelhead." *Transactions of the American Fisheries Society* 143 (4): 864–75.

Hoekstra, P.F., T.M. O'Hara, A.T. Fisk, K. Borgå, K.R. Solomon, and D.C. Muir. 2003. "Trophic Transfer of Persistent Organochlorine Contaminants (OCs) within an Arctic Marine Food Web from the Southern Beaufort-Chukchi Seas." *Environmental Pollution* 124 (3): 509–22. http://dx.doi.org/10.1016/S0269-7491(02)00482-7.

Hoffmann, R.C. 2005. "A Brief History of Aquatic Resource Use in Medieval Europe." *Helgoland Marine Research* 59 (1): 22–30. http://dx.doi.org/10.1007/s10152-004-0203-5.

Høgåsen, H.R., and E. Brun. 2003. "Risk of Inter-River Transmission of *Gyrodactylus salaris* by Migrating Atlantic Salmon Smolts, Estimated by Monte Carlo

Simulation." *Diseases of Aquatic Organisms* 57 (3): 247–54. http://dx.doi.org/10.3354/dao057247.

Hoggart, S.P., R.A. Francis, and M.A. Chadwick. 2012. "Macroinvertebrate Richness on Flood Defence Walls of the Tidal River Thames." *Urban Ecosystems* 15 (2): 327–46. http://dx.doi.org/10.1007/s11252-011-0221-4.

Holling, C.S. 1973. "Resilience and Stability of Ecological Systems." *Annual Review of Ecology and Systematics* 4 (1): 1–23. http://dx.doi.org/10.1146/annurev.es.04.110173.000245.

Holmlund, C.M., and M. Hammer. 2004. "Effects of Fish Stocking on Ecosystem Services: An Overview and Case Study Using the Stockholm Archipelago." *Environmental Management* 33 (6): 799–820. http://dx.doi.org/10.1007/s00267-004-0051-8.

Honda, K., T. Arai, S. Kobayashi, Y. Tsuda, and K. Miyashita. 2012. "Migratory Patterns of Exotic Brown Trout *Salmo trutta* in South Western Hokkaido, Japan, on the Basis of Otolith Sr:Ca Ratios and Acoustic Telemetry." *Journal of Fish Biology* 80 (2): 408–26. http://dx.doi.org/10.1111/j.1095-8649.2011.03183.x.

Hood, W.G. 2002. "Application of Landscape Allometry to Restoration of Tidal Channels." *Restoration Ecology* 10 (2): 213–22. http://dx.doi.org/10.1046/j.1526-100X.2002.02034.x.

–. 2004a. "Likely Scaling of Basin Area with Some Marine Riparian Functions." In *Proceedings of the DFO/PSAT Sponsored Marine Riparian Experts Workshop, Tsawwassen, BC, February 17–18, 2004* (Canadian Manuscript Report of Fisheries and Aquatic Sciences, No. 2680), ed. J.P. Lemieux, J. Brennan, M. Farrell, C.D. Levings, and D. Myers, 55–56. Ottawa: Fisheries and Oceans Canada.

–. 2004b. "Indirect Environmental Effects of Dikes on Estuarine Tidal Channels: Thinking Outside of the Dike for Habitat Restoration and Monitoring." *Estuaries* 27 (2): 273–82. http://dx.doi.org/10.1007/BF02803384.

–. 2012. "Beaver in Tidal Marshes: Dam Effects on Low-Tide Channel Pools and Fish Use of Estuarine Habitat." *Wetlands* 32 (3): 401–10. http://dx.doi.org/10.1007/s13157-012-0294-8.

Hosack, G.R., B.R. Dumbauld, J.L. Ruesink, and D.A. Armstrong. 2006. "Habitat Associations of Estuarine Species: Comparisons of Intertidal Mudflat, Seagrass (*Zostera marina*), and Oyster (*Crassostrea gigas*) Habitats." *Estuaries and Coasts* 29 (6): 1150–60. http://dx.doi.org/10.1007/BF02781816.

Howes, B.L., J.W.H. Dacey, and D.D. Goehringer. 1986. "Factors Controlling the Growth Form of *Spartina alterniflora*: Feedbacks between Above-Ground Production, Sediment Oxidation, Nitrogen and Salinity." *Journal of Ecology* 74 (3): 881–98. http://dx.doi.org/10.2307/2260404.

Hubbell, S.P. 2001. *The Unified Neutral Theory of Biodiversity and Biogeography*. Princeton, NJ: Princeton University Press.

Hubley, P.B., P.G. Amiro, A.J.F. Gibson, G.L. Lacroix, and A.M. Redden. 2008. "Survival and Behaviour of Migrating Atlantic Salmon (*Salmo salar* L.) Kelts in River, Estuarine, and Coastal Habitat." *ICES Journal of Marine Science* 65 (9): 1626–34.

Hughes, J.E., L.A. Deegan, M.J. Weaver, and J.E. Costa. 2002. "Regional Application of an Index of Estuarine Biotic Integrity Based on Fish Communities." *Estuaries and Coasts* 25 (2): 250–63. http://dx.doi.org/10.1007/BF02691312.

Hugman, S.J., A.R. O'Donnell, and G. Mance. 1984. "A Survey of Estuarine Oxygen Concentrations in Relation to the Passage of Migratory Salmonids." WRC Environment, http://aquaticcommons.org/8068/.

Hume, T.M., T. Snelder, M. Weatherhead, and R. Liefting. 2007. "A Controlling Factor Approach to Estuary Classification." *Ocean and Coastal Management* 50 (11–12): 905–29. http://dx.doi.org/10.1016/j.ocecoaman.2007.05.009.

Hutching, J.A. 2011. "Old Wine in New Bottles: Reaction Norms in Salmonid Fishes." *Heredity* 106 (3): 421–37. http://dx.doi.org/10.1038/hdy.2010.166.

Hutchinson, I. 1982. "Vegetation-Environment Relations in a Brackish Marsh, Lulu Island, Richmond, B.C." *Canadian Journal of Botany* 60 (4): 452–62. http://dx.doi.org/10.1139/b82-061.

Hutchison, M.J., and M. Iwata. 1997. "A Comparative Analysis of Aggression in Migratory and Non-Migratory Salmonids." *Environmental Biology of Fishes* 50 (2): 209–15. http://dx.doi.org/10.1023/A:1007327400284.

Hvidsten, N.A., and R.A. Lund. 1988. "Predation on Hatchery-Reared and Wild Smolts of Atlantic Salmon, *Salmo salar* L., in the Estuary of the River Orkla, Norway." *Journal of Fish Biology* 33 (1): 121–26. http://dx.doi.org/10.1111/j.1095-8649.1988.tb05453.x.

Hvidsten, N.A., J.A. Knutsen, E. Torstensen, D. Danielson, and J. Gjoseter. 2000. *Konsekvenser av havneutbygging for laksesmolt fra Numedalslågen, Rapport no. 66* [Consequences of port development for salmon smolts from Numedalslågen, Report No. 66, in Norwegian with English summary]. Trondheim: Norwegian Institute for Nature Research.

Hyslop, E.J. 1980. "Stomach Contents Analysis – A Review of Methods and Their Application." *Journal of Fish Biology* 17 (4): 411–29. http://dx.doi.org/10.1111/j.1095-8649.1980.tb02775.x.

Ims, R.A. 1990. "On the Adaptive Value of Reproductive Synchrony as a Predator-Swamping Strategy." *American Naturalist* 136 (4): 485–98. http://dx.doi.org/10.1086/285109.

International Council for the Exploration of the Sea. 2015. *Report of the Working Group on North Atlantic Salmon (WGNAS), 17–26 March, Moncton, Canada*. ICES CM 2015/ACOM:09. Copenhagen: ICES.

Iriarte, J.A., G.A. Lobos, and F.M. Jaksic. 2005. "Invasive Vertebrate Species in Chile and Their Monitoring by Governmental Agencies." *Revista Chilena de Historia Natural* (Valparaiso, Chile) 78 (1): 143–52.

Irvine, J.R., M. O'Neill, L. Godbout, and J. Schnute. 2013. "Effects of Smolt Release Timing and Size on the Survival of Hatchery-Origin Coho Salmon in the Strait of Georgia." *Progress in Oceanography* 115: 111–18. http://dx.doi.org/10.1016/j.pocean.2013.05.014.

Ishikawa, S. 2004. "Floodplain Vegetation in the Lower Course of the Shimanto River." *Aquabiology* 26 (6): 516–21.

Ivankov, V.N., E.E. Borisovets, and O.A. Rutenko. 2003. "Ecological and Geographical Divergence and Interpopulation Variability of the Masu Salmon *Oncorhynchus masou* as Illustrated by Populations from Primorye and Sakhalin." *Russian Journal of Marine Biology* 29 (1): 12–17. http://dx.doi.org/10.1023/A:1022867502125.

Ivanova, A.N., N.N. Tarasov, K.L. Pusankov, L.V. Ivanova, and E.N. Pusankova. 2001. "Oil Spill Impact on Pacific Salmon (G. *Oncorhynchus*) of Northwestern

Sakhalin (Tengi River Basin as a Pattern)." In *Proceedings of the Hyperspectral Remote Sensing of the Ocean: Conference Volume 4154*, ed. R.J. Fouin, H. Kawamura, and M. Kishino, 124–35. Bellingham, WA: Society of Photo-Optical Instrumentation Engineers.

Iwata, M., and S. Komatsu. 1984. "Importance of Estuarine Residence for Adaptation of Chum Salmon (*Oncorhynchus keta*) Fry to Seawater." *Canadian Journal of Fisheries and Aquatic Sciences* 41 (5): 744–49. http://dx.doi.org/10.1139/f84-086.

Jackson, D.A., P.R. Peres-Neto, and J.D. Olden. 2001. "What Controls Who Is Where in Freshwater Fish Communities: The Roles of Biotic, Abiotic, and Spatial Factors." *Canadian Journal of Fisheries and Aquatic Sciences* 58 (1): 157–70.

Jacobs, S., O. Beauchard, E. Struyf, T. Cox, T. Maris, and P. Meire. 2009. "Restoration of Tidal Freshwater Vegetation Using Controlled Reduced Tide (CRT) along the Schelde Estuary (Belgium)." *Estuarine, Coastal and Shelf Science* 85 (3): 368–76. http://dx.doi.org/10.1016/j.ecss.2009.09.004.

James, G.D., and M.J. Unwin. 1996. "Diet of Chinook Salmon (*Oncorhynchus tshawytscha*) in Canterbury Coastal Waters, New Zealand." *New Zealand Journal of Marine and Freshwater Research* 30 (1): 69–78. http://dx.doi.org/10.1080/00288330.1996.9516697.

Jamieson, G.S., and C.O. Levings. 2001. "Marine Protected Areas in Canada – Implications for Both Conservation and Fisheries Management." *Canadian Journal of Fisheries and Aquatic Sciences* 58 (1): 1–19.

Jarry, M., P. Davaine, and E. Beall. 1998. "A Matrix Model to Study the Colonization by Brown Trout of a Virgin Ecosystem in the Kerguelen Islands." *Acta Biotheoretica* 46 (3): 253–72. http://dx.doi.org/10.1023/A:1001789211146.

Jassby, A.D., W.J. Kimmerer, S.G. Monismith, C.A. Armor, J.E. Cloern, T.M. Powell, Jerry R. Schubel, and Timothy J. Vendlinski. 1995. "Isohaline Position as a Habitat Indicator for Estuarine Populations." *Ecological Applications* 5 (1): 272–89. http://dx.doi.org/10.2307/1942069.

Jelks, H.L., S.J. Walsh, N.M. Burkhead, S. Contreras-Balderas, E. Diaz-Pardo, D.A. Hendrickson, John Lyons, Nicholas E. Mandrak, Frank McCormick, Joseph S. Nelson, et al. 2008. "Conservation Status of Imperiled North American Freshwater and Diadromous Fishes." *Fisheries* (Bethesda, MD) 33 (8): 372–407. http://dx.doi.org/10.1577/1548-8446-33.8.372.

Jellyman, D.J., G.J. Glova, P.M. Sagar, and J.R.E. Sykes. 1997. "Spatio-Temporal Distribution of Fish in the Kakanui River Estuary, South Island, New Zealand." *New Zealand Journal of Marine and Freshwater Research* 31 (1): 103–18. http://dx.doi.org/10.1080/00288330.1997.9516748.

Jenkins, T.M. Jr., S. Diehl, K.W. Kratz, and S.D. Cooper. 1999. "Effects of Population Density on Individual Growth of Brown Trout in Streams." *Ecology* 80 (3): 941–56. http://dx.doi.org/10.1890/0012-9658(1999)080[0941:EOPDOI]2.0.CO;2.

Jensen, A.J., and B.O. Johnsen. 1986. "Different Adaption Strategies of Atlantic Salmon (*Salmo salar*) Populations to Extreme Climates with Special Reference to Some Cold Norwegian Rivers." *Canadian Journal of Fisheries and Aquatic Sciences* 43 (5): 980–84. http://dx.doi.org/10.1139/f86-120.

Jensen, J.L.A., and A.H. Rikardsen. 2008. "Do Northern Riverine Anadromous Arctic Char *Salvelinus alpinus* and Sea Trout *Salmo trutta* Overwinter in Estuarine and Marine Waters?" *Journal of Fish Biology* 73 (7): 1810–18. http://dx.doi.org/10.1111/j.1095-8649.2008.02042.x.

Jensen, Ø., T. Dempster, E.B. Thorstad, I. Uglem, and A. Fredheim. 2010. "Escapes of Fishes from Norwegian Sea-Cage Aquaculture: Causes, Consequences and Prevention." *Aquaculture Environment Interactions* 1 (1): 71–83. http://dx.doi.org/10.3354/aei00008.

Jepsen, N., E. Holthe, and F. Økland. 2006. "Observations of Predation on Salmon and Trout Smolts in a River Mouth." *Fisheries Management and Ecology* 13 (5): 341–41. http://dx.doi.org/10.1111/j.1365-2400.2006.00509.x.

Jobling, M. 1989. "Et godt oppdrettsfor – sett fra fiskens side" [A good rearing environment – seen from the fish's view]. In *Proceedings of AquaNor '89, 5th International Conference and Exhibition of Fishfarming Techniques and Equipment*, 27–34. Trondheim: Norwegian Fish Farmers.

Jobling, M., A.M. Arnesen, B.M. Baardvik, J.S. Christiansen, and E.H. Jørgensen. 1995. "Monitoring Feeding Behaviour and Food Intake: Methods and Applications." *Aquaculture Nutrition* 1 (3): 131–43. http://dx.doi.org/10.1111/j.1365-2095.1995.tb00037.x.

Johnsen, B.O., and A.J. Jensen. 1991. "The *Gyrodactylus* Story in Norway." *Aquaculture* (Amsterdam) 98 (1): 289–302. http://dx.doi.org/10.1016/0044-8486(91)90393-L.

Johnsen, B.O., J.V. Arnekleiv, L. Asplin, B.T. Barlaup, T.F. Næsje, B.O. Rosseland, et al. 2011. "Hydropower Development – Ecological Effects." In *Atlantic Salmon Ecology*, ed. Ø. Aas, S. Einum, A. Klemetsen, and J. Skurdal, 351–376. Oxford: Wiley-Blackwell Publishing.

Johnson, G.E., N.K. Sather, J.R. Skalski, and D.J. Teel. 2014. "Application of Diversity Indices to Quantify Early Life-History Diversity for Chinook Salmon." *Ecological Indicators* 38: 170–80. http://dx.doi.org/10.1016/j.ecolind.2013.11.005.

Johnson, J., T. Johnson, and T. Copeland. 2012. "Defining Life Histories of Precocious Male Parr, Minijack, and Jack Chinook Salmon Using Scale Patterns." *Transactions of the American Fisheries Society* 141 (6): 1545–56. http://dx.doi.org/10.1080/00028487.2012.705256.

Johnson, L.L., G.M. Ylitalo, M.R. Arkoosh, A.N. Kagley, C. Stafford, J.L. Bolton, Jon Buzitis, Bernadita F. Anulacion, and Tracy K. Collier. 2007a. "Contaminant Exposure in Outmigrant Juvenile Salmon from Pacific Northwest Estuaries of the United States." *Environmental Monitoring and Assessment* 124 (1–3): 167–94. http://dx.doi.org/10.1007/s10661-006-9216-7.

Johnson, L.L., G.M. Ylitalo, C.A. Sloan, B.F. Anulacion, A.N. Kagley, M.R. Arkoosh, Tricia A. Lundrigan, Kim Larson, Mark Siipola, and Tracy K. Collier. 2007b. "Persistent Organic Pollutants in Outmigrant Juvenile Chinook Salmon from the Lower Columbia Estuary, USA." *Science of the Total Environment* 374 (2–3): 342–66. http://dx.doi.org/10.1016/j.scitotenv.2006.11.051.

Johnson, S.C., R.B. Blaylock, J. Elphick, and K.D. Hyatt. 1996. "Disease Induced by the Sea Louse (*Lepeophtheirus salmonis*) (Copepoda: Caligidae) in Wild Sockeye Salmon (*Oncorhynchus nerka*) Stocks of Alberni Inlet, British Columbia." *Canadian Journal of Fisheries and Aquatic Sciences* 53 (12): 2888–97. http://dx.doi.org/10.1139/f96-226.

Johnsson, J.I., J. Höjesjö, and I.A. Fleming. 2001. "Behavioural and Heart Rate Responses to Predation Risk in Wild and Domesticated Atlantic Salmon." *Canadian Journal of Fisheries and Aquatic Sciences* 58 (4): 788–94. http://dx.doi.org/10.1139/f01-025.

Jones, P.D. 2006. "Water Quality and Fisheries in the Mersey Estuary, England: A Historical Perspective." *Marine Pollution Bulletin* 53 (1–4): 144–54. http://dx.doi.org/10.1016/j.marpolbul.2005.11.025.

Jonsson, B., and N. Jonsson. 1993. "Partial Migration: Niche Shift versus Sexual Maturation in Fishes." *Reviews in Fish Biology and Fisheries* 3 (4): 348–65. http://dx.doi.org/10.1007/BF00043384.

–. 2006. "Cultured Atlantic Salmon in Nature: A Review of Their Ecology and Interaction with Wild Fish." *ICES Journal of Marine Science* 63 (7): 1162–81. http://dx.doi.org/10.1016/j.icesjms.2006.03.004.

–. 2011. *Ecology of Atlantic Salmon and Brown Trout: Habitat as a Template for Life Histories*. Fish and Fisheries Series, vol. 33. New York: Springer. http://dx.doi.org/10.1007/978-94-007-1189-1.

Jonsson, B., and J. Ruud-Hansen. 1985. "Water Temperature as the Primary Influence on Timing of Seaward Migrations of Atlantic Salmon (*Salmo salar*) Smolts." *Canadian Journal of Fisheries and Aquatic Sciences* 42 (3): 593–95. http://dx.doi.org/10.1139/f85-076.

Jonsson, B., N. Jonsson, E. Brodtkorb, and P.J. Ingebrigtsen. 2001. "Life History Traits of Brown Trout Vary with the Size of Small Streams." *Functional Ecology* 15 (3): 310–17. http://dx.doi.org/10.1046/j.1365-2435.2001.00528.x.

Jutila, E., E. Jokikokko, and E. Ikonen. 2009. "Post Smolt Migration of Atlantic Salmon, *Salmo salar* L., from the Simojoki River to the Baltic Sea." *Journal of Applied Ichthyology* 25 (2): 190–94. http://dx.doi.org/10.1111/j.1439-0426.2009.01212.x.

Jutila, E., E. Jokikokko, and M. Julkunen. 2005. "The Smolt Run and Postsmolt Survival of Atlantic Salmon, *Salmo salar* L., in Relation to Early Summer Water Temperatures in the Northern Baltic Sea." *Ecology of Freshwater Fish* 14 (1): 69–78. http://dx.doi.org/10.1111/j.1600-0633.2005.00079.x.

Kaeriyama, M., and R.R. Edpalina. 2008. "Evaluation of the Biological Interaction between Wild and Hatchery Population for Sustainable Fisheries Management of Pacific Salmon." In *Stock Enhancement and Sea Ranching: Developments, Pitfalls and Opportunities*, 2nd ed., ed. K.M. Leber, S. Kitada, H.L. Blankenship, and T. Svåsand, 245–59. Oxford: Blackwell Publishing.

Kakuda, S. 1973. "Marine Fishes from the Estuary of the Ashida River." *Journal of the Faculty of Fisheries and Animal Husbandry* 12 (1): 83–87.

Kålås, J.A., T.G. Heggberget, P.A. Bjørn, and O. Reitan. 1993. "Feeding Behaviour and Diet of Goosanders (*Mergus merganser*) in Relation to Salmonid Seaward Migration." *Aquatic Living Resources* 6 (1): 31–38. http://dx.doi.org/10.1051/alr:1993003.

Kalish, J.M. 1990. "Use of Otolith Microchemistry to Distinguish the Progeny of Sympatric Anadromous and Non-Anadromous Salmonids." *Fish Bulletin* 88 (4): 657–66.

Karas, N. 2002. *Brook Trout*. New York: Lyons Press.

Kareiva, P., and M. Marvier. 2012. "What Is Conservation Science?" *BioScience* 62 (11): 962–69. http://dx.doi.org/10.1525/bio.2012.62.11.5.

Kareiva, P., S. Watts, R. McDonald, and T. Boucher. 2007. "Domesticated Nature: Shaping Landscapes and Ecosystems for Human Welfare." *Science* 316 (5833): 1866–69. http://dx.doi.org/10.1126/science.1140170.

Karpenko, V.I. 1990. *Main Achievements and Perspectives of Research into the Early Sea Life Period of Kamchatkan Salmon Species*. Canadian Translation of Fisheries and Aquatic Sciences No. 5494. Ottawa: Fisheries and Oceans Canada. [Original paper presented at the Meeting of Soviet and Canadian Scientists on Joint Research in the North Pacific, Khabarovsk, USSR, 20–26 August 1988]

–. 1998. *The Early Sea of Pacific Salmons* [in Russian]. Moscow: VNIRO Publishing.

–. 2003. *Review of Russian Marine Investigations of Juvenile Pacific Salmon*. North Pacific Anadromous Fish Commission Bulletin No. 3, 69–88. Vancouver: North Pacific Anadromous Fish Commission.

Karpenko, V.I., and E.T. Nikolaeva. 1989. "Daily Feeding Rhythm and Rations of Juvenile Chum Salmon, *Oncorhynchus keta*, during River and Early Marine Part of Their Life Cycle" [in Russian]. *Voprosy Ikhtiologii* 29 (4): 75–82.

Karppinen, P., T.S. Mäkinen, J. Erkinaro, V.V. Kostin, R.V. Sadkovskij, A.I. Lupandin, and M. Kaukoranta. 2002. "Migratory and Route-Seeking Behaviour of Ascending Atlantic Salmon in the Regulated River Tuloma." *Hydrobiologia* 483 (1): 23–30. http://dx.doi.org/10.1023/A:1021386319633.

Karr, J.R. 1981. "Assessment of Biotic Integrity Using Fish Communities." *Fisheries* (Bethesda, MD) 6 (6): 21–27. http://dx.doi.org/10.1577/1548-8446(1981)006 <0021:AOBIUF>2.0.CO;2.

Kashiwai, M. 1995. *History of Carrying Capacity Concept as an Index of Ecosystem Productivity* (Review). Bulletin of the Hokkaido National Fisheries Research Institute No. 59, 81–100. Hokkaido: Japan Fisheries Research Agency.

Kask, B.A., T.J. Brown, and C.D. McAllister. 1988. "Nearshore Epibenthos of the Campbell River Estuary and Discovery Passage, 1983, in Relation to Juvenile Chinook Diets." Canadian Technical Report of Fisheries and Aquatic Sciences, No. 1616. Nanaimo, BC: Department of Fisheries and Oceans.

Kawamura, H., M. Miyamoto, M. Nagata, and K. Hirano. 1998. *Interaction between Chum Salmon and Fat Greenling Juveniles in the Coastal Sea of Japan Off Northern Hokkaido*. North Pacific Anadromous Fish Commission Bulletin No. 1, 412–418. Vancouver: North Pacific Anadromous Fish Commission.

Kawamura, K., M. Furukawa, M. Kubota, and Y. Harada. 2012. "Effects of Stocking Hatchery Fish on the Phenotype of Indigenous Populations in the Amago Salmon *Oncorhynchus masou ishikawae* in Japan." *Journal of Fish Biology* 81 (1): 94–109. http://dx.doi.org/10.1111/j.1095-8649.2012.03315.x.

Kazakov, R.V. 1994. "Juvenile Atlantic Salmon in the Varzuga River (White Sea Basin)." *Journal of Fish Biology* 49 (3): 467–77. http://dx.doi.org/10.1111/j.1095-8649.1994.tb01329.x.

Keefer, M.L., C.A. Peery, and C.C. Caudill. 2008. "Migration Timing of Columbia River Spring Chinook Salmon: Effects of Temperature, River Discharge, and Ocean Environment." *Transactions of the American Fisheries Society* 137 (4): 1120–33. http://dx.doi.org/10.1577/T07-008.1.

Keeley, E.R., and J.W.A. Grant. 2001. "Prey Size of Salmonid Fishes in Streams, Lakes, and Oceans." *Canadian Journal of Fisheries and Aquatic Sciences* 58 (6): 1122–32. http://dx.doi.org/10.1139/f01-060.

Kelly, D.W., and J.T.A. Dick. 2005. "Introduction of the Nonindigenous Amphipod *Gammarus pulex* (L.) Alters Population Dynamics and Diet of Juvenile Trout *Salmo trutta* L." *Freshwater Biology* 50 (1): 127–40. http://dx.doi.org/10.1111/j.1365-2427.2004.01315.x.

Kelly, D.W., R.A. Paterson, C.R. Townsend, R. Poulin, and D.M. Tompkins. 2009. "Parasite Spillback: A Neglected Concept in Invasion Ecology?" *Ecology* 90 (8): 2047–56. http://dx.doi.org/10.1890/08-1085.1.

Kennedy, D., and R. Bouchard. 1983. *Sliammon Life, Sliammon Lands*. Vancouver: Talonbooks.

Kennedy, G.J.A., and W.W. Crozier. 1997. "What Is the Value of a Wild Salmon Smolt?" *Fisheries Management and Ecology* 4 (2): 103–10. http://dx.doi.org/10.1046/j.1365-2400.1997.d01-169.x.

Kennish, M., R.J. Livingston, D.G. Raffaelli, and K. Reise. 2008. "Environmental Futures of Estuaries." In *Aquatic Ecosystems: Trends and Global Prospects*, ed. N. Polunin, 188–206. Cambridge: Cambridge University Press. http://dx.doi.org/10.1017/CBO9780511751790.018.

Kent, M. 2011. *Infectious Diseases and Potential Impacts on Survival of Fraser River Sockeye Salmon*. Cohen Commission Technical Report, 1: 1–58. http://epe.lac-bac.gc.ca/100/206/301/pco-bcp/commissions/cohen/cohen_commission/LOCALHOS/EN/INDEX.HTM.

Kimmerer, W.J. 2002. "Effects of Freshwater Flow on Abundance of Estuarine Organisms: Physical Effects or Trophic Linkages?" *Marine Ecology Progress Series* 243: 39–55. http://dx.doi.org/10.3354/meps243039.

–. 2004. "Open Water Processes of the San Francisco Estuary: From Physical Forcing to Biological Responses." *San Francisco Estuary and Watershed Science* 2 (1): 1–142. http://dx.doi.org/10.15447/sfews.2004v2iss1art1.

Kimmerer, W.J., D.D. Murphy, and P.L. Angermeier. 2005. "A Landscape-Level Model for Ecosystem Restoration in the San Francisco Estuary and Watershed." *San Francisco Estuary and Watershed Science* 3 (1), Article 2. http://escholarship.org/uc/item/5846s8qg .

Kimmerer, W.J., L. Brown, S. Culberson, P. Moyle, M. Nobriga, and J. Tompson. 2008. "Aquatic Ecosystems." In *The State of Bay-Delta Science: 2008*, ed. M. Healey, 73–101. Sacramento, CA: CALFED Science Program.

Kinne, O., ed. 1984. *Diseases of Marine Animals. Vol. 4, part 1: Introduction, Pisces*. Hamburg: Biologische Anstalt Helgoland.

Kinnison, M.T., M.J. Unwin, and T.P. Quinn. 2008. "Eco evolutionary vs. Habitat Contributions to Invasion in Salmon: Experimental Evaluation in the Wild." *Molecular Ecology* 17 (1): 405–14. http://dx.doi.org/10.1111/j.1365-294X.2007.03495.x.

Kirwan, M.L., and A.B. Murray. 2008. "Ecological and Morphological Response of Brackish Tidal Marshland to the Next Century of Sea Level Rise: Westham Island, British Columbia." *Global and Planetary Change* 60 (3–4): 471–86. http://dx.doi.org/10.1016/j.gloplacha.2007.05.005.

Kistritz, R.U., K.J. Hall, and I. Yesaki. 1983. "Productivity, Detritus Flux, and Nutrient Cycling in a *Carex lyngbyei* Tidal Marsh." *Estuaries* 6 (3): 227–36. http://dx.doi.org/10.2307/1351514.

Kitano, S. 2004. "Ecological Impacts of Rainbow, Brown and Brook Trout in Japanese Inland Waters." *Global Environmental Research* 8 (1): 41–50.

Kjelson, M.A., P.F. Raquel, and F.W. Fisher. 1982. "Life History of Fall-Run Juvenile Chinook Salmon, *Oncorhynchus tshawytscha*, in the Sacramento–San Joaquin Estuary, California." In *Estuarine Comparisons: Proceedings of the Sixth Biennial International Estuarine Research Conference, Gleneden Beach, Oregon, November 1–6, 1981*, ed. V.S. Kennedy, 393–411. New York: Academic Press. http://dx.doi.org/10.1016/B978-0-12-404070-0.50029-6.

Klein, C.J., A. Chan, L. Kircher, A.J. Cundiff, N. Gardner, Y. Hrovat, A. Scholz, B.E. Kendall, and S. Airamé. 2008. "Striking a Balance between Biodiversity Conservation and Socioeconomic Viability in the Design of Marine Protected Areas." *Conservation Biology* 22 (3): 691–700. http://dx.doi.org/10.1111/j.1523-1739.2008.00896.x.

Klemetsen, A., P.A. Amundsen, J.B. Dempson, B. Jonsson, N. Jonsson, M.F. O'Connell, and E. Mortensen. 2003. "Atlantic Salmon *Salmo salar* L., Brown trout *Salmo trutta* L. and Arctic charr *Salvelinus alpinus* (L.), a Review of Aspects of Their Life Histories." *Ecology of Freshwater Fish* 12 (1): 1–59. http://dx.doi.org/10.1034/j.1600-0633.2003.00010.x.

Kneib, R.T. 1997. "The Role of Tidal Marshes in the Ecology of Estuarine Nekton." In *Oceanography and Marine Biology: An Annual Review, Volume 1*, ed. R.N. Gibson and M. Barnes, 163–220. London: Informa.

Knudsen, E.E., and J.H. Michael, eds. 2009. *Pacific Salmon Environmental and Life History Models: Advancing Science for Sustainable Salmon in the Future*. American Fisheries Society Symposium 71. Bethesda, MD: American Fisheries Society.

Knudsen, F.R., C.B. Schreck, S.M. Knapp, P.S. Enger, and O. Sand. 1997. "Infrasound Produces Flight and Avoidance Responses in Pacific Juvenile Salmonids." *Journal of Fish Biology* 51 (4): 824–29. http://dx.doi.org/10.1111/j.1095-8649.1997.tb02002.x.

Kocik, J.F., J.P. Hawkes, T.F. Sheehan, P.A. Music, and K.F. Beland. 2009. "Assessing Estuarine and Coastal Migration and Survival of Wild Atlantic Salmon Smolts from the Narraguagus River, Maine Using Ultrasonic Telemetry." In *Challenges for Diadromous Fishes in a Dynamic Global Environment* (American Fisheries Society Symposium 69), ed. A.J. Haro, K.L. Smith, R.A. Rulifson, C.M. Moffitt, R.J. Klauda, M.J. Dadswell, et al., 293–310. Bethesda, MD: American Fisheries Society.

Koed, A., H. Baktoft, and B.D. Bak. 2006. "Causes of Mortality of Atlantic Salmon (*Salmo salar*) and Brown Trout (*Salmo trutta*) Smolts in a Restored River and Its Estuary." *River Research and Applications* 22 (1): 69–78. http://dx.doi.org/10.1002/rra.894.

Koljonen, M.L., H. Jansson, T. Paaver, O. Vasin, and J. Koskiniemi. 1999. "Phylogeographic Lineages and Differentiation Pattern of Atlantic Salmon (*Salmo salar*) in the Baltic Sea with Management Implications." *Canadian Journal of Fisheries and Aquatic Sciences* 56 (10): 1766–80. http://dx.doi.org/10.1139/f99-104.

Kolpakov, N.V., E.I. Barabanshchikov, and A.Y. Chepurnoi. 2010. "Species Composition, Distribution, and Biological Conditions of Nonindigenous Fishes in the Estuary of the Razdol'naya River (Peter the Great Bay, Sea of Japan)." *Russian Journal of Biological Invasions* 1 (2): 87–94. Translated from Russian; originally published in *Rossiiskii Zhurnal Biologicheskikh Invasii* 2: 47–57 (2008). http://dx.doi.org/10.1134/S2075111710020062.

Kolpakov, N.V., and P.G. Milovankin. 2010. "Distribution and Seasonal Changes in Fish Abundance in the Estuary of the Razdol'naya River (Peter the Great Bay), Sea of Japan." *Journal of Ichthyology* 50 (6): 445–59. Translated from Russian; originally published in *Voprosy Ikhtiologii* 50 (4): 495–509.

Kondolf, G.M., A.J. Boulton, S. O'Daniel, G.C. Poole, F.J. Rahel, E.H. Stanley, et al. 2006. "Process-Based Ecological River Restoration: Visualizing Three-Dimensional Connectivity and Dynamic Vectors to Recover Lost Linkages." *Ecology and Society* 11 (2): 5. http://www.ecologyandsociety.org/vol11/iss2/art5/.

Korman, J., B. Bravender, and C.D. Levings. 1997. *Utilization of the Campbell River Estuary by Juvenile Chinook Salmon* (Oncorhynchus tshawytscha) *in 1994*. Canadian Technical Report of Fisheries and Aquatic Sciences, No. 2169. Ottawa: Fisheries and Oceans Canada.

Korstrom, J.S., and I.K. Birtwell. 2006. "Effects of Suspended Sediment on the Escape Behavior and Cover-Seeking Response of Juvenile Chinook Salmon in Freshwater." *Transactions of the American Fisheries Society* 135 (4): 1006–16. http://dx.doi.org/10.1577/T05-194.1.

Korsu, K., and A. Huusko. 2009. "Propagule Pressure and Initial Dispersal as Determinants of Establishment Success of Brook Trout (*Salvelinus fontinalis*, Mitchill 1814)." *Aquatic Invasions* 4 (4): 619–26. http://dx.doi.org/10.3391/ai.2009.4.4.8.

Koski, K.V. 2009. "The Fate of Coho Salmon Nomads: The Story of an Estuarine-Rearing Strategy Promoting Resilience." *Ecology and Society* 14 (1): 4.

Koski, M., and B.M. Johnson. 2002. "Functional Response of Kokanee Salmon (*Oncorhynchus nerka*) to Daphnia at Different Light Levels." *Canadian Journal of Fisheries and Aquatic Sciences* 59 (4): 707–16. http://dx.doi.org/10.1139/f02-045.

Kostachuk, R.A., and J.L. Luternauer. 2004. "Sedimentary Processes and Their Environmental Significance: Lower Main Channel, Fraser River Estuary." In *Fraser River Delta, British Columbia: Issues of an Urban Estuary*, vol. 567, ed. B.J. Groulx, D.C. Mosher, J.L. Luternauer, and D.E. Bilderback, 81–92. Geological Survey of Canada Bulletin. Ottawa: Natural Resources Canada.

Kotler, B.P., and R.D. Holt. 1989. "Predation and Competition: The Interaction of Two Types of Species Interactions." *Oikos* 54 (2): 256–60. http://dx.doi.org/10.2307/3565279.

Koval, M.V., and S.L. Gorin. 2013. "Influence of the Conditions in the Hairuzova and Belogolovaya Estuaries (Western Kamchatka) on Total Pacific Salmon Abundance." In *3rd International Workshop on Migration and Survival Mechanisms of Juvenile Salmon and Steelhead in Ocean Ecosystems* (North Pacific Anadromous Fish Commission Technical Report No. 9), ed. N.D. Davis and C. Chan, 222–27. Vancouver: North Pacific Anadromous Fish Commission.

Kowalik, Z. 2004. "Tide Distribution and Tapping into Tidal Energy." *Oceanologia* 46 (3): 291–331.

Kraabøl, M., S.I. Johnsen, J. Museth, and O.T. Sandlund. 2009. "Conserving Iteroparous Fish Stocks in Regulated Rivers: The Need for a Broader Perspective!" *Fisheries Management and Ecology* 16 (4): 337–40. http://dx.doi.org/10.1111/j.1365-2400.2009.00666.x.

Kramer, D.L. 1987. "Dissolved Oxygen and Fish Behavior." *Environmental Biology of Fishes* 18 (2): 81–92. http://dx.doi.org/10.1007/BF00002597.

Kravtsova, V.I., and N.S. Mit'kinykh. 2011. "Mouths of World Rivers in the Atlas of Space Images." *Water Resources* 38 (1): 1–17. http://dx.doi.org/10.1134/S0097807811010064.

Krentz, L.K. 2007. "Habitat Use, Movement, and Life History Variation of Coastal Cutthroat Trout *Oncorhynchus clarkii clarkii* in the Salmon River Estuary, Oregon." MS thesis, Oregon State University.

Krkošek, M. 2010. "Sea Lice and Salmon in Pacific Canada: Ecology and Policy." *Frontiers in Ecology and the Environment* 8 (4): 201–9. http://dx.doi.org/10.1890/080097.

Krkošek, M., C.W. Revie, P.G. Gargan, O.T. Skilbrei, B. Finstad, and C.D. Todd. 2013. "Impact of Parasites on Salmon Recruitment in the Northeast Atlantic Ocean." *Proceedings of the Royal Society B: Biological Sciences* 280 (1750): 20122359.

Kroglund, F., B. Finstad, S.O. Stefansson, T.O. Nilsen, T. Kristensen, B.O. Rosseland, H.C. Teien, and B. Salbu. 2007. "Exposure to Moderate Acid Water and Aluminum Reduces Atlantic Salmon Post-Smolt Survival." *Aquaculture* (Amsterdam) 273 (2–3): 360–73. http://dx.doi.org/10.1016/j.aquaculture.2007.10.018.

Kruzynski, G.M., and I.K. Birtwell. 1994. "A Predation Bioassay to Quantify the Ecological Significance of Sublethal Responses of Juvenile Chinook Salmon (*Oncorhynchus tshawytscha*) to the Antisapstain Fungicide TCMTB." *Canadian Journal of Fisheries and Aquatic Sciences* 51 (8): 1780–90. http://dx.doi.org/10.1139/f94-180.

Kukulka, T., and D.A. Jay. 2003. "Impacts of Columbia River Discharge on Salmonid Habitat. 2. Changes in Shallow-Water Habitat." *Journal of Geophysical Research* 108 (C9): 3294. http://dx.doi.org/10.1029/2003JC001829.

Kwak, T.J., and J.B. Zedler. 1997. "Food Web Analysis of Southern California Coastal Wetlands Using Multiple Stable Isotopes." *Oecologia* 110 (2): 262–77. http://dx.doi.org/10.1007/s004420050159.

L'Abée-Lund, J.H. 1994. "Effect of Smolt Age, Sex and Environmental Conditions on Sea Age at First Maturity of Anadromous Brown Trout, *Salmo trutta*, in Norway." *Aquaculture* (Amsterdam) 121 (1–3): 65–71. http://dx.doi.org/10.1016/0044-8486(94)90008-6.

Labenia, J.S., D.H. Baldwin, B.L. French, J.W. Davis, and N.L. Scholz. 2007. "Behavioral Impairment and Increased Predation Mortality in Cutthroat Trout Exposed to Carbaryl." *Marine Ecology Progress Series* 329: 1–11. http://dx.doi.org/10.3354/meps329001.

Lackey, R.T., D.H. Lach, and S.L. Duncan. 2006. "The Challenge of Restoring Wild Salmon." In *Salmon 2100: The Future of Wild Pacific Salmon*, ed. R.T. Lackey, D.H. Lach, and S.L. Duncan, 1–11. Bethesda, MD: American Fisheries Society.

Lake, R.G., and S.G. Hinch. 1999. "Acute Effects of Suspended Sediment Angularity on Juvenile Coho Salmon (*Oncorhynchus kisutch*)." *Canadian Journal of Fisheries and Aquatic Sciences* 56 (5): 862–67. http://dx.doi.org/10.1139/f99-024.

Lalli, C.M., and T.R. Parsons. 1997. *Biological Oceanography: An Introduction*. 2nd ed. Oxford: Butterworth-Heinemann.

Lambelet, M., M. Rehkämper, T. van de Flierdt, Z. Xue, K. Kreissig, B. Coles, Don Porcelli, and Per Andersson. 2013. "Isotopic Analysis of Cd in the Mixing Zone of Siberian Rivers with the Arctic Ocean – New Constraints on Marine Cd Cycling and the Isotope Composition of Riverine Cd." *Earth and Planetary Science Letters* 361: 64–73. http://dx.doi.org/10.1016/j.epsl.2012.11.034.

Lappalainen, J., and J. Soininen. 2006. "Latitudinal Gradients in Niche Breadth and Position – Regional Patterns in Freshwater Fish." *Naturwissenschaften* 93 (5): 246–50. http://dx.doi.org/10.1007/s00114-006-0093-2.

Lara, A., R. Villalba, and R. Urrutia. 2008. "A 400-Year Tree-Ring Record of the Puelo River Summer–Fall Streamflow in the Valdivian Rainforest Eco-region, Chile." *Climatic Change* 86 (3–4): 331–56. http://dx.doi.org/10.1007/s10584-007-9287-7.

Largier, J.L., J.T. Hollibaugh, and S.V. Smith. 1997. "Seasonally Hypersaline Estuaries in Mediterranean-Climate Regions." *Estuarine, Coastal and Shelf Science* 45 (6): 789–97. http://dx.doi.org/10.1006/ecss.1997.0279.

Larkin, P.A. 1956. "Interspecific Competition and Population Control in Freshwater Fish." *Journal of the Fisheries Board of Canada* 13 (3): 327–42. http://dx.doi.org/10.1139/f56-022.

Larsen, P.F., E.E. Nielsen, A. Koed, D.S. Thomsen, P.A. Olsvik, and V. Loeschcke. 2008. "Interpopulation Differences in Expression of Candidate Genes for Salinity Tolerance in Winter Migrating Anadromous Brown Trout (*Salmo trutta*)." *BioMed Central Genetics* 9: 12–20.

Larson, M.R., M.G. Foreman, C.D. Levings, and M.R. Tarbotton. 2003. "Dispersion of Discharged Ship Ballast Water in Vancouver Harbour, Juan de Fuca Strait, and Offshore of the Washington Coast." *Journal of Environmental Engineering and Science* 2 (3): 163–76. http://dx.doi.org/10.1139/s03-014.

Larsson, S., and I. Berglund. 2005. "The Effect of Temperature on the Energetic Growth Efficiency of Arctic Charr (*Salvelinus alpinus* L.) from Four Swedish Populations." *Journal of Thermal Biology* 30 (1): 29–36. http://dx.doi.org/10.1016/j.jtherbio.2004.06.001.

Larsson, U., R. Elmgren, and F. Wulff. 1983. "Eutrophication and the Baltic Sea." *Ambio* 14: 9–14.

Lasne, E., B. Bergerot, S. Lek, and P. Laffaille. 2007. "Fish Zonation and Indicator Species for the Evaluation of the Ecological Status of Rivers: Example of the Loire Basin (France)." *River Research and Applications* 23 (8): 877–90. http://dx.doi.org/10.1002/rra.1030.

Launey, S., G. Brunet, R. Guyomard, and P. Davaine. 2010. "Role of Introduction History and Landscape in the Range Expansion of Brown Trout (*Salmo trutta* L.) in the Kerguelen Islands." *Journal of Heredity* 101 (3): 270–83. http://dx.doi.org/10.1093/jhered/esp130.

Lavery, S., and B. Donovan. 2005. "Flood Risk Management in the Thames Estuary Looking Ahead 100 Years." *Philosophical Transactions of the Royal Society A: Mathematical, Physical and Engineering Sciences* 363 (1831): 1455–74. http://dx.doi.org/10.1098/rsta.2005.1579.

Layzer, J.A. 2012. "The Purpose and Politics of Ecosystem-Based Management." In *Sustainability Science: The Emerging Paradigm and the Urban Environment*, ed. M.P. Weinstein and R.E. Turner, 177–97. New York: Springer. http://dx.doi.org/10.1007/978-1-4614-3188-6_9.

Lecklin, T., R. Ryömä, and S. Kuikka. 2011. "A Bayesian Network for Analyzing Biological Acute and Long-Term Impacts of an Oil Spill in the Gulf of Finland."

Marine Pollution Bulletin 62 (12): 2822–35. http://dx.doi.org/10.1016/j.marpolbul.2011.08.045.

Leman, V.N., and V.V. Chebanova. 2005. "New Data on the Ecology of Juveniles of the Western Kamchatka King Salmon *Oncorhynchus tshawytscha* in the Riverine and Estuarine Life Periods." *Journal of Ichthyology* 45: 318–27. Translated from Russian; originally published in *Voprosy Ikhthiologii* 45: 395–404.

Lenormand, S., J.J. Dodson, and A. Menard. 2004. "Seasonal and Ontogenetic Patterns in the Migration of Anadromous Brook Charr (*Salvelinus fontinalis*)." *Canadian Journal of Fisheries and Aquatic Sciences* 61 (1): 54–67. http://dx.doi.org/10.1139/f03-137.

Leopold, L.B., M.G. Wolman, and J.P. Miller. 1964. *Fluvial Processes in Geomorphology*. San Francisco: William H. Freeman and Sons. Repr. New York: Dover Publications, 1995.

Lester, R.J.G., and K. MacKenzie. 2009. "The Use and Abuse of Parasites as Stock Markers for Fish." *Fisheries Research* 97 (1–2): 1–2. http://dx.doi.org/10.1016/j.fishres.2008.12.016.

Levings, C.D. 1976. "River Diversion and Intertidal Benthos at the Squamish River Delta, British Columbia." In *Fresh Water on the Sea* (Proceedings Symposium on Influence of Freshwater Outflow on the Biological Processes in Fjords and Coastal Waters, April 22–25, 1974, Geilo, Norway), ed. S. Skreslet, R. Leinebo, J.B.L. Mathews, E. Sakshaug, 193–202. Oslo: Association of Norwegian Oceanographers.

–. 1980a. *Vertical Distribution and Abundance of Epibenthos and Macro-Zooplankton in the Lower Fraser Estuary*. Canadian Data Report, Fisheries and Aquatic Sciences No. 241. Nanaimo, BC: Fisheries and Oceans Canada.

–. 1980b. "Consequences of Training Walls and Jetties for Aquatic Habitats at two British Columbia Estuaries." *Coastal Engineering* 4: 111–36. http://dx.doi.org/10.1016/0378-3839(80)90010-1.

–. 1982. *Short Term Use of a Low Tide Refuge in a Sandflat by Juvenile Chinook* (Oncorhynchus tshawytscha), *Fraser River Estuary*. Canadian Technical Report of Fisheries and Aquatic Sciences, No. 1111. West Vancouver: Fisheries and Oceans Canada.

–. 1984. "Commentary: Progress in Attempts to Test the Null Hypothesis that Juvenile Salmonids Aren't Dependent on Estuaries." In *The Influence of Ocean Conditions on the Production of Salmonids in the North Pacific: A Workshop, November 8–10, 1983, Newport, Oregon*, ed. W.G. Pearcy, 287–96. Corvallis: Oregon State University Press.

–. 1985. "Juvenile Salmonid Use of Habitats Altered by a Coal Port in the Fraser River Estuary, British Columbia." *Marine Pollution Bulletin* 16 (6): 248–54. http://dx.doi.org/10.1016/0025-326X(85)90510-7.

–. 1993. "Requirements for Genetic Data on Adaptations to Environment and Habitats of Salmonids." In *Proceedings of NATO Advanced Study Institute: Genetic Conservation of Salmonid Fishes, Moscow, Idaho, June 23–29, 1992*, ed. J.G. Cloud and G.H. Thorgaard, 49–66. New York: Plenum Press. http://dx.doi.org/10.1007/978-1-4615-2866-1_4.

–. 1994a. "Feeding Behaviour of Juvenile Salmon and Significance of Habitat during Estuary and Early Sea Phase." *Nordic Journal of Freshwater Research* 69: 7–16.

–. 1994b. "Life on the Edge: Structural and Functional Aspects of Chinook and Coho Rearing Habitats on the Margins of the Lower Fraser River." In *Salmon Ecosystem Restoration: Myth and Reality. Proceedings of the 1994 Northeast Pacific Chinook and Coho Salmon Workshop, Eugene, Oregon, November 7– 10, 1994,* ed. M. Keefe and P. Lawson, 139–47. Corvallis: Oregon Chapter, American Fisheries Society.

–. 1994c. "Science and Management Needed to Maintain Salmon Production in Estuaries of the Northeast Pacific." In *Proceedings of the Joint Symposium of Estuarine Research Federation and Estuarine and Coastal Sciences Association, September 14–18, 1992, Plymouth, England,* ed. K. Dyer and R.J. Orth, 417–21. Fredensborg, Denmark: Olsen and Olsen.

–. 2000a. "An Overview Assessment of Compensation and Mitigation Techniques Used to Assist Fish Habitat Management in British Columbia Estuaries." In *Sustainable Fisheries Management: Pacific Salmon,* ed. E.E. Knudsen, C.R. Steward, D.D. MacDonald, J.E. Williams, and D.W. Reiser, 341–47. Boca Raton, FL: CRC Press.

–. 2000b. "Critical Environmental Information for Management Plans in Pacific Estuaries." In *Proceedings from PACON 1999, Russian Academy of Sciences, June 23–25, 1999, Moscow, Russia,* 337–43. Honolulu: PACON International.

–. 2004. "Knowledge of Fish Ecology and Its Application to Habitat Management." In *Fraser River Delta, British Columbia: Issues of an Urban Estuary,* vol. 567 (Geological Survey of Canada Bulletin), ed. B.J. Groulx, D.C. Mosher, J.L. Luternauer, and D.E. Bilderback, 81–92. Ottawa: Natural Resources Canada.

Levings, C.D., I.K. Birtwell, and G.E. Piercey. 2003. *Beach Seine Catch Data from Sechelt Inlet and Agamemnon Channel, British Columbia.* Canadian Data Report of Fisheries and Aquatic Sciences No. 1110. West Vancouver: Fisheries and Oceans Canada.

Levings, C.D., and D. Bouillon. 1997. "Criteria for Evaluating the Survival Value of Estuaries for Salmonids." In *Estuarine and Ocean Survival of Northeastern Pacific Salmon* (NOAA Technical Memorandum NMFS-NWFSC 29), ed. R.L. Emmett and M.H. Schiewe, 159–68. Washington, DC: US Department of Commerce.

–. 2008. "Scaling Salmonid Life History by Habitat Area: A Conceptual Approach to Estimating Estuarine Conservation Needs." In *Reconciling Fisheries with Conservation:* Proceedings of the Fourth World Fisheries Congress, ed. J.L. Nielsen, J.J. Dodson, K. Friedland, T.R. Hamon, J. Musick, and E. Verspoor, 1597–1604. Bethesda, MD: American Fisheries Society.

Levings, C.D., D.E. Boyle, and T.R. Whitehouse. 1995. "Distribution and Feeding of Juvenile Pacific Salmon in Freshwater Tidal Creeks of the Lower Fraser River, British Columbia." *Fisheries Management and Ecology* 2 (4): 299–308. http://dx.doi.org/10.1111/j.1365-2400.1995.tb00121.x.

Levings, C.D., K. Conlin, and B. Raymond. 1991. "Intertidal Habitats Used by Juvenile Chinook Salmon (*Oncorhynchus tshawytscha*) Rearing in the North Arm of the Fraser River Estuary." *Marine Pollution Bulletin* 22 (1): 20–26. http://dx.doi.org/10.1016/0025-326X(91)90440-4.

Levings, C.D., N.A. Hvidsten, and B.O. Johnsen. 1994. "Feeding of Atlantic Salmon (*Salmo salar* L.) Post-Smolts in a Fjord, Central Norway." *Canadian Journal of Zoology* 72 (5): 834–39. http://dx.doi.org/10.1139/z94-113.

Levings, C.D., and G.S. Jamieson. 2001. *Marine and Estuarine Riparian Habitats and Their Role in Coastal Ecosystems.* CSAS Research Document 2001/109. Ottawa: Canadian Science Advisory Secretariat.

Levings, C.D., and J.S. Macdonald. n.d. "Chinook Salmon and Estuarine Ecosystems: Final Results of a Transfer Experiment to Help Evaluate Estuary Dependency." Unpublished manuscript. West Vancouver: Department of Fisheries and Oceans, Centre for Aquaculture and Environmental Research.

–. 1991. "Rehabilitation of Estuarine Fish Habitat at Campbell River, British Columbia." In *American Fisheries Symposium 10: Proceedings of the 1988 Fisheries Bioengineering Symposium,* ed. J. Colt and R.J. White, 176–90. Bethesda, MD: American Fisheries Society.

Levings, C.D., C.D. McAllister, and B.C. Chang. 1986. "Differential Use of the Campbell River Estuary, British Columbia, by Wild and Hatchery Reared Juvenile Chinook Salmon (*Oncorhynchus tshawytscha*)." *Canadian Journal of Fisheries and Aquatic Sciences* 43 (7): 1386–97. http://dx.doi.org/10.1139/f86-172.

Levings, C.D., and N.G. McDaniel. 1976. *Industrial Disruption of Invertebrate Communities on Beaches in Howe Sound, British Columbia.* Fisheries and Marine Services Research and Development Directorate Technical Report No. 663. West Vancouver: Pacific Environment Institute.

Levings, C.D., and D.J.H. Nishimura. 1997. "Created and Restored Marshes in the Lower Fraser River, British Columbia: A Summary of Their Functioning as Fish Habitat." *Water Quality Research Journal of Canada* 32 (3): 599–618.

Levings, C.D., and T.G. Northcote. 2004. "Effects of Forestry on Estuarine Ecosystems Supporting Fishes." In *Fishes and Forestry – Worldwide Watershed Interactions and Management,* ed. T.G. Northcote and G.F. Hartman, 320–35. Oxford: Blackwell Science. http://dx.doi.org/10.1002/9780470995242.ch15.

Levings, C.D., C.D. McAllister, J.S. Macdonald, T.J. Brown, M.S. Kotyk, and B.A. Kask. 1989. "Chinook Salmon (*Oncorhynchus tshawytscha*) and Estuarine Habitat: A Transfer Experiment Can Help Evaluate Estuary Dependency." In *Proceedings of the National Workshop on Effects of Habitat Alteration on Salmonid Stocks* (Canadian Special Publication of Fisheries and Aquatic Sciences No. 105), ed. C.D. Levings, L.B. Holtby, and M.A. Henderson, 116–22. Ottawa: Fisheries and Oceans Canada.

Levings, C.D., A. Ervik, P. Johannessen, and J. Aure. 1995. "Ecological Criteria Used to Help Site Fish Farms in Fjords." *Estuaries* 18 (1): 81–90. http://dx.doi.org/10.2307/1352284.

Levings, C.D., M.S. North, G.E. Piercey, G.S. Jamieson, and B.D. Smiley. 1999. "Mapping Nearshore and Intertidal Marine Habitats with Remote Sensing and GPS: The Importance of Spatial and Temporal Scales." In *Proceedings of the Annual Meeting of the Marine Technology Society and IEEE, Seattle, Washington. Vol. 3: Riding the Crest into the 21st Century,* ed. J. Bruce and C. Brancart, 1249–55. Seattle: Institute of Electrical and Electronics Engineers. http://dx.doi.org/10.1109/OCEANS.1999.800170.

Levings, C.D., K.L. Barry, J.A. Grout, G.E. Piercey, A.D. Marsden, A.P. Coombs, and B. Mossop. 2004. "Effects of Acid Mine Drainage on the Estuarine Food Web, Britannia Beach, Howe Sound, British Columbia, Canada." *Hydrobiologia* 525 (1–3): 185–202. http://dx.doi.org/10.1023/B:HYDR.0000038866.20304.3d.

Levy, D.A., and A.D. Cadenhead. 1995. "Selective Tidal Stream Transport of Adult Sockeye Salmon (*Oncorhynchus nerka*) in the Fraser River Estuary." *Canadian Journal of Fisheries and Aquatic Sciences* 52 (1): 1–12. http://dx.doi.org/10.1139/f95-001.

Levy, D.A., and C.D. Levings. 1978. *A Description of the Fish Community of the Squamish River Estuary, British Columbia: Relative Abundance, Seasonal Changes, and Feeding Habits of Salmonids.* Canadian Manuscript Report of Fisheries and Marine Service No. 1475. West Vancouver: Fisheries and Marine Service.

Levy, D.A., and T.G. Northcote. 1982. "Juvenile Salmon Residency in a Marsh Area of the Fraser River Estuary." *Canadian Journal of Fisheries and Aquatic Sciences* 39 (2): 270–76. http://dx.doi.org/10.1139/f82-038.

Levy, D.A., T.G. Northcote, and R.M. Barr. 1982. *Effects of Estuarine Log Storage on Juvenile Salmon.* Technical Report 26. Vancouver: Westwater Research Centre.

Levy, D.A., T.G. Northcote, and G.J. Birch. 1979. *Juvenile Salmon Utilization of Tidal Channels in the Fraser River Estuary, British Columbia.* Technical Report 23. Vancouver: Westwater Research Centre.

Lim, P.G., and C.D. Levings. 1973. *Distribution and Biomass of Intertidal Vascular Plants on the Squamish Delta.* Fisheries Research Board of Canada Manuscript Report Series, No. 1219. West Vancouver: Pacific Environment Institute.

Lima, A.L., J.B. Hubeny, C.M. Reddy, J.W. King, K.A. Hughen, and T.I. Eglinton. 2005. "High-Resolution Historical Records from Pettaquamscutt Riverbasin Sediments. ^{210}Pb and Varve Chronologies Validate Record of ^{137}Cs Released by the Chernobyl Accident." *Geochimica et Cosmochimica Acta* 69 (7): 1803–12. http://dx.doi.org/10.1016/j.gca.2004.10.009.

Limburg, K.E., and J.R. Waldman. 2009. "Dramatic Declines in North Atlantic Diadromous Fishes." *BioScience* 59 (11): 955–65. http://dx.doi.org/10.1525/bio.2009.59.11.7.

Lincoln, R.J., G.A. Boxshall, and P.F. Clark. 1987. *A Dictionary of Ecology, Evolution, and Systematics.* 1st ed. London: Cambridge University Press.

Lindley, S.T., and M.S. Mohr. 2003. "Modeling the Effect of Striped Bass (*Morone saxatilis*) on the Population Viability of Sacramento River Winter-Run Chinook Salmon (*Oncorhynchus tshawytscha*)." *Fish Bulletin* 101 (2): 321–31.

Lindley, S.T., R.S. Schick, E. Mora, P.B. Adams, J.J. Anderson, S. Greene, et al. 2007. "Framework for Assessing Viability of Threatened and Endangered Chinook Salmon and Steelhead in the Sacramento–San Joaquin Basin." *San Francisco Estuary and Watershed Science* 5 (1): 1–26.

Lindley, S.T., C.B. Grimes, M.S. Mohr, W. Peterson, J. Stein, J.T. Anderson, et al. 2009. *What Caused the Sacramento River Fall Chinook Stock Collapse?* NOAA Technical Memorandum NMFS, NOAA-TM-NMFS-SWFSC-447. Washington, DC: US Department of Commerce.

List, G.F., P.B. Mirchandani, M.A. Turnquist, and K.G. Zografos. 1991. "Modeling and Analysis for Hazardous Materials Transportation: Risk Analysis, Routing/Scheduling and Facility Location." *Transportation Science* 25 (2): 100–14.

Lobry, J., L. Mourand, E. Rochard, and P. Elie. 2003. "Structure of the Gironde Estuarine Fish Assemblages: A Comparison of European Estuaries Perspective." *Aquatic Living Resources* 16 (2): 47–58. http://dx.doi.org/10.1016/S0990-7440(03)00031-7.

Lobry, J., V. David, S. Pasquaud, M. Lepage, B. Sautour, and E. Rochard. 2008. "Diversity and Stability of an Estuarine Trophic Network." *Marine Ecology Progress Series* 358: 13–25. http://dx.doi.org/10.3354/meps07294.

Locke, A., J.M. Hanson, G.J. Klassen, S.M. Richardson, and C.I. Aubé. 2003. "The Damming of the Petitcodiac River: Species, Populations, and Habitats Lost." *Northeastern Naturalist* 10 (1): 39–54. http://dx.doi.org/10.1656/1092-6194 (2003)010[0039:TDOTPR]2.0.CO;2.

London Rivers Action Plan, The. 2015. "The London Rivers Action Plan: A Tool to Help Restore Rivers for People and Nature." January 2009. http://www.therrc.co.uk/lrap/lplan.pdf.

Lorenzen, K. 2008. "Understanding and Managing Enhancement Fisheries Systems." *Reviews in Fisheries Science* 16 (1–3): 10–23. http://dx.doi.org/10.1080/10641260701790291.

Lorz, H.W., and B.P. McPherson. 1976. "Effects of Copper or Zinc in Freshwater on the Adaptation to Sea Water and ATPase Activity, and the Effects of Copper on Migratory Disposition of Coho Salmon (*Oncorhynchus kisutch*)." *Journal of the Fisheries Research Board of Canada* 33 (9): 2023–30.

Lott, M.A. 2004. "Habitat-Specific Feeding Ecology of Ocean-Type Juvenile Chinook Salmon in the Lower Columbia River." MS thesis, University of Washington.

Lotze, H.K., H.S. Lenihan, B.J. Bourque, R.H. Bradbury, R.G. Cooke, M.C. Kay, et al. 2006. "Depletion, Degradation, and Recovery Potential of Estuaries and Coastal Seas." *Science* 312 (5781): 1806–9. http://dx.doi.org/10.1126/science.1128035.

Lower Columbia Estuary Partnership. 2015. Who We Are. http://www.estuarypartnership.org.

Lubchenco, J., S.R. Palumbi, S.D. Gaines, and S. Andelman. 2003. "Plugging a Hole in the Ocean: The Emerging Science of Marine Reserves." *Ecological Applications* 13 (sp1): 3–7. http://dx.doi.org/10.1890/1051-0761(2003)013[0003:PAHITO]2.0.CO;2.

Luck, M., N. Maumenee, D. Whited, J. Lucotch, S. Chilcote, M. Lorang, Daniel Goodman, Kyle McDonald, John Kimball, and Jack Stanford. 2010. "Remote Sensing Analysis of Physical Complexity of North Pacific Rim Rivers to Assist Wild Salmon Conservation." *Earth Surface Processes and Landforms* 35 (11): 1330–43. http://dx.doi.org/10.1002/esp.2044.

Luo, Pingping, Bin He, Kaoru Takara, Bam HN Razafindrabe, Daniel Nover, and Yosuke Yamashiki. 2011. "Spatiotemporal trend analysis of recent river water quality conditions in Japan." *Journal of Environmental Monitoring* 13 (10): 2819–29.

Lyse, A.A., S.O. Stefansson, and A. Ferno. 1998. "Behaviour and Diet of Sea Trout Post-Smolts in a Norwegian Fjord System." *Journal of Fish Biology* 52 (5): 923–36. http://dx.doi.org/10.1111/j.1095-8649.1998.tb00593.x.

MacCrimmon, H.R., and B.L. Gots. 1979. "World Distribution of Atlantic Salmon, *Salmo salar*." *Journal of the Fisheries Research Board of Canada* 36 (4): 422–57. http://dx.doi.org/10.1139/f79-062.

Macdonald, A.L. 1984. "Seasonal Use of Nearshore Intertidal Habitats by Juvenile Pacific Salmon on the Delta Front of the Fraser River Estuary, British Columbia." MSc thesis, University of Victoria.

Macdonald, J.S., I.K. Birtwell, and G.M. Kruzynski. 1987. "Food and Habitat Utilization by Juvenile Salmonids in the Campbell River Estuary." *Canadian Journal of Fisheries and Aquatic Sciences* 44 (6): 1233–46. http://dx.doi.org/10.1139/f87-146.

Macdonald, J.S., and B.D. Chang. 1993. "Seasonal Use by Fish of Nearshore Areas in an Urbanized Coastal Inlet in Southwestern British Columbia." *Northwest Science* 67 (2): 63–77.

Macdonald, J.S., C.D. Levings, C.D. McAllister, U.H.M. Fagerlund, and J.R. McBride. 1988. "A Field Experiment to Test the Importance of Estuaries for Chinook Salmon (*Oncorhynchus tshawytscha*) Survival: Short Term Results." *Canadian Journal of Fisheries and Aquatic Sciences* 45 (8): 1366–77. http://dx.doi.org/10.1139/f88-160.

Macdonald, R.W. 2000. "Arctic Estuaries and Ice: A Positive Negative Estuarine Couple." In *The Freshwater Budget of the Arctic Ocean*, ed. E.L. Lewis, E.P. Jones, P. Lemke, T. Prouse, and P. Wadhams, 383–407. Boston: Kluwer. http://dx.doi.org/10.1007/978-94-011-4132-1_17.

Mace, P.M. 1983. "Predator-Prey Functional Responses and Predation by Staghorn Sculpins *Leptocottus armatus* on Chum Salmon, *Oncorhynchus keta*." PhD dissertation, University of British Columbia.

MacFarlane, R.B. 2010. "Energy Dynamics and Growth of Chinook Salmon (*Oncorhynchus tshawytscha*) from the Central Valley of California during the Estuarine Phase and First Ocean Year." *Canadian Journal of Fisheries and Aquatic Sciences* 67 (10): 1549–65. http://dx.doi.org/10.1139/F10-080.

MacFarlane, R.B., and E.C. Norton. 2002. "Physiological Ecology of Juvenile Chinook Salmon (*Oncorhynchus tshawytscha*) at the Southern End of Their Distribution, the San Francisco Estuary and Gulf of the Farallones, California." *Fish Bulletin* 100 (2): 244–57.

MacIsaac, H.J., L.F. Herborg, and J.R. Muirhead. 2007. "Modeling Biological Invasions of Inland Waters." In *Biological Invaders in Inland Waters: Profiles, Distribution, and Threats*, ed. F. Gherardi, 347–68. Dordrecht, Netherlands: Springer. http://dx.doi.org/10.1007/978-1-4020-6029-8_18.

MacKenzie, W.D., D. Remington, and J. Shaw. 2000. *Estuaries on the North Coast of British Columbia: A Reconnaissance Survey of Selected Sites*. Unpublished draft. Victoria: Ministry of Environment, Lands and Parks/Ministry of Forests. https://www.for.gov.bc.ca/HRE/becweb/Downloads/Downloads_Wetlands/North%20Coast%20Estuaries.pdf.

Macklin, M.G., K.A. Hudson-Edwards, and E.J. Dawson. 1997. "The Significance of Pollution from Historic Metal Mining in the Pennine Ore Fields on River Sediment Contaminant Fluxes to the North Sea." *Science of the Total Environment* 194–95: 391–97. http://dx.doi.org/10.1016/S0048-9697(96)05378-8.

Maclaurin, J., and K. Sterelny. 2008. *What Is Biodiversity?* Chicago: University of Chicago Press. http://dx.doi.org/10.7208/chicago/9780226500829.001.0001.

MacLean, S.A., E.M. Caldarone, and J.M. St. Onge-Burns. 2008. "Estimating Recent Growth Rates of Atlantic Salmon Smolts Using RNA-DNA Ratios from Nonlethally Sampled Tissues." *Transactions of the American Fisheries Society* 137 (5): 1279–84. http://dx.doi.org/10.1577/T07-254.1.

Magnhagen, C. 1988. "Predation Risk and Foraging in Juvenile Pink (*Oncorhynchus gorbuscha*) and Chum Salmon (*O. keta*)." *Canadian Journal of Fisheries and Aquatic Sciences* 45 (4): 592–96. http://dx.doi.org/10.1139/f88-072.

Magnusson, A., and R. Hilborn. 2003. "Estuarine Influence on Survival Rates of Coho (*Oncorhynchus kisutch*) and Chinook Salmon (*Oncorhynchus tshawytscha*) Released from Hatcheries on the US Pacific Coast." *Estuaries* 26 (4): 1094–1103. http://dx.doi.org/10.1007/BF02803366.

Maier, G.O., and C.A. Simenstad. 2009. "The Role of Marsh-Derived Macrodetritus to the Food Webs of Juvenile Chinook Salmon in a Large Altered Estuary." *Estuaries and Coasts* 32 (5): 984–98. http://dx.doi.org/10.1007/s12237-009-9197-1.

Maltby, L.L., A.C. Paetzold, and P.H. Warren. 2010. "Sustaining Industrial Activity and Ecological Quality: The Potential Role of an Ecosystem Services Approach." In *Ecology of Industrial Pollution*, ed. L.C. Batty and K.B. Hallberg, 327–44. Cambridge: Cambridge University Press. http://dx.doi.org/10.1017/CBO9780511805561.017.

Marchand, J., I. Codling, P. Drake, M. Elliot, L. Pihl, and J. Rebelo. 2002. "Environmental Quality of Estuaries." In *Fishes in Estuaries*, ed. M. Elliott and K. Hemingway, 322–409. Oxford: Blackwell Science. http://dx.doi.org/10.1002/9780470995228.ch7.

Marchetti, M.P., P.B. Moyle, and R. Levine. 2004. "Alien Fishes in California Watersheds: Characteristics of Successful and Failed Invaders." *Ecological Applications* 14 (2): 587–96. http://dx.doi.org/10.1890/02-5301.

Marchetti, M.P., T. Light, J. Feliciano, T. Armstrong, Z. Hogan, J. Viers, and P.B. Moyle. 2001. "Homogenization of California's Fish Fauna through Abiotic Change." In *Biotic Homogenization*, ed. J.L. Lockwood and M.L. McKinney, 259–78. London: Kluwer Academic/Plenum Publishers. http://dx.doi.org/10.1007/978-1-4615-1261-5_13.

Marín, V.H., L.E. Delgado, and P. Bachmann. 2008. "Conceptual PHES-System Models of the Aysén Watershed and Fjord (Southern Chile), Testing a Brainstorming Strategy." *Journal of Environmental Management* 88 (4): 1109–18. http://dx.doi.org/10.1016/j.jenvman.2007.05.012.

Marine, K.R., and J.J. Cech Jr. 2004. "Effects of High Water Temperature on Growth, Smoltification, and Predator Avoidance in Juvenile Sacramento River Chinook Salmon." *North American Journal of Fisheries Management* 24 (1): 198–210. http://dx.doi.org/10.1577/M02-142.

Marmorek, D., and C. Peters. 2001. "Finding a PATH toward Scientific Collaboration: Insights from the Columbia River Basin." *Conservation Ecology* 5 (2): 8. http://www.ecologyandsociety.org/vol5/iss2/art8/.

Marshall, S., and M. Elliott. 1997. "A Comparison of Univariate and Multivariate Numerical and Graphical Techniques for Determining Inter and Intraspecific Feeding Relationships in Estuarine Fish." *Journal of Fish Biology* 51 (3): 526–45. http://dx.doi.org/10.1111/j.1095-8649.1997.tb01510.x.

Marston, R.B. 1904. "The Increase of Fish-Destroying Birds and Seals. Part I. Birds." In *The Twentieth Century Monthly Review. Vol. 55: January-June 1904*, ed. J. Knowles, 107. New York: Leonard Scott Publications.

Martin, F., R.D. Hedger, J.J. Dodson, L. Fernandes, D. Hatin, F. Caron, and F.G. Whoriskey. 2009. "Behavioural Transition during the Estuarine Migration of

Wild Atlantic Salmon (*Salmo salar* L.) Smolt." *Ecology of Freshwater Fish* 18 (3): 406–17. http://dx.doi.org/10.1111/j.1600-0633.2009.00357.x.

Marty, G.D., S.M. Saksida, and T.J. Quinn. 2010. "Relationship of Farm Salmon, Sea Lice, and Wild Salmon Populations." *Proceedings of the National Academy of Sciences of the United States of America* 107 (52): 22599–604. http://dx.doi.org/10.1073/pnas.1009573108.

Maser, C., and J.R. Sedell. 1994. *From the Forest to the Sea: The Ecology of Wood in Streams, Rivers, Estuaries, and Oceans.* Delray Beach, FL: St. Lucie Press.

Matern, S.A., P.B. Moyle, and L.C. Pierce. 2002. "Native and Alien Fishes in a California Estuarine Marsh: Twenty-One Years of Changing Assemblages." *Transactions of the American Fisheries Society* 131 (5): 797–816. http://dx.doi.org/10.1577/1548-8659(2002)131<0797:NAAFIA>2.0.CO;2.

Mather, M.E. 1998. "The Role of Context-Specific Predation in Understanding Patterns Exhibited by Anadromous Salmon." *Canadian Journal of Fisheries and Aquatic Sciences* 55 (S1): 232–46. http://dx.doi.org/10.1139/d98-002.

Mathieson, S., A. Cattrijsse, M.J. Costa, P. Drake, M. Elliott, J. Gardner, and J. Marchand. 2000. "Fish Assemblages of European Tidal Marshes: A Comparison Based on Species, Families and Functional Guilds." *Marine Ecology Progress Series* 204: 225–42. http://dx.doi.org/10.3354/meps204225.

Mathur, D., W.H. Bason, E.J. Purdy Jr., and C.A. Silver. 1985. "A Critique of the In Stream Flow Incremental Methodology." *Canadian Journal of Fisheries and Aquatic Sciences* 42 (4): 825–31. http://dx.doi.org/10.1139/f85-105.

Matishov, G.G., and E.G. Berestovskii. 2010. "Problems of Preserving the Diversity of Salmon in Russia's North and Far East." *Herald of the Russian Academy of Sciences* 80 (1): 69–73. Translated from Russian; originally published in *Vestnik Rossiiskoi Akademii Nauk* 80 (1): 52–56. http://dx.doi.org/10.1134/S1019331610010090.

Maximenkov, V.V. 2007. *Feeding by and Trophic Interactions of Juvenile Fishes Inhabiting River Estuaries and Coastal Waters of Kamchatka.* Publication No. 278 [in Russian]. Petropavlovsk-Kamchatsky: Kamchatka Research Institute of Fisheries and Oceanography.

Mayama, H., and Y. Ishida. 2003. *Japanese Studies on the Early Ocean Life of Juvenile Salmon.* North Pacific Anadromous Fish Commission Bulletin No. 3. Vancouver: North Pacific Anadromous Fish Commission.

McCabe, G.T., W.D. Muir, R.L. Emmett, and J.T. Durkin. 1983. "Interrelationships between Juvenile Salmonids and Nonsalmonid Fish in the Columbia River Estuary." *Fish Bulletin* 81 (4): 815–26.

McCarthy, I.D., and S. Waldron. 2000. "Identifying Migratory *Salmo trutta* Using Carbon and Nitrogen Stable Isotope Ratios." *Rapid Communications in Mass Spectrometry* 14 (15): 1325–31. http://dx.doi.org/10.1002/1097-0231(20000815)14:15<1325::AID-RCM980>3.0.CO;2-A.

McClure, M.M., S.M. Carlson, T.J. Beechie, G.R. Pess, J.C. Jorgensen, S.M. Sogard, et al. 2008. "Evolutionary Consequences of Habitat Loss for Pacific Anadromous Salmonids." *Evolutionary Applications* 1 (2): 300–18. http://dx.doi.org/10.1111/j.1752-4571.2008.00030.x.

McCormick, S.D. 1994. "Ontogeny and Evolution of Salinity Tolerance in Anadromous Salmonids: Hormones and Heterochrony." *Estuaries* 17 (1): 26–33. http://dx.doi.org/10.2307/1352332.

—. 2013. "Smolt Physiology and Endocrinology." In *Fish Physiology. Vol. 32: Euryhaline Fishes*, ed. S.D. McCormick, A.P. Farrell, and C.J. Brauner, 199–252. Amsterdam: Elsevier.

McCormick, S.D., and R.J. Naiman. 1984. "Osmoregulation in the Brook Trout, *Salvelinus fontinalis* – II. Effects of Size, Age and Photoperiod on Seawater Survival and Ionic Regulation." *Comparative Biochemistry and Physiology. Part A, Physiology* 79 (1): 17–28. http://dx.doi.org/10.1016/0300-9629(84)90704-7.

McCormick, S.D., L.P. Hansen, T.P. Quinn, and R.L. Saunders. 1998. "Movement, Migration, and Smolting of Atlantic Salmon (*Salmo salar*)." *Canadian Journal of Fisheries and Aquatic Sciences* 55 (S1): 77–92. http://dx.doi.org/10.1139/d98-011.

McCormick, S.D., M.F. O'Dea, A.M. Moeckel, D.T. Lerner, and B.T. Björnsson. 2005. "Endocrine Disruption of Parr-Smolt Transformation and Seawater Tolerance of Atlantic Salmon by 4-Nonylphenol and 17β-estradiol." *General and Comparative Endocrinology* 142 (3): 280–88. http://dx.doi.org/10.1016/j.ygcen.2005.01.015.

McCormick, S.D., D.T. Lerner, M.Y. Monette, K. Nieves-Puigdoller, J.T. Kelly, and B.T. Björnsson. 2009. "Taking It with You When You Go: How Perturbations to the Freshwater Environment, Including Temperature, Dams, and Contaminants, Affect Marine Survival of Salmon." In *Proceedings of Challenges for Diadromous Fishes in a Dynamic Global Environment, Symposium 69*, ed. A.J. Haro, K.L. Smith, R.A. Rulifson, C.M. Moffitt, R.J. Klauda, M.J. Dadswell, et al., 195–214. Bethesda, MD: American Fisheries Society.

McDowall, R.M. 1994. "The Origin of New Zealand's Chinook Salmon, *Oncorhynchus tshawytscha*." *Marine Fisheries Review* 56 (1): 1–10.

—. 1995. "Seasonal Pulses in Migrations of New Zealand Diadromous Fish and the Potential Impacts of River Mouth Closure." *New Zealand Journal of Marine and Freshwater Research* 29 (4): 517–26. http://dx.doi.org/10.1080/00288330.1995.9516684.

—. 2006. "Crying Wolf, Crying Foul, or Crying Shame: Alien Salmonids and a Biodiversity Crisis in the Southern Cool-Temperate Galaxioid Fishes?" *Reviews in Fish Biology and Fisheries* 16 (3–4): 233–422. http://dx.doi.org/10.1007/s11160-006-9017-7.

McDowall, R.M., R.M. Allibone, and W.L. Chadderton. 2001. "Issues for the Conservation and Management of Falkland Islands Freshwater Fishes." *Aquatic Conservation: Marine Freshwater Ecosystems* 11 (6): 473–86. http://dx.doi.org/10.1002/aqc.499.

McInerney, J.E. 1964. "Salinity Preference: An Orientation Mechanism in Salmon Migration." *Journal of the Fisheries Board of Canada* 21 (5): 995–1018. http://dx.doi.org/10.1139/f64-092.

McKenzie, J.R., B. Parsons, A.C. Seitz, R.K. Kopf, M.G. Mesa, and Q. Phelps, eds. 2012. *Advances in Fish Tagging and Marking Technology*. American Fisheries Society Symposium 76. Bethesda, MD: American Fisheries Society.

McKinley, A.C., L. Ryan, M.A. Coleman, N.A. Knott, G. Clark, M.D. Taylor, and E.L. Johnston. 2011. "Putting Marine Sanctuaries into Context: A Comparison of Estuary Fish Assemblages over Multiple Levels of Protection and Modification." *Aquatic Conservation: Marine and Freshwater Ecosystems* 21 (7): 636–48. http://dx.doi.org/10.1002/aqc.1223.

McLusky, D.S., and M. Elliott. 2004. *The Estuarine Ecosystem: Ecology, Threats, and Management*. Oxford: Oxford University Press. http://dx.doi.org/10.1093/acprof: oso/9780198525080.001.0001.

–. 2007. "Transitional Waters: A New Approach, Semantics or Just Muddying the Waters?" *Estuarine, Coastal and Shelf Science* 71 (3–4): 359–63. http://dx.doi.org/10.1016/j.ecss.2006.08.025.

McMahon, T.E., and G.F. Hartman. 1988. "Variation in the Degree of Silvering of Wild Coho Salmon, *Oncorhynchus kisutch*, Smolts Migrating Seaward from Carnation Creek, British Columbia." *Journal of Fish Biology* 32 (6): 825–33. http://dx.doi.org/10.1111/j.1095-8649.1988.tb05426.x.

McMahon, T.E., and L.B. Holtby. 1992. "Behaviour, Habitat Use, and Movements of Coho Salmon (*Oncorhynchus kisutch*) Smolt during Seaward Migration." *Canadian Journal of Fisheries and Aquatic Sciences* 49 (7): 1478–85. http://dx.doi.org/10.1139/f92-163.

McNicol, R.E., and S.E. MacLellan. 2010. "Accuracy of Using Scales to Age Mixed-Stock Chinook Salmon of Hatchery Origin." *Transactions of the American Fisheries Society* 139 (3): 727–34. http://dx.doi.org/10.1577/T09-033.1.

McPhail, J.D. 2007. *Freshwater Fishes of British Columbia*. Edmonton: University of Alberta Press.

Menzies, C.R., and C. Butler. 2006. "Introduction – Understanding Traditional Ecological Knowledge." In *Traditional Ecological Knowedge and Natural Resource Management*, ed. C.R. Menzies, 1–17. Lincoln: University of Nebraska Press.

Merrell, T.R., and K.V. Koski. 1978. "Habitat Values of Coastal Wetlands for Pacific Coast Salmonids." In *Proceedings of Symposium on Wetland Functions and Values: The State of Our Understanding*, ed. P.E. Greeson, J.R. Clark, and J.E. Clark, 256–66. Minneapolis: American Water Resources Association.

Methven, D.A., R.L. Haedrich, and G.A. Rose. 2001. "The Fish Assemblage of a Newfoundland Estuary: Diel, Monthly and Annual Variation." *Estuarine, Coastal and Shelf Science* 52 (6): 669–87. http://dx.doi.org/10.1006/ecss.2001.0768.

Meybeck, M. 2003. "Global Analysis of River Systems: From Earth System Controls to Anthropocene Syndromes." *Philosophical Transactions of the Royal Society of London: Biological Sciences* 358 (1440): 1935–55. http://dx.doi.org/10.1098/rstb.2003.1379.

Michener, R.H., and D.M. Schell. 1994. "Stable Isotope Ratios as Tracers in Marine Aquatic Food Webs." In *Stable Isotopes in Ecology and Environmental Science*, ed. K. Lajtha and R.H. Michener, 138–57. Oxford: Blackwell Scientific Publications.

Middlemas, S.J., T.R. Barton, J.D. Armstrong, and P.M. Thompson. 2006. "Functional and Aggregative Responses of Harbour Seals to Changes in Salmonid Abundance." *Proceedings of the Royal Society of London: Biological Sciences* 273 (1583): 193–98. http://dx.doi.org/10.1098/rspb.2005.3215.

Middlemas, S.J., D.C. Stewart, S. Mackay, and J.D. Armstrong. 2009. "Habitat Use and Dispersal of Post-Smolt Sea Trout *Salmo trutta* in a Scottish Sea Loch System." *Journal of Fish Biology* 74 (3): 639–51. http://dx.doi.org/10.1111/j.1095-8649.2008.02154.x.

Middlemas, S.J., R.J. Fryer, D. Tulett, and J.D. Armstrong. 2013. "Relationship between Sea Lice Levels on Sea Trout and Fish Farm Activity in Western Scotland." *Fisheries Management and Ecology* 20 (1): 68–74. http://dx.doi.org/10.1111/fme.12010.

Mikhailov, V.N., and S.L. Gorin. 2012. "New Definitions, Regionalization, and Typification of River Mouth Areas and Estuaries as Their Parts." *Water Resources* 39 (3): 247–60. http://dx.doi.org/10.1134/S0097807812030050.

Miller, B.A., and S. Sadro. 2003. "Residence Time and Seasonal Movements of Juvenile Coho Salmon in the Ecotone and Lower Estuary of Winchester Creek, South Slough, Oregon." *Transactions of the American Fisheries Society* 132 (3): 546–59. http://dx.doi.org/10.1577/1548-8659(2003)132<0546:RTASMO>2.0.CO;2.

Miller, J.A. 2011. "Effects of Water Temperature and Barium Concentration on Otolith Composition along a Salinity Gradient: Implications for Migratory Reconstructions." *Journal of Experimental Marine Biology and Ecology* 405 (1–2): 42–52. http://dx.doi.org/10.1016/j.jembe.2011.05.017.

Miller, J.A., A. Gray, and J. Merz. 2010. "Quantifying the Contribution of Juvenile Migratory Phenotypes in a Population of Chinook Salmon *Oncorhynchus tshawytscha*." *Marine Ecology Progress Series* 408: 227–40. http://dx.doi.org/10.3354/meps08613.

Miller, J.A., and C.A. Simenstad. 1997. "A Comparative Assessment of a Natural and Created Estuarine Slough as Rearing Habitat for Juvenile Chinook and Coho Salmon." *Estuaries* 20 (4): 792–806. http://dx.doi.org/10.2307/1352252.

Miller, J.A., D.J. Teel, A. Baptista, C.A. Morgan, and Michael Bradford. 2013. "Disentangling Bottom-up and Top-down Effects on Survival during Early Ocean Residence in a Population of Chinook Salmon (*Oncorhynchus tshawytscha*)." *Canadian Journal of Fisheries and Aquatic Sciences* 70 (4): 617–29. http://dx.doi.org/10.1139/cjfas-2012-0354.

Miller, K.M., S. Li, K.H. Kaukinen, N. Ginther, E. Hammill, J.M. Curtis, D.A. Patterson, T. Sierocinski, L. Donnison, P. Pavlidis, et al. 2011. "Genomic Signatures Predict Migration and Spawning Failure in Wild Canadian Salmon." *Science* 331 (6014): 214–17. http://dx.doi.org/10.1126/science.1196901.

Miller, R.J., and E.L. Brannon. 1982. "The Origin and Development of Life History Patterns in Pacific Salmonidae." In *Proceedings of Salmon and Trout Migratory Behavior Symposium*, ed. E.L. Brannon and E.O. Salo, 296–309. Seattle: University of Washington.

Milliman, J.D., and K.L. Farnsworth. 2011. *River Discharge to the Coastal Ocean: A Global Synthesis*. Cambridge: Cambridge University Press. http://dx.doi.org/10.1017/CBO9780511781247.

Milliman, J.D., and J.P.M. Syvitski. 1992. "Geomorphic/Tectonic Control of Sediment Discharge to the Ocean: The Importance of Small Mountainous Rivers." *Journal of Geology* 100 (5): 525–44. http://dx.doi.org/10.1086/629606.

Mills, D.H. 1989. *Ecology and Management of Atlantic Salmon*. London: Chapman and Hall.

Mills, E.L. 1969. "The Community Concept in Marine Zoology, with Comments on Continua and Instability in Some Marine Communities: A Review." *Journal of the Fisheries Board of Canada* 26 (6): 1415–28. http://dx.doi.org/10.1139/f69-132.

Milne, G. 1936. "Normal Erosion as a Factor in Soil Profile Development." *Nature* 138 (3491): 548–49. http://dx.doi.org/10.1038/138548c0.

Mina, M.V. 1991. "Problems of Protection of Fish Faunas in the USSR." *Netherlands Journal of Zoology* 42 (2): 200–13. http://dx.doi.org/10.1163/156854291X00289.

Mitchell, E. 2003. *Fly Rodding Estuaries: How to Fish Salt Ponds, Coastal Rivers and Tidal Creeks*. Mechanicsburg, PA: Stackpole Books.

Moksness, E., E. Dahl, and J.G. Støttrup, eds. 2013. *Global Challenges in Integrated Coastal Zone Management*. Oxford: Wiley-Blackwell Publishing. http://dx.doi.org/10.1002/9781118496480.

Montgomery, D.R. 2000. "Coevolution of the Pacific Salmon and Pacific Rim Topography." *Geology* 28 (12): 1107–10. http://dx.doi.org/10.1130/0091-7613(2000)28<1107:COTPSA>2.0.CO;2.

Montgomery, W.L., S.D. McCormick, R.J. Naiman, F.G. Whoriskey Jr., and G.A. Black. 1983. "Spring Migratory Synchrony of Salmonid, Catostomid, and Cyprinid Fishes in Rivière á la Truite, Québec." *Canadian Journal of Zoology* 61 (11): 2495–2502. http://dx.doi.org/10.1139/z83-331.

Moore, A., A.P. Scott, N. Lower, I. Katsiadaki, and L. Greenwood. 2003. "The Effects of 4-Nonylphenol and Atrazine on Atlantic Salmon (*Salmo salar* L.) Smolts." *Aquaculture* (Amsterdam) 222 (1–4): 253–63. http://dx.doi.org/10.1016/S0044-8486(03)00126-1.

Moore, M., B.A. Berejikian, and E.P. Tezak. 2013. "A Floating Bridge Disrupts Seaward Migration and Increases Mortality of Steelhead Smolts in Hood Canal, Washington State." *PLoS One* 8 (9): e73427. http://dx.doi.org/10.1371/journal.pone.0073427.

Moran, P., D.J. Teel, M.A. Banks, T.D. Beacham, M.R. Bellinger, S.M. Blankenship, John R. Candy, John Carlos Garza, Jon E. Hess, Shawn R. Narum, et al. 2013. "Divergent Life-History Races Do Not Represent Chinook Salmon Coast-wide: The Importance of Scale in Quaternary Biogeography." *Canadian Journal of Fisheries and Aquatic Sciences* 70 (3): 415–35. http://dx.doi.org/10.1139/cjfas-2012-0135.

Morgan, C.A., J.R. Cordell, and C.A. Simenstad. 1997. "Sink or Swim? Copepod Population Maintenance in the Columbia River Estuarine Turbidity-Maxima Region." *Marine Biology* 129 (2): 309–17. http://dx.doi.org/10.1007/s002270050171.

Morgan, J.D., and G.K. Iwama. 1998. "Salinity Effects on Oxygen Consumption, Gill Na^+, K^+, -ATPase and Ion Regulation in Juvenile Coho Salmon." *Journal of Fish Biology* 53 (5): 1110–19.

Morin, R., and J.J. Dodson. 1986. "The Ecology of Fishes in James Bay, Hudson Bay and Hudson Strait." In *Canadian Inland Seas* (Elsevier Oceanography Series, vol. 44), ed. I.P. Martini, 293–326. Amsterdam: Elsevier Science Publishers. http://dx.doi.org/10.1016/S0422-9894(08)70908-5.

Morita, K. 2001. "The Growth History of Anadromous White-Spotted Charr in Northern Japan: A Comparison between River and Sea Life." *Journal of Fish Biology* 59 (6): 1556–65. http://dx.doi.org/10.1111/j.1095-8649.2001.tb00220.x.

Morita, K., S. Yamamoto, and N. Hoshino. 2000. "Extreme Life History Change of White-Spotted Char (*Salvelinus leucomaenis*) after Damming." *Canadian Journal of Fisheries and Aquatic Sciences* 57 (6): 1300–6. http://dx.doi.org/10.1139/f00-050.

Morita, K., T. Saito, Y. Miyakoshi, M.A. Fukuwaka, T. Nagasawa, and M. Kaeriyama. 2006. "A Review of Pacific Salmon Hatchery Programmes on Hokkaido Island, Japan." *ICES Journal of Marine Science* 63 (7): 1353–63. http://dx.doi.org/10.1016/j.icesjms.2006.03.024.

Moriyama, S., F.G. Ayson, and H. Kawauchi. 2000. "Growth Regulation by Insulin-like Growth Factor-I in Fish." *BioScience, Biotechnology, and Biochemistry* 64 (8): 1553–62. http://dx.doi.org/10.1271/bbb.64.1553.

Morley, S.A., J.D. Toft, and K.M. Hanson. 2012. "Ecological Effects of Shoreline Armoring on Intertidal Habitats of a Puget Sound Urban Estuary." *Estuaries and Coasts* 35 (3): 774–84. http://dx.doi.org/10.1007/s12237-012-9481-3.

Morris, M.R., D.J. Fraser, A.J. Heggelin, F.G. Whoriskey, J.W. Carr, S.F. O'Neil, and J.A. Hutchings. 2008. "Prevalence and Recurrence of Escaped Farmed Atlantic Salmon (*Salmo salar*) in Eastern North American Rivers." *Canadian Journal of Fisheries and Aquatic Sciences* 65 (12): 2807–26. http://dx.doi.org/10.1139/F08-181.

Mörth, C.M., C. Humborg, H. Eriksson, Å. Danielsson, M. Rodriguez Medina, S. Löfgren, et al. 2007. "Modeling Riverine Nutrient Transport to the Baltic Sea: A Large-Scale Approach." *AMBIO: A Journal of the Human Environment* 36 (2): 124–33. http://dx.doi.org/10.1579/0044-7447(2007)36[124:MRNTTT]2.0.CO;2.

Moser, M.L., A.F. Olson, and T.P. Quinn. 1991. "Riverine and Estuarine Migratory Behavior of Coho Salmon (*Oncorhynchus kisutch*) Smolts." *Canadian Journal of Fisheries and Aquatic Sciences* 48 (9): 1670–78. http://dx.doi.org/10.1139/f91-198.

Mote, P.W., and E.P. Salathé. 2010. "Future Climate in the Pacific Northwest." *Climatic Change* 102 (1–2): 29–50. http://dx.doi.org/10.1007/s10584-010-9848-z.

Mothersill, C., C. Bucking, R.W. Smith, N. Agnihotri, A. O'Neill, M. Kilemade, and C.B. Seymour. 2006. "Communication of Radiation-Induced Stress or Bystander Signals between Fish in Vivo." *Environmental Science and Technology* 40 (21): 6859–64. http://dx.doi.org/10.1021/es061099y.

Moyle, P.B. 1999. "Effects of Invading Species on Freshwater and Estuarine Ecosystems." In *Invasive Species and Biodiversity Management*, ed. O.T. Sandlund, P.J. Schei, and A. Viken, 177–91. Dordrecht, Netherlands: Kluwer Academic. http://dx.doi.org/10.1007/978-94-011-4523-7_12.

–. 2014. "Novel Aquatic Ecosystems: The New Reality for Streams in California and Other Mediterranean Climate Regions." *River Research and Applications* 30 (10): 1335–44. http://dx.doi.org/10.1002/rra.2709.

Moyle, P.B., J.R. Lund, W.A. Bennett, and W.E. Fleenor. 2010. "Habitat Variability and Complexity in the Upper San Francisco Estuary." *San Francisco Estuary and Watershed Science* 8 (3): 1–24.

Munro, J.A., and W.A. Clemens. 1937. *The American Merganser in British Columbia and Its Relation to the Fish Population.* Biological Board of Canada Bulletin 55. Ottawa: Biological Board of Canada.

Munsch, S.H., J.R. Cordell, J.D. Toft, and E.E. Morgan. 2014. "Effects of Seawalls and Piers on Fish Assemblages and Juvenile Salmon Feeding Behavior." *North American Journal of Fisheries Management* 34 (4): 814–27. http://dx.doi.org/10.1080/02755947.2014.910579.

Murakami, M., Y. Oonishi, and H. Kunishi. 1985. "A Numerical Simulation of the Distribution of Water Temperature and Salinity in the Seto Inland Sea." *Journal of the Oceanographical Society of Japan* 41 (4): 213–24. http://dx.doi.org/10.1007/BF02109271.

Muraoka, K., K. Amano, and J. Miwa. 2011. "Effects of Suspended Solids Concentration and Particle Size on Survival and Gill Structure in Fish." In *Proceedings of the 34th World Congress of the International Association for Hydro-Environment Research and Engineering: 33rd Hydrology and Water Resources Symposium and 10th Conference on Hydraulics in Water Engineering*, ed. E.M. Valentine, C.J. Apelt, J. Ball, H. Chanson, R. Cox, R. Ettema, et al., 2893–2900. Barton, Australia: ACT Engineers. http://search.informit.com.au/document Summary;dn=359462576937081;res=IELENG.

Murphy, M.L. 1984. "Primary Production and Grazing in Freshwater and Intertidal Reaches of a Coastal Stream, Southeast Alaska." *Limnology and Oceanography* 29 (4): 805–15. http://dx.doi.org/10.4319/lo.1984.29.4.0805.

–. 1985. "Die-offs of Pre-Spawn Adult Pink Salmon and Chum Salmon in Southeastern Alaska." *North American Journal of Fisheries Management* 5 (2B): 302–8. http://dx.doi.org/10.1577/1548-8659(1985)5<302:DOPAPS>2.0.CO;2.

Murphy, M.L., S.W. Johnson, and D.J. Csepp. 2000. "A Comparison of Fish Assemblages in Eelgrass and Adjacent Subtidal Habitats near Craig, Alaska." *Alaska Fishery Research Bulletin* 7: 11–21.

Murphy, M.L., J.F. Thedinga, and K.V. Koski. 1988. "Size and Diet of Juvenile Pacific Salmon during Seaward Migration through a Small Estuary in Southeastern Alaska USA." *Fish Bulletin* 86 (2): 213–22.

Murphy, M.L., J. Heifetz, J.F. Thedinga, S.W. Johnson, and K.V. Koski. 1989. "Habitat Utilization by Juvenile Pacific Salmon (*Oncorhynchus*) in the Glacial Taku River, Southeast Alaska." *Canadian Journal of Fisheries and Aquatic Sciences* 46 (10): 1677–85. http://dx.doi.org/10.1139/f89-213.

Myers, K.W. 1978. "Comparative Analysis of Stomach Contents of Cultured and Wild Juvenile Salmonids in Yaquina Bay, Oregon." In *Fish Food Habits Studies: Proceedings of the 2nd Pacific Northwest Technical Workshop, Maple Valley, Washington, October 10–13, 1978*, ed. S.J. Lipovsky and C.A. Simenstad, 155–62. Seattle: Washington Sea Grant.

Myers, K.W., and H.F. Horton. 1982. "Temporal Use of an Oregon Estuary by Hatchery and Wild Juvenile Salmon." In *Estuarine Comparisons*, ed. V.S. Kennedy, 377–92. New York: Academic Press. http://dx.doi.org/10.1016/B978-0-12-404070-0.50028-4.

Nagasawa, K. 2004. "Sea Lice, *Lepeophtheirus salmonis* and *Caligus orientalis* (Copepoda: Caligidae), of Wild and Farmed Fish in Sea and Brackish Waters of Japan and Adjacent Regions: A Review." *Zoological Studies* (Taipei) 43 (2): 173–78.

Nagata, M., and M. Kaeriyama. 2003. "Salmonid Status and Conservation in Japan." In *Proceedings of Speaking for the Salmon: The World Summit on Salmon, June 10–13, 2003*, ed. P. Gallaugher and L. Wood, 89–98. Burnaby, BC: Simon Fraser University.

Nagrodski, A., G.D. Raby, C.T. Hasler, M.K. Taylor, and S.J. Cooke. 2012. "Fish Stranding in Freshwater Systems: Sources, Consequences, and Mitigation." *Journal of Environmental Management* 103: 133–41. http://dx.doi.org/10.1016/j.jenvman.2012.03.007.

Naiman, R.J., J.R. Alldredge, D.A. Beauchamp, P.A. Bisson, J. Congleton, C.J. Henny, N. Huntly, R. Lamberson, C. Levings, E.N. Merrill, et al. 2012. "Developing a Broader Scientific Foundation for River Restoration: Columbia River

Food Webs." *Proceedings of the National Academy of Sciences of the United States of America* 109 (52): 21201–7. http://dx.doi.org/10.1073/pnas.1213408109.

Naiman, R.J., and J.R. Sibert. 1978. "Transport of Nutrients and Carbon from the Nanaimo River to Its Estuary." *Limnology and Oceanography* 23 (6): 1183–93. http://dx.doi.org/10.4319/lo.1978.23.6.1183.

Naiman, R.J., and J.J. Latterell. 2005. "Principles for Linking Fish Habitat to Fisheries Management and Conservation." *Journal of Fish Biology* 67 (sB): 166–85. http://dx.doi.org/10.1111/j.0022-1112.2005.00921.x.

Nakamura, F., T. Sudo, S. Kameyama, and M. Jitsu. 1997. "Influences of Channelization on Discharge of Suspended Sediment and Wetland Vegetation in Kushiro Marsh, Northern Japan." *Geomorphology* 18 (3–4): 279–89. http://dx.doi.org/10.1016/S0169-555X(96)00031-1.

Naman, S.W., and C.S. Sharpe. 2012. "Predation by Hatchery Yearling Salmonids on Wild Subyearling Salmonids in the Freshwater Environment: A Review of Studies, Two Case Histories, and Implications for Management." *Environmental Biology of Fishes* 94 (1): 21–28. http://dx.doi.org/10.1007/s10641-011-9819-x.

National Oceanic and Atmospheric Administration. 2014a. "Welcome to Estuaries." http://oceanservice.noaa.gov/education/kits/estuaries/lessons/estuaries_tutorial.pdf.

–. 2014b. "National Estuarine Research Reserves System." http://nerrs.noaa.gov/.

Natural Capital Project. 2014. "The Natural Capital Project." http://www.naturalcapitalproject.org/InVEST.html.

Nature Trust of British Columbia. 2013. "The Pacific Estuary Conservation Program." http://www.naturetrust.bc.ca/about-us/partners/programs/.

Naughton, G.P., M.L. Keefer, T.S. Clabough, M.A. Jepson, S.R. Lee, C.A. Peery, C.C. Caudill, and Michael Bradford. 2011. "Influence of Pinniped-Caused Injuries on the Survival of Adult Chinook Salmon (*Oncorhynchus tshawytscha*) and Steelhead Trout (*Oncorhynchus mykiss*) in the Columbia River Basin." *Canadian Journal of Fisheries and Aquatic Sciences* 68 (9): 1615–24. http://dx.doi.org/10.1139/f2011-064.

Navodaru, I., M. Staras, and I. Cernisencu. 2001. "The Challenge of Sustainable Use of the Danube Delta Fisheries, Romania." *Fisheries Management and Ecology* 8 (4 5): 323–32. http://dx.doi.org/10.1046/j.1365-2400.2001.00257.x.

Neill, S.R., and J.M. Cullen. 1974. "Experiments on Whether Schooling by Their Prey Affects the Hunting Behaviour of Cephalopods and Fish Predators." *Journal of Zoology* 172 (4): 549–69. http://dx.doi.org/10.1111/j.1469-7998.1974.tb04385.x.

Neilson, J.D., G.H. Geen, and D. Bottom. 1985. "Estuarine Growth of Juvenile Chinook Salmon (*Oncorhynchus tshawytscha*) as Inferred from Otolith Microstructure." *Canadian Journal of Fisheries and Aquatic Sciences* 42 (5): 899–908. http://dx.doi.org/10.1139/f85-114.

Nelson, J.S. 1994. *Fishes of the World*. 3rd ed. New York: John Wiley and Sons.

Netboy, A. 1980. *Salmon – The World's Most Harassed Fish*. Tulsa, OK: Winchester Press.

Newman, K.B., and J. Rice. 2002. "Modeling the Survival of Chinook Salmon Smolts Outmigrating through the Lower Sacramento River System." *Journal of the American Statistical Association* 97 (460): 983–93. http://dx.doi.org/10.1198/016214502388618771.

Newson, M., D. Sear, and C. Soulsby. 2012. "Incorporating Hydromorphology in Strategic Approaches to Managing Flows for Salmonids." *Fisheries Management and Ecology* 19 (6): 490–99. http://dx.doi.org/10.1111/j.1365-2400.2011.00822.x.

Ng, S.H., C.G. Artieri, I.E. Bosdet, R. Chiu, R.G. Danzmann, W.S. Davidson, Moira M. Ferguson, Christopher D. Fjell, Bjorn Hoyheim, Steven J.M. Jones, et al. 2005. "A Physical Map of the Genome of Atlantic Salmon, *Salmo salar*." *Genomics* 86 (4): 396–404. http://dx.doi.org/10.1016/j.ygeno.2005.06.001.

Nicholas, J.W. 1997. "On the Nature of Data and Their Role in Salmon Conservation." In *Pacific Salmon and Their Ecosystems: Status and Future Options*, ed. D.J. Stouder, P.A. Bisson, and R.J. Naiman, 53–60. New York: Chapman and Hall. http://dx.doi.org/10.1007/978-1-4615-6375-4_6.

Nicholas, J.W., and D.G. Hankin. 1988. *Chinook Salmon Populations in Oregon Coastal Basins: Description of Life Histories and Assessment of Recent Trends in Run Strengths*. Corvallis: Oregon Department of Fish and Wildlife.

Nichols, K.M., A.F. Edo, P.A. Wheeler, and G.H. Thorgaard. 2008. "The Genetic Basis of Smoltification-Related Traits in *Oncorhynchus mykiss*." *Genetics* 179 (3): 1559–75. http://dx.doi.org/10.1534/genetics.107.084251.

Nicolas, D., J. Lobry, O. Le Pape, and P. Boët. 2010. "Functional Diversity in European Estuaries: Relating the Composition of Fish Assemblages to the Abiotic Environment." *Estuarine, Coastal and Shelf Science* 88 (3): 329–38. http://dx.doi.org/10.1016/j.ecss.2010.04.010.

Nielsen, J.L., G.T. Ruggerone, and C.E. Zimmerman. 2013. "Adaptive Strategies and Life History Characteristics in a Warming Climate: Salmon in the Arctic?" *Environmental Biology of Fishes* 96 (10–11): 1187–226. http://dx.doi.org/10.1007/s10641-012-0082-6.

Niemelä, E., T.S. Makinen, K. Moen, E. Hassinen, J. Erkinaro, M. Länsman, and M. Julkunen. 2000. "Age, Sex Ratio and Timing of the Catch of Kelts and Ascending Atlantic Salmon in the Subarctic River Teno." *Journal of Fish Biology* 56 (4): 974–85. http://dx.doi.org/10.1111/j.1095-8649.2000.tb00886.x.

Niemi, G.J., and M.E. McDonald. 2004. "Application of Ecological Indicators." *Annual Review of Ecology Evolution and Systematics* 35 (1): 89–111. http://dx.doi.org/10.1146/annurev.ecolsys.35.112202.130132.

Niksirat, H., and A. Abdoli. 2009. "On the Status of the Critically Endangered Caspian Brown Trout, *Salmo trutta caspius*, during Recent Decades in the Southern Caspian Sea Basin (Osteichthyes: Salmonidae)." *Zoology in the Middle East* 46 (1): 55–60. http://dx.doi.org/10.1080/09397140.2009.10638328.

Nilssen, K.J., O.A. Gulseth, M. Iversen, and R. Kjùl. 1997. "Summer Osmoregulatory Capacity of the World's Northernmost Living Salmonid." *American Journal of Physiology* 272 (3 Pt 2): R743–R749.

Nislow, K. 2010. "Riparian Management: Alternative Paradigms and Implications for Wild Salmon." In *Salmonid Fisheries: Freshwater Habitat Management*, ed. P. Kemp, 164–82. Oxford: Wiley-Blackwell Publishing. http://dx.doi.org/10.1002/9781444323337.ch7.

Nobriga, M.L., F. Feyrer, R.D. Baxter, and M. Chotkowski. 2005. "Fish Community Ecology in an Altered River Delta: Spatial Patterns in Species Composition, Life History Strategies, and Biomass." *Estuaries* 28 (5): 776–85. http://dx.doi.org/10.1007/BF02732915.

Noga, E.J. 2010. *Fish Disease: Diagnosis and Treatment.* Oxford: Wiley-Blackwell Publishing. http://dx.doi.org/10.1002/9781118786758.

Nordeng, H. .2009. "Char Ecology: Natal Homing in Sympatric Populations of Anadromous Arctic Char *Salvelinus alpinus* (L.), Roles of Pheromone Recognition." *Ecology of Freshwater Fish* 18 (1): 41–51. http://dx.doi.org/10.1111/j.1600-0633.2008.00320.x.

North Atlantic Salmon Conservation Organization. 2009. *Protection, Restoration and Enhancement of Salmon Habitat: Focus Area Report, Norway.* http://www.nasco.int/pdf/far_habitat/HabitatFAR_Norway.pdf.

Northcote, T.G., R.S. Gregory, and C. Magnhagen. 2007. *Contrasting Space and Food Use among Three Species of Juvenile Pacific Salmon* (Oncorhynchus) *Cohabiting Tidal Marsh Channels of a Large Estuary.* Canadian Technical Report of Fisheries and Aquatic Sciences, No. 2759. Ottawa: Fisheries and Oceans Canada.

Northcote, T.G., N.T. Johnston, and K. Tsumura. 1979. *Feeding Relationships and Food Web Structure of the Lower Fraser River Fishes.* Technical Report No. 16. Vancouver: Westwater Research Centre.

Novomodnyi, G.V., and V.A. Belyaev. 2002. "Predation by Lamprey Smolts *Lampetra japonica* as a Main Cause of Amur Chum Salmon and Pink Salmon Mortality in the Early Sea Period of Life." In *Proceedings of 2002 Joint Meeting on Causes of Marine Mortality of Salmon in the North Pacific and North Atlantic Oceans and in the Baltic Sea* (North Pacific Anadromous Fish Commission Technical Report No. 4), ed. Y. Ishida and M. Windsor, 81–82. Vancouver: North Pacific Anadromous Fish Commission.

Novomodnyi, G., P. Sharov, and S. Zolotukhin. 2004. *Amur Fish: Wealth and Crisis.* Vladivostok: World Wildlife Federation – Russian Far East.

Null, R.E., K.S. Niemelä, and S.F. Hamelberg. 2013. "Post-Spawn Migrations of Hatchery-Origin *Oncorhynchus mykiss* Kelts in the Central Valley of California." *Environmental Biology of Fishes* 96 (2–3): 341–53. http://dx.doi.org/10.1007/s10641-012-0075-5.

O'Hara, G. 2007. "Water Management Aspects of the Britannia Mine Remediation Project, British Columbia, Canada." *Mine Water and the Environment* 26 (1): 46–54. http://dx.doi.org/10.1007/s10230-007-0148-4.

O'Malley, K.G., M.J. Ford, and J.J. Hard. 2010. "Clock Polymorphism in Pacific Salmon: Evidence for Variable Selection along a Latitudinal Gradient." *Proceedings of the Royal Society B: Biological Sciences* 277 (1701): 3703–14.

O'Neal, S.L., and J.A. Stanford. 2011. "Partial Migration in a Robust Brown Trout Population of a Patagonian River." *Transactions of the American Fisheries Society* 140 (3): 623–35. http://dx.doi.org/10.1080/00028487.2011.585577.

Odenkirk, J., and S. Owens. 2007. "Expansion of a Northern Snakehead Population in the Potomac River System." *Transactions of the American Fisheries Society* 136 (6): 1633–39. http://dx.doi.org/10.1577/T07-025.1.

Odum, W.E. 1970. "Insidious Alteration of the Estuarine Environment." *Transactions of the American Fisheries Society* 99 (4): 836–47. http://dx.doi.org/10.1577/1548-8659(1970)99<836:IAOTEE>2.0.CO;2.

–. 1988. "Comparative Ecology of Tidal Freshwater and Salt Marshes." *Annual Review of Ecology and Systematics* 19 (1): 147–76. http://dx.doi.org/10.1146/annurev.es.19.110188.001051.

Ohji, M., T. Arai, and N. Miyazaki. 2007. "Comparison of Organotin Accumulation in the Masu Salmon (*Oncorhynchus masou*) Accompanying Migratory Histories." *Estuarine, Coastal and Shelf Science* 72 (4): 721–31. http://dx.doi.org/10.1016/j.ecss.2006.12.004.

Ohji, M., H. Harino, and T. Arai. 2011. "Differences in Organotin Accumulation in Relation to Life History in the White-Spotted Charr *Salvelinus leucomaenis*." *Marine Pollution Bulletin* 62 (2): 318–26. http://dx.doi.org/10.1016/j.marpolbul.2010.10.008.

Okada, S., and A. Taniguchi. 1971. "Size Relationship between Salmon Juveniles in Shore Waters and Their Prey Animals." *Bulletin of the Faculty of Fisheries, Hokkaido University* 22 (1): 30–36.

Olden, J.D., and D.A. Jackson. 2002. "A Comparison of Statistical Approaches for Modelling Fish Species Distributions." *Freshwater Biology* 47 (10): 1976–95. http://dx.doi.org/10.1046/j.1365-2427.2002.00945.x.

Olson, A.F., and T.P. Quinn. 1993. "Vertical and Horizontal Movements of Adult Chinook Salmon *Oncorhynchus tshawytscha* in the Columbia River Estuary." *Fish Bulletin* 91 (1): 171–78.

Ono, K., and C.A. Simenstad. 2014. "Reducing the Effect of Overwater Structures on Migrating Juvenile Salmon: An Experiment with Light." *Ecological Engineering* 71: 180–89. http://dx.doi.org/10.1016/j.ecoleng.2014.07.010.

Orsi, J.A., and A.C. Wertheimer. 1995. "Marine Vertical Distribution of Juvenile Chinook and Coho Salmon in Southeastern Alaska." *Transactions of the American Fisheries Society* 124 (2): 159–69. http://dx.doi.org/10.1577/1548-8659(1995)124<0159:MVDOJC>2.3.CO;2.

Otte, G., and C.D. Levings. 1975. *Distribution of Macro Invertebrate Communities on a Mud Flat Influenced by Sewage, Fraser River Estuary, British Columbia*. Technical Report No. 476. West Vancouver: Research and Development Directorate, Pacific Environment Institute.

Otto, R.G. 1971. "Effects of Salinity on the Survival and Growth of Pre-Smolt Coho Salmon (*Oncorhynchus kisutch*)." *Journal of the Fisheries Board of Canada* 28 (3): 343–49. http://dx.doi.org/10.1139/f71-046.

Pacific North Coast Integrated Management Area. 2014. "Pacific North Coast Integrated Management Area." http://www.pncima.org/.

Pacific Northwest Aquatic Monitoring Partnership. 2014. "Pacific Northwest Aquatic Monitoring Partnership." http://www.pnamp.org/.

Pacific Northwest Hatchery Scientific Review Group. 2009. "Columbia River Hatchery Reform System-wide Report." http://www.hatcheryreform.us/hrp/reports/system/welcome_show.action.

Pacific Salmon Commission. 2008. *Twenty-First Annual Report 2005/2006*. Vancouver: Pacific Salmon Commission.

Paling, E.I., M. Fonseca, M.M. van Katwijk, and M. van Keulen. 2009. "Seagrass Restoration." In *Coastal Wetlands: An Integrated Ecosystems Approach, Seagrass Restoration*, ed. G.M.E. Perillo, E. Wolanski, D.R. Cahoon, and M. Brinson, 687–713. Amsterdam: Elsevier.

Palm, R.C., D.B. Powell, A. Skillman, and K. Godtfredsen. 2003. "Immunocompetence of Juvenile Chinook Salmon against *Listonella anguillarum* Following Dietary Exposure to Polycyclic Aromatic Hydrocarbons." *Environmental Toxicology and Chemistry* 22 (12): 2986–94. http://dx.doi.org/10.1897/02-561.

Palmer, M.A. 2009. "Reforming Watershed Restoration: Science in Need of Application and Applications in Need of Science." *Estuaries and Coasts* 32 (1): 1–17. http://dx.doi.org/10.1007/s12237-008-9129-5.

Panov, V.E., A.F. Alimov, S.M. Golubkov, M.I. Orlova, and I.V. Telesh. 2008. "Environmental Problems and Challenges for Coastal Zone Management in the Neva Estuary (Eastern Gulf of Finland)." In *Ecology of Baltic Coastal Waters. Vol. 197: Ecological Studies*, ed. U. Schiewer, 171–84. New York: Springer.

Pardo, I., and P.D. Armitage. 1997. "Species Assemblages as Descriptors of Mesohabitats." *Hydrobiologia* 344 (1): 111–28. http://dx.doi.org/10.1023/A:1002958412237.

Parker, R.R. 1971. "Size Selective Predation among Juvenile Salmonid Fishes in a British Columbia Inlet." *Journal of the Fisheries Research Board of Canada* 28 (10): 1503–10. http://dx.doi.org/10.1139/f71-231.

Parks Canada Agency. 2000. *"Unimpaired for Future Generations"? Protecting Ecological Integrity with Canada's National Parks. Vol.1: A Call to Action. Report of the Panel on the Ecological Integrity of Canada's National Parks*. Catalogue No. R62–323/2000–1. Ottawa: Minister of Public Works and Government Services.

Parrish, J.D., D.P. Braun, and R.S. Unnasch. 2003. "Are We Conserving What We Say We Are? Measuring Ecological Integrity within Protected Areas." *BioScience* 53 (9): 851–60. http://dx.doi.org/10.1641/0006-3568(2003)053[0851:AWCWWS]2.0.CO;2.

Parsons, T.R. 1982. "The Future of Controlled Ecosystem Enclosure Experiments." In *Marine Mesocosms: Biological and Chemical Research in Experimental Ecosystems*, ed. G.D. Grice and M.R. Reeve, 411–18. New York: Springer-Verlag. http://dx.doi.org/10.1007/978-1-4612-5645-8_30.

Pascual, M., P. Bentzen, C. Riva Rossi, G. Mackey, M.T. Kinnison, and R. Walker. 2001. "First Documented Case of Anadromy in a Population of Introduced Rainbow Trout in Patagonia, Argentina." *Transactions of the American Fisheries Society* 130 (1): 53–67. http://dx.doi.org/10.1577/1548-8659(2001)130<0053:FDCOAI>2.0.CO;2.

Pasquaud, S., J. Lobry, and P. Elie. 2007. "Facing the Necessity of Describing Estuarine Ecosystems: A Review of Food Web Ecology Study Techniques." *Hydrobiologia* 588 (1): 159–72. http://dx.doi.org/10.1007/s10750-007-0660-3.

Paszkowski, C.A., and B.L. Olla. 1985. "Foraging Behavior of Hatchery-Produced Coho Salmon (*Oncorhynchus kisutch*) Smolts on Live Prey." *Canadian Journal of Fisheries and Aquatic Sciences* 42 (12): 1915–21. http://dx.doi.org/10.1139/f85-237.

Pavey, S.A., J.L. Nielsen, R.H. MacKas, T.R. Hamon, and F. Breden. 2010. "Contrasting Ecology Shapes Juvenile Lake-Type and Riverine Sockeye Salmon." *Transactions of the American Fisheries Society* 139 (5): 1584–94. http://dx.doi.org/10.1577/T09-182.1.

Pavlov, D.S., and K.A. Savvaitova. 2008. "On the Problem of Ratio of Anadromy and Residence in Salmonids (Salmonidae)." *Journal of Ichthyology* 48 (9): 778–91. Translated from Russian; originally published in *Voprosy Ikhtiologii* 48 (6): 81–824. http://dx.doi.org/10.1134/S0032945208090099.

Pavlovskaya, L.P. 1995. "Fishery in the Lower Amu-Darya under the Impact of Irrigated Agriculture." In *Inland Fisheries under the Impact of Irrigated Agriculture: Central Asia* (Food and Agriculture Organization Fisheries Circular No. 894), ed. T. Petr, 42–57. Rome: Food and Agriculture Organization of the United Nations.

Pearcy, W.G. 1992. *Ocean Ecology of North Pacific Salmonids*. Seattle: Washington Sea Grant.

Pearcy, W.G., T. Nishiyama, T. Fujii, and K. Masuda. 1984. "Diel Variations in the Feeding Habits of Pacific Salmon Caught in Gill Nets during a 24-Hour Period in the Gulf of Alaska." *Fish Bulletin* 82 (2): 391–99.

Pearcy, W.G., C.D. Wilson, A.W. Chung, and J.W. Chapman. 1989. "Distribution, and Production of Juvenile Chum Salmon, *Oncorhynchus keta*, in Netarts Bay, Oregon." *Fish Bulletin* 87 (3): 553–68.

Peeler, E.J., and M.A. Thrush. 2004. "Qualitative Analysis of the Risk of Introducing *Gyrodactylus salaris* into the United Kingdom." *Diseases of Aquatic Organisms* 62 (1–2): 103–13. http://dx.doi.org/10.3354/dao062103.

Pennell, W., and B.A. Barton, eds. 1996. *Principles of Salmonid Culture*. Amsterdam: Elsevier.

Perrier, C., G. Evanno, J. Belliard, R. Guyomard, and J.L. Baglinière. 2010. "Natural Recolonization of the Seine River by Atlantic Salmon (*Salmo salar*) of Multiple Origins." *Canadian Journal of Fisheries and Aquatic Sciences* 67 (1): 1–4. http://dx.doi.org/10.1139/F09-190.

Perrings, C., A. Duraiappah, A. Larigauderie, and H. Mooney. 2011. "The Biodiversity and Ecosystem Services Science-Policy Interface." *Science* 331 (6021): 1139–40. http://dx.doi.org/10.1126/science.1202400.

Perry, R.W., J.R. Skalski, P.L. Brandes, P.T. Sandstrom, A.P. Klimley, A. Ammann, and B. MacFarlane. 2010. "Estimating Survival Migration Route Probabilities of Juvenile Chinook Salmon in Sacramento River Estuary." *North American Journal of Fisheries Management* 30 (1): 142–56. http://dx.doi.org/10.1577/M08-200.1.

Persson, L., and A.M. De Roos. 2006. "Food Dependent Individual Growth and Population Dynamics in Fishes." *Journal of Fish Biology* 69 (sc): 1–20. http://dx.doi.org/10.1111/j.1095-8649.2006.01269.x.

Persson, P., K. Sundell, B.T. Björnsson, and H. Lundqvist. 1998. "Calcium Metabolism and Osmoregulation during Sexual Maturation of River Running Atlantic Salmon." *Journal of Fish Biology* 52 (2): 334–49. http://dx.doi.org/10.1111/j.1095-8649.1998.tb00801.x.

Pess, G.R., T.P. Quinn, S.R. Gephard, and R. Saunders. 2014. "Re-colonization of Atlantic and Pacific Rivers by Anadromous Fishes: Linkages between Life History and the Benefits of Barrier Removal." *Reviews in Fish Biology and Fisheries* 24 (3): 881–900. http://dx.doi.org/10.1007/s11160-013-9339-1.

Peterson, D.P., and K.D. Fausch. 2003. "Testing Population Level Mechanisms of Invasion by a Mobile Vertebrate: A Simple Conceptual Framework for Salmonids in Streams." *Biological Invasions* 5 (3): 239–59. http://dx.doi.org/10.1023/A:1026155628599.

Peterson, W.T., R.D. Brodeur, and W.G. Pearcy. 1982. "Food Habits of Juvenile Salmon in the Oregon Coastal Zone, June 1979." *Fisheries Bulletin* 80 (4): 841–51.

Petitcodiac Watershed Alliance. 2014. "Petitcodiac Fish Recovery Coalition." http://petitcodiacwatershed.org/about-the-alliance/.

Piccolo, M.C., and G.M. Perillo. 1999. "The Argentina Estuaries: A Review." In *Estuaries of South America*, ed. G.M. Perillo, M.C. Piccolo, and M. Pino-Quivira, 101–32. Berlin: Springer. http://dx.doi.org/10.1007/978-3-642-60131-6_6.

Pickard, G.L., and W.J. Emery. 1990. *Descriptive Physical Oceanography: An Introduction*. 5th ed. Oxford: Pergamon Press.

Pickard, G.L., and B.R. Stanton. 1980. "Pacific Fjords – A Review of Their Water Characteristics." In *Fjord Oceanography: Proceedings from NATO Workshop on Fjord Oceanography, Sidney, Canada, June 4–9, 1979*, ed. H.J. Freeland, D. M. Farmer, and C. D. Levings, 1–51. New York: Plenum Press.

Pitcher, T.J., and J.K. Parrish. 1993. "Functions of Shoaling Behaviour in Teleosts." In *Behaviour of Teleost Fishes*, 2nd ed., ed. T.J. Pitcher, 363–439. London: Chapman and Hall. http://dx.doi.org/10.1007/978-94-011-1578-0_12.

Poff, N.L., J.D. Allan, M.B. Bain, J.R. Karr, K.L. Prestegaard, B.D. Richter, Richard E. Sparks, and Julie C. Stromberg. 1997. "The Natural Flow Regime." *BioScience* 47 (11): 769–84. http://dx.doi.org/10.2307/1313099.

Pokhrel, D., and T. Viraraghavan. 2004. "Treatment of Pulp and Paper Mill Wastewater – A Review." *Science of the Total Environment* 333 (1–3): 37–58. http://dx.doi.org/10.1016/j.scitotenv.2004.05.017.

Polis, G.A., W. Anderson, and R.D. Holt. 1997. "Towards an Integration of Landscape Ecology and Food Web Ecology: The Dynamics of Spatially Subsidized Food Webs." *Annual Review of Ecology and Systematics* 28 (1): 289–316. http://dx.doi.org/10.1146/annurev.ecolsys.28.1.289.

Polivka, K.M. 2005. "Resource Matching across Habitats Is Limited by Competition at Patch Scales in an Estuarine-Opportunist Fish." *Canadian Journal of Fisheries and Aquatic Sciences* 62 (4): 913–24. http://dx.doi.org/10.1139/f04-235.

Pomeroy, W.M., and C.D. Levings. 1980. "Association and Feeding Relationships between *Eogammarus confervicolus* (Amphipoda, Gammaridae) and Benthic Algae on Sturgeon and Roberts Banks, Fraser River Estuary." *Canadian Journal of Fisheries and Aquatic Sciences* 37 (1): 1–10. http://dx.doi.org/10.1139/f80-001.

Poole, W.R., C.J. Byrne, M.G. Dillane, K.F. Whelan, and P.G. Gargan. 2002. "The Irish Sea Trout Enhancement Program: A Review of the Broodstock and Ova Production Programmes." *Fisheries Management and Ecology* 9 (6): 315–28. http://dx.doi.org/10.1046/j.1365-2400.2002.00315.x.

Pope, K.L., S.E. Lochmann, and M.K. Young. 2010. "Methods for Assessing Fish Populations." In *Inland Fisheries Management in North America*, 3rd ed., ed. M.C. Quist and W.A. Hubert, 325–53. Bethesda, MD: American Fisheries Society.

Port of London Authority. 2015. "About London VTS." http://www.pla.co.uk/Safety/About-London-VTS.

Potter, E.C.E., and M.G. Pawson. 1991. *Gill Netting*. Laboratory Leaflet 69. Lowestoft, UK: Ministry of Agriculture, Fisheries and Food, Directorate of Fisheries Research.

Poulin, R., and K.N. Mouritsen. 2006. "Climate Change, Parasitism and the Structure of Intertidal Ecosystems." *Journal of Helminthology* 80 (2): 183–91. http://dx.doi.org/10.1079/JOH2006341.

Power, G. 1969. *The Salmon of Ungava Bay*. Technical Paper 22. Calgary: Arctic Institute of North America.

Power, G., and G. Shooner. 1966. "Juvenile Salmon in the Estuary and Lower Nabisipi River and Some Results of Tagging." *Journal of the Fisheries Board of Canada* 23 (7): 947–61. http://dx.doi.org/10.1139/f66-088.

Powles, H., M. Bradford, R. Bradford, W.G. Doubleday, S. Innes, and C.D. Levings. 2000. "Assessing and Protecting Endangered Marine Species." *ICES Journal of Marine Science* 57 (3): 669–76.

Prato, T. 2003. "Multiple-Attribute Evaluation of Ecosystem Management for the Missouri River System." *Ecological Economics* 45 (2): 297–309. http://dx.doi.org/10.1016/S0921-8009(03)00077-6.

Priede, I.G., J.F. de L.G. Solbé, J.E. Nott, K.T. O'Grady, and D. Cragg-Hine. 1988. "Behaviour of Adult Atlantic Salmon, *Salmo salar* L., in the Estuary of the River Ribble in Relation to Variations in Dissolved Oxygen and Tidal Flow." *Journal of Fish Biology* 33 (Suppl. sA): 133–39. http://dx.doi.org/10.1111/j.1095-8649.1988.tb05567.x.

Pritchard, D.W. 1952. "Estuarine Hydrography." In *Advances in Geophysics*, ed. H.E. Landsberg, 243–80. New York: Academic Press.

Prunet, P., M.T. Cairns, S. Winberg, and T.G. Pottinger. 2008. "Functional Genomics of Stress Responses in Fish." *Reviews in Fisheries Science* 16 (Suppl. 1): 157–66. http://dx.doi.org/10.1080/10641260802341838.

Quinn, T.P. 2005. *The Behaviour and Ecology of Pacific Salmon and Trout*. Seattle: University of Washington Press.

Quinn, T.P., and M.J. Unwin. 1993. "Variation in Life History Patterns among New Zealand Chinook Salmon (*Oncorhynchus tshawytscha*) Populations." *Canadian Journal of Fisheries and Aquatic Sciences* 50 (7): 1414–21. http://dx.doi.org/10.1139/f93-162.

Quiñones, R.M., and T.J. Mulligan. 2005. "Habitat Use by Juvenile Salmonids in the Smith River Estuary, California." *Transactions of the American Fisheries Society* 134 (5): 1147–58. http://dx.doi.org/10.1577/T04-092.1.

Raat, A.J.P. 1988. *Synopsis of Biological Data on the Northern Pike*, Esox lucius Linnaeus, 1758. Food and Agriculture Organization, Fisheries Synopsis No. 30. Rome: Food and Agriculture Organization of the United Nations.

Rachlin, J.W., B.E. Warkentine, and A. Pappantoniou. 1989. "The Use of Niche Breadth and Proportional Similarity in Feeding to Stipulate Resource Utilization Strategies in Fish." *Journal of Freshwater Ecology* 5 (1): 103–12. http://dx.doi.org/10.1080/02705060.1989.9665218.

Rahr, G., and X. Augerot. 2006. "A Proactive Sanctuary Strategy to Anchor and Restore High-Priority Wild Salmon Ecosystems." In *Salmon 2100: The Future of Wild Pacific Salmon*, ed. R.T. Lackey, D.H. Lach, and S.L. Duncan, 465–89. Bethesda, MD: American Fisheries Society.

Ramsar Convention Secretariat. 2014. "About the Ramsar Convention and Its Mission." http://www.ramsar.org/about/the-ramsar-convention-and-its-mission.

Ramstad, K.M., C.A. Woody, G.K. Sage, and F.W. Allendorf. 2004. "Founding Events Influence Genetic Population Structure of Sockeye Salmon (*Oncorhynchus nerka*) in Lake Clark, Alaska." *Molecular Ecology* 13 (2): 277–90. http://dx.doi.org/10.1046/j.1365-294X.2003.2062.x.

Rand, P.S., B.A. Berejikian, A. Bidlack, D. Bottom, J. Gardner, M. Kaeriyama, Rich Lincoln, Mitsuhiro Nagata, Todd N. Pearsons, Michael Schmidt, et al. 2012. "Ecological Interactions between Wild and Hatchery Salmonids and Key Recommendations for Research and Management Actions in Selected Regions of the North Pacific." *Environmental Biology of Fishes* 94 (1): 343–58. http://dx.doi.org/10.1007/s10641-012-9988-2.

Randall, D.J., and P.A. Wright. 1995. "Circulation and Gas Transfer." In *Physiological Ecology of Pacific Salmon*, ed. C. Groot, L. Margolis, and W.C. Clarke, 441–58. Vancouver: UBC Press.

Rasmussen, C., C.O. Ostberg, D.R. Clifton, J.L. Holloway, and R.J. Rodriguez. 2003. "Identification of a Genetic Marker that Discriminates Ocean-Type and Stream-Type Chinook Salmon in the Columbia River Basin." *Transactions of the American Fisheries Society* 132 (1): 131–42. http://dx.doi.org/10.1577/1548-8659(2003)132<0131:IOAGMT>2.0.CO;2.

Raubenheimer, D. 2011. "Towards a Quantitative Nutritional Ecology: The Right-Angled Mixture Triangle." *Ecological Monographs* 81 (3): 407–27. http://dx.doi.org/10.1890/10-1707.1.

Raven, S.J., and J.C. Coulson. 2001. "Effects of Cleaning a Tidal River of Sewage on Gull Numbers: A Before-and-After Study of the River Tyne, Northeast England." *Bird Study* 48 (1): 48–58. http://dx.doi.org/10.1080/00063650109461202.

Ravine, D. 2007. *Manawatu River Estuary Ramsar Management Plan 2007–2012. Horizons Regional Council, Department of Conservation and Horowhenua District Council.* Foxton, New Zealand: Manawatu Estuary Trust. https://www.horizons.govt.nz/assets/horizons/Images/Council/Regional%20Council/Cncl%2010%20Feb%2009/09-01%20Annex%20A%20Manawatu%20Estuary%20Management%20Plan.pdf.

Rechisky, E.L., D.W. Welch, A.D. Porter, M.C. Jacobs-Scott, P.M. Winchell, and J.L. McKern. 2012. "Estuarine and Early-Marine Survival of Transported and In-River Migrant Snake River Spring Chinook Salmon Smolts." *Scientific Reports* 2: 448. http://dx.doi.org/10.1038/srep00448.

Reimchen, T.E. 1998. "Nocturnal Foraging Behaviour of Black Bears, *Ursus americanus*, on Moresby Island, British Columbia." *Canadian Field Naturalist* 112 (3): 446–50.

Reimers, P.E. 1968. "Social Behavior among Juvenile Fall Chinook Salmon." *Journal of the Fisheries Board of Canada* 25 (9): 2005–8. http://dx.doi.org/10.1139/f68-179.

–. 1973. "The Length of Residence of Juvenile Fall Chinook Salmon in Sixes River, Oregon." *Research Reports of the Fish Commission of Oregon* 4 (2): 3–42.

Renkawitz, M.D., T.F. Sheehan, and G.S. Goulette. 2012. "Swimming Depth, Behavior, and Survival of Atlantic Salmon Postsmolts in Penobscot Bay, Maine." *Transactions of the American Fisheries Society* 141 (5): 1219–29. http://dx.doi.org/10.1080/00028487.2012.688916.

Reynoldson, T.B., J. Culp, R. Lowell, and J.S. Richardson. 2005. "Fraser River Basin." In *Rivers of North America*, ed. A.C. Benke and C.E. Cushing, 696–732. Amsterdam: Elsevier Academic Press. http://dx.doi.org/10.1016/B978-012088253-3/50018-3.

Rice, C.A. 2006. "Effects of Shoreline Modification on a Northern Puget Sound Beach: Microclimate and Embryo Mortality in Surf Smelt (*Hypomesus pretiosus*)." *Estuaries and Coasts* 29 (1): 63–71. http://dx.doi.org/10.1007/BF02784699.

Rice, C.A., L.L. Johnson, P. Roni, B.E. Feist, W.G. Hood, L.M. Tear, and C.A. Simenstad. 2005. "Monitoring Rehabilitation in Temperate North American Estuaries." In *Monitoring Stream and Watershed Restoration*, ed. P. Roni, 165–204. Bethesda, MD: American Fisheries Society.

Rice, J.A. 1993. "Forecasting Abundance from Habitat Measures Using Non-Parametric Density Estimate Methods." *Canadian Journal of Fisheries and Aquatic Sciences* 50 (8): 1690–98. http://dx.doi.org/10.1139/f93-190.

Rich, H.B., T.P. Quinn, M.D. Scheuerell, and D.E. Schindler. 2009. "Climate and Intraspecific Competition Control the Growth and Life History of Juvenile Sockeye Salmon (*Oncorhynchus nerka*) in Iliamna Lake, Alaska." *Canadian Journal of Fisheries and Aquatic Sciences* 66 (2): 238–46. http://dx.doi.org/10.1139/F08-210.

Rich, W.H., and H.B. Holmes. 1929. "Experiments in Marking Young Chinook Salmon on the Columbia River, 1916 to 1927." *Bulletin of the United States Bureau of Fisheries* 44: 215–64.

Richardson, J., and D.A. Dixon. 2004. "Modeling the Hydraulic Zone of Influence of Connecticut Yankee Nuclear Power Plant's Cooling Water Intake Structure." *American Fisheries Society Monograph* 9: 513–24.

Richardson, J.S., T.J. Lissimore, M.C. Healey, and T.G. Northcote. 2000. "Fish Communities of the Lower Fraser River (Canada) and a 21-Year Contrast." *Environmental Biology of Fishes* 59 (2): 125–40. http://dx.doi.org/10.1023/A:1007681332484.

Richter, A., and S.A. Kolmes. 2005. "Maximum Temperature Limits for Chinook, Coho, and Chum Salmon, and Steelhead Trout in the Pacific Northwest." *Reviews in Fisheries Science* 13 (1): 23–49. http://dx.doi.org/10.1080/10641260590885861.

Richter, B.D., S. Postel, C. Revenga, T. Scudder, B. Lehner, A. Churchill, and M. Chow. 2010. "Lost in Development's Shadow: The Downstream Human Consequences of Dams." *Water Alternatives* 3 (2): 14–42.

Ricker, W.E. 1975. *Computation and Interpretation of Biological Statistics of Fish Populations*. Bulletin of Fisheries Research Board of Canada No. 191. Ottawa: Department of the Environment, Fisheries and Marine Services.

Ricketts, E.F., J. Calvin, and J.W. Hedgpeth. 1992. *Between Pacific Tides*. 5th ed. Revised by D.W. Phillips. Palo Alto, CA: Stanford University Press.

Ricklefs, R.E. 2008. "Disintegration of the Ecological Community." *American Naturalist* 172 (6): 741–50. http://dx.doi.org/10.1086/593002.

Riddell, B.E., and W. Leggett. 1981. "Evidence of an Adaptive Basis for Geographic Variation in Body Morphology and Time of Downstream Migration of Juvenile Atlantic Salmon (*Salmo salar*)." *Canadian Journal of Fisheries and Aquatic Sciences* 38 (3): 308–20. http://dx.doi.org/10.1139/f81-042.

Rieman, B.E., C.L. Smith, R.J. Naiman, G.T. Ruggerone, C.C. Wood, N. Huntly, Erik N. Merrill, J. Richard Alldredge, Peter A. Bisson, James Congleton, et al. 2015. "A Comprehensive Approach for Habitat Restoration in the Columbia Basin." *Fisheries* (Bethesda, MD) 40 (3): 124–35. http://dx.doi.org/10.1080/03632415.2015.1007205.

Rikardsen, A.H., P.A. Amundsen, P.A. Bjørn, and M. Johansen. 2000. "Comparison of Growth, Diet and Food Consumption of Sea Run and Lake Dwelling Arctic Charr." *Journal of Fish Biology* 57 (5): 1172–88. http://dx.doi.org/10.1111/j.1095-8649.2000.tb00479.x.

Rikardsen, A.H., M. Haugland, P.A. Bjørn, B. Finstad, R. Knudsen, J.B. Dempson, J.C. Holst, N.A. Hvidsten, and M. Holm. 2004. "Geographical Differences in Marine Feeding of Atlantic Salmon Post-Smolts in Norwegian Fjords." *Journal of Fish Biology* 64 (6): 1655–79. http://dx.doi.org/10.1111/j.0022-1112.2004.00425.x.

Rikardsen, A.H., O.H. Diserud, J.M. Elliott, J.B. Dempson, J. Sturlaugsson, and A.J. Jensen. 2007. "The Marine Temperature and Depth Preferences of Arctic Charr (*Salvelinus alpinus*) and Sea Trout (*Salmo trutta*), as Recorded by Data Storage Tags." *Fisheries Oceanography* 16 (5): 436–47. http://dx.doi.org/10.1111/j.1365-2419.2007.00445.x.

Riva Rossi, C.M.R., C.P. Lessa, and M.A. Pascual. 2004. "The Origin of Introduced Rainbow Trout (*Oncorhynchus mykiss*) in the Santa Cruz River, Patagonia, Argentina, as Inferred from Mitochondrial DNA." *Canadian Journal of Fisheries and Aquatic Sciences* 61 (7): 1095–1101. http://dx.doi.org/10.1139/f04-056.

Roberge, J.M., and P. Angelstam. 2004. "Usefulness of the Umbrella Species Concept as a Conservation Tool." *Conservation Biology* 18 (1): 76–85. http://dx.doi.org/10.1111/j.1523-1739.2004.00450.x.

Roberts, R.J., ed. 2012. *Fish Pathology*. Oxford: Wiley-Blackwell Publishing. http://dx.doi.org/10.1002/9781118222942.

Robitaille, J.A., Y. Cote, G. Schooner, and G. Hayeur. 1986. "Growth and Maturation Patterns of Atlantic Salmon *Salmo salar* in the Koksoak River, Ungava, Québec." In *Salmonid Age at Maturity* (Canadian Special Publication of Fisheries and Aquatic Sciences No. 89), ed. D.J. Meerburg, 62–69. Ottawa: National Research Council Press.

Roby, D.D., D.E. Lyons, D.P. Craig, K. Collis, and G.H. Visser. 2003. "Quantifying the Effect of Predators on Endangered Species Using a Bioenergetics Approach: Caspian Terns and Juvenile Salmonids in the Columbia River Estuary." *Canadian Journal of Zoology* 81 (2): 250–65. http://dx.doi.org/10.1139/z02-242.

Rodríguez, J.P. 2001. "Exotic Species Introductions into South America: An Underestimated Threat?" *Biodiversity and Conservation* 10 (11): 1983–96. http://dx.doi.org/10.1023/A:1013151722557.

Roegner, G.C., J.A. Needoba, and A.M. Baptista. 2011a. "Coastal Upwelling Supplies Oxygen-Depleted Water to the Columbia River Estuary." *PLoS One* 6 (4): e18672. http://dx.doi.org/10.1371/journal.pone.0018672.

Roegner, G.C., C. Seaton, and A.M. Baptista. 2011b. "Climatic and Tidal Forcing of Hydrography and Chlorophyll Concentrations in the Columbia River Estuary." *Estuaries and Coasts* 34 (2): 281–96. http://dx.doi.org/10.1007/s12237-010-9340-z.

Roegner, G.C., H.L. Diefenderfer, A.B. Borde, R.M. Thom, E.M. Dawley, A.H. Whiting, et al. 2009. *Protocols for Monitoring Habitat Restoration Projects in the Lower Columbia River and Estuary*. NOAA Technical Memorandum NMFS-NWFSC-97. Washington, DC: US Department of Commerce.

Roegner, G.C., E.W. Dawley, M. Russell, A. Whiting, and D.J. Teel. 2010. "Juvenile Salmonid Use of Reconnected Tidal Freshwater Wetlands in Grays River,

Lower Columbia River Basin." *Transactions of the American Fisheries Society* 139 (4): 1211–32. http://dx.doi.org/10.1577/T09-082.1.

Roegner, G.C., R. McNatt, D.J. Teel, and D.L. Bottom. 2012. "Distribution, Size, and Origin of Juvenile Chinook Salmon in Shallow-Water Habitats of the Lower Columbia River and Estuary, 2002–2007." *Marine and Coastal Fisheries* 4 (1): 450–72. http://dx.doi.org/10.1080/19425120.2012.675982.

Rogers, D.E. 2001. *Estimates of Annual Salmon Runs from the North Pacific, 1951–2001*. University of Washington Report No. SAFS-UW-0115. Seattle: University of Washington.

Romanuk, T.N., and C.D. Levings. 2003. "Associations between Arthropods and the Supralittoral Ecotone: Dependence of Aquatic and Terrestrial Taxa on Riparian Vegetation." *Environmental Entomology* 32 (6): 1343–53. http://dx.doi.org/10.1603/0046-225X-32.6.1343.

–. 2005. "Stable Isotope Analysis of Trophic Position and Terrestrial vs. Marine Carbon Sources for Juvenile Pacific Salmonids in Nearshore Marine Habitats." *Fisheries Management and Ecology* 12 (2): 113–21. http://dx.doi.org/10.1111/j.1365-2400.2004.00432.x.

–. 2010. "Reciprocal Subsidies and Food Web Pathways Leading to Chum Salmon Fry in a Temperate Marine-Terrestrial Ecotone." *PLoS One* 5 (4): e10073. http://dx.doi.org/10.1371/journal.pone.0010073.

Rombouts, I., G. Beaugrand, X. Fizzala, F. Gaill, S.P.R. Greenstreet, S. Lamare, F. Le Loc'h, A. McQuatters-Gollop, B. Mialet, N. Niquil, et al. 2013. "Food Web Indicators under the Marine Strategy Framework Directive: From Complexity to Simplicity?" *Ecological Indicators* 29: 246–54. http://dx.doi.org/10.1016/j.ecolind.2012.12.021.

Romer, J.D., C.A. Leblanc, S. Clements, J.A. Ferguson, M.L. Kent, D. Noakes, and C.B. Schreck. 2013. "Survival and Behavior of Juvenile Steelhead Trout (*Oncorhynchus mykiss*) in Two Estuaries in Oregon, USA." *Environmental Biology of Fishes* 96 (7): 849–63. http://dx.doi.org/10.1007/s10641-012-0080-8.

Rosenfeld, J.S., and S. Boss. 2001. "Fitness Consequences of Habitat Use for Juvenile Cutthroat Trout: Energetic Costs and Benefits in Pools and Riffles." *Canadian Journal of Fisheries and Aquatic Sciences* 58 (3): 585–93. http://dx.doi.org/10.1139/f01-019.

Roslyi, Y.S., and G.V. Novomodnyi. 1996. "Elimination of the Young of the Genus *Oncorhynchus* from the Amur River by the Arctic Lamprey *Lampetra japonica* and by Other Predatory Fish in the Early Marine Period of Life." *Journal of Ichthyology* 36 (1): 46–50. Translated from Russian; originally published in *Voprosy Ikthiologii* 36 (1): 50–54.

Ross, P.S., C. Kennedy, L.K. Shelley, K.B. Tierney, D.A. Patterson, W.L. Fairchild, R.W. Macdonald, and Karen Kidd. 2013. "The Trouble with Salmon: Relating Pollutant Exposure to Toxic Effect in Species with Transformational Life Histories and Lengthy Migrations." *Canadian Journal of Fisheries and Aquatic Sciences* 70 (8): 1252–64. http://dx.doi.org/10.1139/cjfas-2012-0540.

Rounsefell, G.A. 1958. "Anadromy in North American Salmonidae." *Fisheries Bulletin* 58: 171–85.

Rowan, D.J., and J.B. Rasmussen. 1996. "Measuring the Bioenergetic Cost of Fish Activity In Situ Using a Globally Dispersed Radiotracer (^{137}Cs)." *Canadian*

Journal of Fisheries and Aquatic Sciences 53 (4): 734–45. http://dx.doi.org/10.1139/f95-046.

Ruckelshaus, M.H., P. Levin, J.B. Johnson, and P.M. Kareiva. 2002. "The Pacific Salmon Wars: What Science Brings to the Challenge of Recovering Species." *Annual Review of Ecology and Systematics* 33 (1): 665–706. http://dx.doi.org/10.1146/annurev.ecolsys.33.010802.150504.

Ruggerone, G.T., and B.M. Connors. 2015. "Productivity and Life History of Sockeye Salmon in Relation to Competition with Pink and Sockeye Salmon in the North Pacific Ocean." *Canadian Journal of Fisheries and Aquatic Sciences* 72 (6): 818–33. http://dx.doi.org/10.1139/cjfas-2014-0134.

Runge, M.C. 2011. "An Introduction to Adaptive Management for Threatened and Endangered Species." *Journal of Fish and Wildlife Management* 2 (2): 220–33. http://dx.doi.org/10.3996/082011-JFWM-045.

Russell, G., S.J. Hawkins, L.C. Evans, H.D. Jones, and G.D. Holmes. 1983. "Restoration of a Disused Dock Basin as a Habitat for Marine Benthos and Fish." *Journal of Applied Ecology* 20 (1): 43–58. http://dx.doi.org/10.2307/2403375.

Ryall, P.E., and C.D. Levings. 1987. *Juvenile Salmon Utilization of Rejuvenated Tidal Channels in the Squamish Estuary, British Columbia*. Canadian Manuscript Report of Fisheries and Aquatic Sciences No. 1904. West Vancouver: Fisheries and Oceans Canada.

Ryan, B.A., S.G. Smith, J.A.M. Butzerin, and J.W. Ferguson. 2003. "Relative Vulnerability to Avian Predation of Juvenile Salmonids Tagged with Passive Integrated Transponders in the Columbia River Estuary, 1998–2000." *Transactions of the American Fisheries Society* 132 (2): 275–88. http://dx.doi.org/10.1577/1548-8659(2003)132<0275:RVTAPO>2.0.CO;2.

Ryder, J.L., J.K. Kenyon, D. Buffett, K. Moore, M. Ceh, and K. Stipec. 2007. *An Integrated Biophysical Assessment of Estuarine Habitats in British Columbia to Assist Regional Conservation Planning*. Technical Report Series No. 476. Delta, BC: Canadian Wildlife Service.

Ryer, C.H., and B.L. Olla. 1991. "Agonistic Behavior in a Schooling Fish: Form, Function and Ontogeny." *Environmental Biology of Fishes* 31 (4): 355–63. http://dx.doi.org/10.1007/BF00002360.

Sahan, E., K. Sabbe, G. Creach, G. Hernandez-Requet, W. Vyverman, L.J. Stal, and G. Muyzer. 2007. "Community Structure and Seasonal Dynamics of Diatom Biofilms and Associated Grazers in Intertidal Mudflats." *Aquatic Microbial Ecology* 47 (3): 253–66. http://dx.doi.org/10.3354/ame047253.

Saito, T., I. Shimizu, J. Seki, and K. Nagasawa. 2009. "Relationship between Zooplankton Abundance and the Early Marine Life History of Juvenile Chum Salmon *Oncorhynchus keta* in Eastern Hokkaido, Japan." *Fisheries Science* 75 (2): 303–16. http://dx.doi.org/10.1007/s12562-008-0040-6.

Salminen, M., E. Erkamo, and J. Salmi. 2001. "Diet of Post-Smolt and One-Sea-Winter Atlantic Salmon in the Bothnian Sea, Northern Baltic." *Journal of Fish Biology* 58 (1): 16–35. http://dx.doi.org/10.1111/j.1095-8649.2001.tb00496.x.

Salvanes, A.G.V. 2001. "Ocean Ranching." In *Encyclopedia of Ocean Sciences*, ed. J. Steele, S. Thorpe, and K. Turekian, 1973–82. London: Academic Press. http://dx.doi.org/10.1006/rwos.2001.0485.

Sanchez, W., and J.M. Porcher. 2009. "Fish Biomarkers for Environmental Monitoring within the Water Framework Directive of the European Union." *Trends in Analytical Chemistry* 28 (2): 150–58. http://dx.doi.org/10.1016/j.trac.2008.10.012.

Sand, O., P.S. Enger, H.E. Karlsen, and F.R. Knudsen. 2001. "Detection of Infrasound in Fish and Behavioral Responses to Intense Infrasound in Juvenile Salmonids and European Silver Eels: A Minireview. In *Behavioral Technologies for Fish Guidance* (American Fisheries Society Symposium 26), ed. C.C. Coutant, 183–93. Bethesda, MD: American Fisheries Society.

Sanderson, B.L., K.A. Barnas, and A.M. Wargo Rub. 2009. "Nonindigenous Species of the Pacific Northwest: An Overlooked Risk to Endangered Salmon?" *BioScience* 59 (3): 245–56. http://dx.doi.org/10.1525/bio.2009.59.3.9.

Sandu, P.G.C., and L. Oprea. 2013. "The Influence of Environmental Abiotic Factors on the Qualitative and Quantitative Structure of Ichthyofauna from Predeltaic Danube Area." *Scientific Papers Animal Science and Biotechnologies* 46 (1): 251–59.

Sano, S., and S. Abe. 1967. "Ecological Study of the Masou Salmon (*Oncorhynchus masou* Brevoort): The Observation on the Smolt in the Coastal Waters." *Science Reports of the Hokkaido Salmon Hatchery* 21:1–9.

Saunders, R., M.A. Hachey, and C.W. Fay. 2006. "Maine's Diadromous Fish Community Past, Present, and Implications for Atlantic Salmon Recovery." *Fisheries* (Bethesda, MD) 31 (11): 537–47. http://dx.doi.org/10.1577/1548-8446 (2006)31[537:MDFC]2.0.CO;2.

Savini, D., A. Occhipinti-Ambrogi, A. Marchini, E. Tricarico, F. Gherardi, S. Olenin, and S. Gollasch. 2010. "The Top 27 Animal Alien Species Introduced into Europe for Aquaculture and Related Activities." *Journal of Applied Ichthyology* 26 (S2): 1–7. http://dx.doi.org/10.1111/j.1439-0426.2010.01503.x.

Sayre, N.F. 2008. "The Genesis, History, and Limits of Carrying Capacity." *Annals of the Association of American Geographers* 98 (1): 120–34. http://dx.doi.org/10.1080/00045600701734356.

Schaal, G., P. Riera, and C. Leroux. 2009. "Trophic Significance of the Kelp *Laminaria digitata* (Lamour.) for the Associated Food Web: A Between-Sites Comparison." *Estuarine, Coastal and Shelf Science* 85 (4): 565–72. http://dx.doi.org/10.1016/j.ecss.2009.09.027.

Scheifele, P.M., S. Andrew, R.A. Cooper, M. Darre, F.E. Musiek, and L. Max. 2005. "Indication of a Lombard Vocal Response in the St. Lawrence River Beluga." *Journal of the Acoustical Society of America* 117 (3 3Pt1): 1486–91. http://dx.doi.org/10.1121/1.1835508.

Schindler, D.E., R. Hilborn, B. Chasco, C.P. Boatright, T.P. Quinn, L.A. Rogers, and M.S. Webster. 2010. "Population Diversity and the Portfolio Effect in an Exploited Species." *Nature* 465 (7298): 609–12. http://dx.doi.org/10.1038/nature09060.

Scholle, J., and B. Schuchardt. 2012. "A Fish-Based Index of Biotic Integrity – FAT-TW an Assessment Tool for Transitional Waters of the Northern German Tidal Estuaries." *Coastline Reports* 18: 1–73.

Schreiber, A., and G. Diefenbach. 2005. "Population Genetics of the European Trout (*Salmo trutta* L.) Migration System in the River Rhine: Recolonization by

Sea Trout." *Ecology of Freshwater Fish* 14 (1): 1–13. http://dx.doi.org/10.1111/ j.1600-0633.2004.00072.x.

Schulte-Oehlmann, U., J. Oehlmann, and F. Keil. 2011. "Before the Curtain Falls: Endocrine-Active Pesticides – A German Contamination Legacy." *Reviews of Environmental Contamination and Toxicology* 213: 137–59. http://dx.doi.org/ 10.1007/978-1-4419-9860-6_5.

Scott, D., J.W. Moore, L.M. Herborg, C. Clarke Murray, and N.R. Serrao. 2013. "A Non-Native Snakehead Fish in British Columbia, Canada: Capture, Genetics, Isotopes, and Policy Consequences." *Management of Biological Invasions* 4 (4): 265–71. http://dx.doi.org/10.3391/mbi.2013.4.4.01.

Scottish Environment Protection Agency. 2014a. "Biodiversity Indicator, Estuarine Fish." http://www.snh.gov.uk/docs/B447222.pdf.

–. 2014b. "Fish Farm Manual, Estuarine Fish." http://www.sepa.org.uk/regulations/ water/aquaculture/fish-farm-manual/.

Scudder, C.G.E. 1989. "The Adaptive Significance of Marginal Populations: A General Perspective." In *Proceedings of the National Workshop on Effects of Habitat Alteration on Salmonid Stocks* (Canadian Special Publication of Fisheries and Aquatic Sciences No. 105), ed. C.D. Levings, L.B. Holtby, and M.A. Henderson, 180–85. Ottawa: Fisheries and Oceans Canada.

Seaman, G.A., L.F. Lowry, and K.J. Frost. 1982. "Foods of Beluga Whales *Delphinapterus leucas* in Western Alaska USA." *Cetology* 44: 1–19.

Semmens, B.X. 2008. "Acoustically Derived Fine-Scale Behaviors of Juvenile Chinook Salmon (*Oncorhynchus tshawytscha*) Associated with Intertidal Benthic Habitats in an Estuary." *Canadian Journal of Fisheries and Aquatic Sciences* 65 (9): 2053–62. http://dx.doi.org/10.1139/F08-107.

Semushin, T.V., and A.P. Novoselov. 2009. "Species Composition of Ichthyofauna of Baidaratskaya Bay of the Kara Sea." *Journal of Ichthyology* 49 (5): 362–75. Translated from Russian; originally published in *Voprosy Ikhtiologii* 49 (3): 304–17. http://dx.doi.org/10.1134/S0032945209050026.

Servos, M.R., K.R. Munkittrick, J.H. Carey, and G.J. Van Der Kraak, eds. 1996. *Environmental Fate and Effects of Pulp and Paper Mill Effluents*. Delray Beach, FL: St. Lucie Press.

Severn Estuary/Môr Hafren European Marine Site. 2009. "Natural England and the Countryside Council for Wales' Advice Given under Regulation 33(2)(a) of the Conservation (Natural Habitats, &c.) Regulations 1994, as Amended. June 2009." http://publications.naturalengland.org.uk/file/3977366.

Shaffer, J.A., P. Crain, B. Winter, M.L. McHenry, C. Lear, and T.J. Randle. 2008. "Nearshore Restoration of the Elwha River through Removal of the Elwha and Glines Canyon Dams: An Overview." *Northwest Science* 82 (sp1): 48–58. http:// dx.doi.org/10.3955/0029-344X-82.S.I.48.

Shaposhnikova, G.K. 1950. "Ryby Amu-Daryi (Fishes of the Amu-Darya)" [in Russian]. *Trudy Zoologicheskogo Instituta Akademii Nauk S.S.S.R.* 9: 16–54.

Shearer, K.D., and T. Åsgård. 1992. "The Effect of Water-Borne Magnesium on the Dietary Magnesium Requirement of the Rainbow Trout (*Oncorhynchus mykiss*)." *Fish Physiology and Biochemistry* 9 (5–6): 387–92. http://dx.doi. org/10.1007/BF02274219.

Sheaves, M. 2001. "Are There Really Few Piscivorous Fishes in Shallow Estuarine Habitats?" *Marine Ecology Progress Series* 222: 279–90. http://dx.doi.org/10.3354/meps222279.

Sherwood, C.R., D.A. Jay, R. Bradford Harvey, P. Hamilton, and C.A. Simenstad. 1990. "Historical Changes in the Columbia River Estuary." *Progress in Oceanography* 25 (1–4): 299–352. http://dx.doi.org/10.1016/0079-6611(90)90011-P.

Shevlyakov, V.A., and V.A. Parensky. 2010. "Traumatization of Kamchatka River Pacific Salmon by Lampreys." *Russian Journal of Marine Biology* 36 (5): 396–400. http://dx.doi.org/10.1134/S106307401005010X.

Shinada, A., N. Shiga, and S. Ban. 1999. "Structure and Magnitude of Diatom Spring Bloom in Funka Bay, Southwestern Hokkaido, Japan, as Influenced by the Intrusion of Coastal Oyashio Water." *Plankton Biology and Ecology* 46: 24–29.

Shine, R., and L. Schwarzkopf. 1992. "The Evolution of Reproductive Effort in Lizards and Snakes." *Evolution* 46 (1): 62–75.

Shipman, H. 2008. *A Geomorphic Classification of Puget Sound Nearshore Landforms*. Puget Sound Nearshore Partnership Report No. 2008–01. Seattle: Seattle District of US Army Corps of Engineers.

Shively, R.S., T.P. Poe, and S.T. Sauter. 1996. "Feeding Response by Northern Squawfish to a Hatchery Release of Juvenile Salmonids in the Clearwater River, Idaho." *Transactions of the American Fisheries Society* 125 (2): 230–36. http://dx.doi.org/10.1577/1548-8659(1996)125<0230:FRBNST>2.3.CO;2.

Shrimpton, J.M., D.A. Patterson, J.G. Richards, S.J. Cooke, P.M. Schulte, S.G. Hinch, and A.P. Farrell. 2005. "Ionoregulatory Changes in Different Populations of Maturing Sockeye Salmon *Oncorhynchus nerka* during Ocean and River Migration." *Journal of Experimental Biology* 208 (21): 4069–78. http://dx.doi.org/10.1242/jeb.01871.

Shurin, J.B., E.T. Borer, E.W. Seabloom, K. Anderson, C.A. Blanchette, B. Broitman, Scott D. Cooper, and Benjamin S. Halpern. 2002. "A Cross Ecosystem Comparison of the Strength of Trophic Cascades." *Ecology Letters* 5 (6): 785–91. http://dx.doi.org/10.1046/j.1461-0248.2002.00381.x.

Sibert, J.R. 1979. "Detritus and Juvenile Salmon Production in the Nanaimo Estuary: II. Meiofauna Available as Food to Juvenile Chum Salmon (*Oncorhynchus keta*)." *Journal of the Fisheries Board of Canada* 36 (5): 497–503. http://dx.doi.org/10.1139/f79-073.

–. 1981. "Intertidal Hyperbenthic Populations in the Nanaimo Estuary." *Marine Biology* 64 (3): 259–65.

Sibert, J.R., and B. Kask. 1978. "Do Fish Have Diets?" In *Proceedings of the 1977 Northeast Pacific Chinook and Coho Salmon Workshop* (Fisheries and Marine Service Canadian Technical Report No.759), ed. B.G. Shepherd and R.M. Ginetz, 48–56. Vancouver: Department of Fisheries and the Environment.

Sibert, J.R., and S. Obreski. 1977. "Frequency Distributions of Food Item Counts in Individual Fish Stomachs." In *Fish Food Habit Studies* (WSG-WO-77-2), ed. C.A. Simenstad and S.J. Sipovsky, 107–14. Seattle: Washington Sea Grant.

Sibert, J.R., T.J. Brown, M.C. Healey, B.A. Kask, and R.J. Naiman. 1977. "Detritus-Based Food Webs: Exploitation by Juvenile Chum Salmon (*Oncorhynchus keta*)." *Science* 196 (4290): 649–50. http://dx.doi.org/10.1126/science.196.4290.649.

Sibly, R.M., T.D. Williams, and M.B. Jones. 2000. "How Environmental Stress Affects Density Dependence and Carrying Capacity in a Marine Copepod." *Journal of Applied Ecology* 37 (3): 388–97. http://dx.doi.org/10.1046/j.1365 -2664.2000.00534.x.

Sigholt, T., and B. Finstad. 1990. "Effect of Low Temperature on Seawater Tolerance in Atlantic Salmon (*Salmo salar*) Smolts." *Aquaculture* (Amsterdam) 84 (2): 167–72. http://dx.doi.org/10.1016/0044-8486(90)90346-O.

Simenstad, C.A., and J.R. Cordell. 2000. "Ecological Assessment Criteria for Restoring Anadromous Salmonid Habitat in Pacific Northwest Estuaries." *Ecological Engineering* 15 (3–4): 283–302. http://dx.doi.org/10.1016/S0925-8574 (00)00082-3.

Simenstad, C.A., K.L. Fresh, and E.O. Salo. 1982. "The Role of Puget Sound and Washington Coastal Estuaries in the Life History of Pacific Salmon: An Unappreciated Function." In *Estuarine Comparisons*, ed. V.S. Kennedy, 343–64. New York: Academic Press. http://dx.doi.org/10.1016/B978-0-12-404070-0. 50026-0.

Simenstad, C.A., D. Reed, and M. Ford. 2006. "When Is Restoration Not? Incorporating Landscape-Scale Processes to Restore Self-Sustaining Ecosystems in Coastal Wetland Restoration." *Ecological Engineering* 26 (1): 27–39. http://dx.doi.org/10.1016/j.ecoleng.2005.09.007.

Simenstad, C.A., and R.C. Wissmar. 1985. "^{13}C Evidence of the Origins and Fates of Organic Carbon in Estuarine and Nearshore Food Webs." *Marine Ecology Progress Series* 22: 141–52. http://dx.doi.org/10.3354/meps022141.

Simenstad, C.A., W.G. Hood, R.M. Thom, D.A. Levy, and D.L. Bottom. 2002. "Landscape Structure and Scale Constraints on Restoring Estuarine Wetlands for Pacific Coast Juvenile Fishes." In *Concepts and Controversies in Tidal Marsh Ecology*, ed. M.P. Weinstein and D.A. Kreeger, 597–630. New York: Springer. http://dx.doi.org/10.1007/0-306-47534-0_28.

Simenstad, C.A., M. Logsdon, K. Fresh, H. Shipman, M. Dethier, and J. Newton. 2006. *Conceptual Model for Assessing Restoration of Puget Sound Nearshore Ecosystems*. Puget Sound Nearshore Partnership Report No. 2006–03. Seattle: Washington Sea Grant Program, University of Washington. http://puget soundnearshore.org/technical_papers/conceptual_model.pdf.

Skalski, J.R., R.A. Buchanan, and J. Griswold. 2009. "Review of Marking Methods and Release-Recapture Designs for Estimating the Survival of Very Small Fish: Examples from the Assessment of Salmonid Fry Survival." *Reviews in Fisheries Science* 17 (3): 391–401. http://dx.doi.org/10.1080/10641260902752199.

Skilbrei, O.T., and V. Wennevik. 2006. "Survival and Growth of Sea-Ranched Atlantic Salmon, *Salmo salar* L., Treated against Sea Lice before Release." *ICES Journal of Marine Science* 63 (7): 1317–25. http://dx.doi.org/10.1016/ j.icesjms.2006.04.012.

Sloan, C.A., B.F. Anulacion, J.L. Bolton, D. Boyd, O.P. Olson, S.Y. Sol, Gina M. Ylitalo, and Lyndal L. Johnson. 2010. "Polybrominated Diphenyl Ethers in Outmigrant Juvenile Chinook Salmon from the Lower Columbia River and Estuary and Puget Sound, Washington." *Archives of Environmental Contamination and Toxicology* 58 (2): 403–14. http://dx.doi.org/10.1007/s00244-009 -9391-y.

Smith, F.G.W., ed. 1974. *Handbook of Marine Science*. Cleveland: CRC Press.

Smith, I.P., and G.W. Smith. 1997. "Tidal and Diel Timing of River Entry by Adult Atlantic Salmon Returning to the Aberdeenshire Dee, Scotland." *Journal of Fish Biology* 50 (3): 463–74. http://dx.doi.org/10.1111/j.1095-8649.1997.tb01942.x.

Smith, M.W., and D.K. Rushton. 1963. "A Study of Barachois Ponds in the Bras d'Or Lake Area of Cape Breton Island, Nova Scotia." *Proceedings of the Nova Scotia Institute of Science* 26: 3–17.

Smith, T.B., and S. Skúlason. 1996. "Evolutionary Significance of Resource Polymorphisms in Fishes, Amphibians, and Birds." *Annual Review of Ecology and Systematics* 27 (1): 111–33. http://dx.doi.org/10.1146/annurev.ecolsys.27.1.111.

Solazzi, M.F., E. Nickelson, and S.L. Johnson. 1991. "Survival, Contribution, and Return of Hatchery Coho Salmon (*Oncorhynchus kisutch*) Released into Freshwater, Estuarine, and Marine Environments." *Canadian Journal of Fisheries and Aquatic Sciences* 48 (2): 248–53. http://dx.doi.org/10.1139/f91-034.

Solomon, D.J. 1973. "Evidence for Pheromone-Influenced Homing by Migrating Atlantic Salmon, *Salmo salar* (L.)." *Nature* 244 (5413): 231–32. http://dx.doi.org/10.1038/244231a0.

Solomon, D.J., and E.C.E. Potter. 1988. "First Results with a New Estuarine Fish Tracking System." *Journal of Fish Biology* 33 (Suppl. sA): 127–32. http://dx.doi.org/10.1111/j.1095-8649.1988.tb05566.x.

Sommer, T.R., W.C. Harrell, and M.L. Nobriga. 2005. "Habitat Use and Stranding Risk of Juvenile Chinook Salmon on a Seasonal Floodplain." *North American Journal of Fisheries Management* 25 (4): 1493–1504. http://dx.doi.org/10.1577/M04-208.1.

Sommer, T.R., M.L. Nobriga, W.C. Harrell, W. Batham, and W.J. Kimmerer. 2001. "Floodplain Rearing of Juvenile Chinook Salmon: Evidence of Enhanced Growth and Survival." *Canadian Journal of Fisheries and Aquatic Sciences* 58 (2): 325–33. http://dx.doi.org/10.1139/f00-245.

Sommer, T.R., Chuck Armor, Randall Baxter, Richard Breuer, Larry Brown, Mike Chotkowski, Steve Culberson, Fredrick Feyrer, Marty Gingras, Bruce Herbold, et al. 2007. "The Collapse of Pelagic Fishes in the Upper San Francisco Estuary." *Fisheries* (Bethesda, MD) 32 (6): 270–77. http://dx.doi.org/10.1577/1548-8446(2007)32[270:TCOPFI]2.0.CO;2.

Soto, D., and F. Jara. 2007. "Using Natural Ecosystem Services to Diminish Salmon-Farming Footprints in Southern Chile." In *Ecological and Genetic Implications of Aquaculture Activities*, ed. T.M. Bert, 459–75. New York: Springer. http://dx.doi.org/10.1007/978-1-4020-6148-6_26.

Soto, D., F. Jara, and C. Moreno. 2001. "Escaped Salmon in the Inner Seas, Southern Chile: Facing Ecological and Social Conflicts." *Ecological Applications* 11 (6): 1750–62. http://dx.doi.org/10.1890/1051-0761(2001)011[1750:ESITIS]2.0.CO;2.

Southwood, T.R.E. 1977. "Habitat, the Templet for Ecological Strategies?" *Journal of Animal Ecology* 46 (2): 336–65. http://dx.doi.org/10.2307/3817.

Spares, A.D., M.J.W. Stokesbury, R.K. O'Dor, and T.A. Dick. 2012. "Temperature, Salinity and Prey Availability Shape the Marine Migration of Arctic Char, *Salvelinus alpinus*, in a Macrotidal Estuary." *Marine Biology* 159 (8): 1633–46. http://dx.doi.org/10.1007/s00227-012-1949-y.

Spilseth, S.A., and C.A. Simenstad. 2011. "Seasonal, Diel, and Landscape Effects on Resource Partitioning between Juvenile Chinook Salmon (*Oncorhynchus tshawytscha*) and Threespine Stickleback (*Gasterosteus aculeatus*) in the Columbia River Estuary." *Estuaries and Coasts* 34 (1): 159–71. http://dx.doi.org/10.1007/s12237-010-9349-3.

Spromberg, J.A., and L.L. Johnson. 2008. "Potential Effects of Freshwater and Estuarine Contaminant Exposure on Lower Columbia River Chinook Salmon (*Oncorhynchus tshawytscha*) Populations." In *Demographic Toxicity Methods in Ecological Risk Assessment*, ed. H.R. Akcakaya, J.D. Stark, and T.S. Bridges, 123–42. Oxford: Oxford University Press.

Stalberg, H.C., R.B. Lauzier, E.A. MacIsaac, M. Porter, and C. Murray. 2009. *Canada's Policy for Conservation of Wild Pacific Salmon: Stream, Lake, and Estuarine Habitat Indicators*. Canadian Manuscript Report of Fisheries and Aquatic Sciences No. 2859. Ottawa: Fisheries and Oceans Canada.

Stanford, J.A., M.S. Lorang, and F.R. Hauer. 2005. "The Shifting Habitat Mosaic of River Ecosystems." *Verhandlungen – Internationale Vereinigung für Theoretische und Angewandte Limnologie* 29 (1): 123–36.

Stanhope, M.J., and C.D. Levings. 1985. "Growth and Production of *Eogammarus confervicolus* (Amphipoda: Anisogammaridae) at a Log Storage Site and in Areas of Undisturbed Habitat within the Squamish Estuary, British Columbia." *Canadian Journal of Fisheries and Aquatic Sciences* 42 (11): 1733–40. http://dx.doi.org/10.1139/f85-217.

Stasko, A.B. 1975. "Progress of Migrating Atlantic Salmon (*Salmo salar*) along an Estuary, Observed by Ultrasonic Tracking." *Journal of Fish Biology* 7 (3): 329–38. http://dx.doi.org/10.1111/j.1095-8649.1975.tb04607.x.

Stein, J.E., T. Hom, T.K. Collier, D.W. Brown, and U. Varanasi. 1995. "Contaminant Exposure and Biochemical Effects in Outmigrant Juvenile Chinook Salmon from Urban and Non-Urban Estuaries of Puget Sound, Washington." *Environmental Toxicology and Chemistry* 14 (6): 1019–29. http://dx.doi.org/10.1002/etc.5620140613.

Stevens, M., G. Rappe, J. Maes, B. van Asten, and F. Ollevier. 2004. "*Micropogonias undulatus* (L.), Another Exotic Arrival in European Waters." *Journal of Fish Biology* 64 (4): 1143–46. http://dx.doi.org/10.1111/j.1095-8649.2004.00369.x.

Stewart, D.C., S.J. Middlemas, and A.F. Youngson. 2006. "Population Structuring in Atlantic Salmon (*Salmo salar*), Evidence of Genetic Influence on the Timing of Smolt Migration in Sub-Catchment Stocks." *Ecology of Freshwater Fish* 15 (4): 552–58. http://dx.doi.org/10.1111/j.1600-0633.2006.00197.x.

Stewart, D.C., G.W. Smith, and A.F. Youngson. 2002. "Tributary-Specific Variation in Timing of Return of Adult Atlantic Salmon (*Salmo salar*) to Fresh Water Has a Genetic Component." *Canadian Journal of Fisheries and Aquatic Sciences* 59 (2): 276–81. http://dx.doi.org/10.1139/f02-011.

Strange, J.S. 2013. "Factors Influencing the Behavior and Duration of Residence of Adult Chinook Salmon in a Stratified Estuary." *Environmental Biology of Fishes* 96 (2–3): 225–43. http://dx.doi.org/10.1007/s10641-012-0004-7.

Strauss, S.Y., J.A. Lau, and S.P. Carroll. 2006. "Evolutionary Responses of Natives to Introduced Species: What Do Introductions Tell Us about Natural Communities?" *Ecology Letters* 9 (3): 357–74. http://dx.doi.org/10.1111/j.1461-0248.2005.00874.x.

Strayer, D.L. 2010. "Alien Species in Fresh Waters: Ecological Effects, Interactions with Other Stressors, and Prospects for the Future." *Freshwater Biology* 55 (S1): 152–74. http://dx.doi.org/10.1111/j.1365-2427.2009.02380.x.

Stringfellow, W., J. Herr, G. Litton, M. Brunell, S. Borglin, J. Hanlon, Carl Chen, Justin Graham, Remie Burks, Randy Dahlgren, et al. 2009. "Investigation of River Eutrophication as Part of a Low Dissolved Oxygen Total Maximum Daily Load Implementation." *Water Science and Technology* 59 (1): 9–14. http://dx.doi.org/10.2166/wst.2009.739.

Sturdevant, M.V., E. Fergusson, N. Hillgruber, C. Reese, J. Orsi, R. Focht, Alex Wertheimer, and Bill Smoker. 2012. "Lack of Trophic Competition among Wild and Hatchery Juvenile Chum Salmon during Early Marine Residence in Taku Inlet, Southeast Alaska." *Environmental Biology of Fishes* 94 (1): 101–16. http://dx.doi.org/10.1007/s10641-011-9899-7.

Sundh, H., T.O. Nilsen, J. Lindström, L. Hasselberg-Frank, S.O. Stefansson, S.D. McCormick, and K. Sundell. 2014. "Development of Intestinal Ion Transporting Mechanisms during Smoltification and Seawater Acclimation in Atlantic Salmon *Salmo salar*." *Journal of Fish Biology* 85 (4): 1227–52. http://dx.doi.org/10.1111/jfb.12531.

Svenning, M.A., R. Borgstrom, T.O. Dehli, G. Moen, R.T. Barrett, T. Pedersen, and W. Vader. 2005. "The Impact of Marine Fish Predation on Atlantic Salmon Smolts (*Salmo salar*) in the Tana Estuary, North Norway, in the Presence of an Alternate Prey, Lesser Sand Eel (*Ammodytes marinus*)." *Fisheries Research* (Amsterdam) 76 (3): 466–74. http://dx.doi.org/10.1016/j.fishres.2005.06.015.

Sverdrup, A., E. Kjellsby, P.G. Krüger, R. Floysand, F.R. Knudsen, P.S. Enger, G. Serck-Hanssen, and K.B. Helle. 1994. "Effects of Experimental Seismic Shock on Vasoactivity of Arteries, Integrity of the Vascular Endothelium and on Primary Stress Hormones of the Atlantic Salmon." *Journal of Fish Biology* 45 (6): 973–95. http://dx.doi.org/10.1111/j.1095-8649.1994.tb01067.x.

Swanson, H.K., K.A. Kidd, J.A. Babaluk, R.J. Wastle, P.P. Yang, N.M. Halden, and J.D. Reist. 2010. "Anadromy in Arctic Populations of Lake Trout (*Salvelinus namaycush*), Otolith Microchemistry, Stable Isotopes, and Comparisons with Arctic Char (*Salvelinus alpinus*)." *Canadian Journal of Fisheries and Aquatic Sciences* 67 (5): 842–53. http://dx.doi.org/10.1139/F10-022.

Symons, P.E.K., and J.D. Martin. 1978. "Discovery of Juvenile Pacific Salmon (Coho) in a Small Coastal Stream of New Brunswick." *Fish Bulletin* 78: 487–89.

Tabor, R.A., G.S. Brown, and V.T. Luiting. 2004. "The Effect of Light Intensity on Sockeye Salmon Fry Migratory Behavior and Predation by Cottids in the Cedar River, Washington." *North American Journal of Fisheries Management* 24 (1): 128–45. http://dx.doi.org/10.1577/M02-095.

Tack, S.L. 1970. "The Summer Distribution and Standing Stock of the Fishes of Izembek Lagoon, Alaska." MS thesis, University of Alaska.

Takami, T. 1998. "Seawater Tolerance of White-Spotted Charr (*Salvelinus leucomaenis*) Related to Water Temperature." *Scientific Reports Hokkaido Fish Hatchery* 52: 11–19.

Tanaka, H., N.X. Tinh, M. Umeda, R. Hirao, E. Pradjoko, A. Mano, and K. Udo. 2012. "Coastal and Estuarine Morphology Changes Induced by the 2011 Great East Japan Earthquake Tsunami." *Coastal Engineering Journal* 54 (01): 1250010–35. http://dx.doi.org/10.1142/S0578563412500106.

Tanner, C.D., J.R. Cordell, J. Rubey, and L.M. Tear. 2002. "Restoration of Freshwater Intertidal Habitat Functions at Spencer Island, Everett, Washington." *Restoration Ecology* 10 (3): 564–76. http://dx.doi.org/10.1046/j.1526-100X.2002.t01-1-02034.x.

Tarbotton, M. and P.C. Harrison. 1996. "A Review of Recent Physical and Biological Development of the Southern Roberts Bank Seagrass System 1950–1994." Prepared for the Roberts Bank Environmental Review Committee. Vancouver: Triton Consultants Ltd., Vancouver.

Taylor, E.B. 1990. "Variability in Agonistic Behavior and Salinity Tolerance between and within Two Populations of Juvenile Chinook Salmon (*Oncorhynchus tshawytscha*) with Contrasting Life Histories." *Canadian Journal of Fisheries and Aquatic Sciences* 47 (11): 2172–80. http://dx.doi.org/10.1139/f90-242.

Teel, D.J., C. Baker, D.R. Kuligowski, T.A. Friesen, and B. Shields. 2009. "Genetic Stock Composition of Subyearling Chinook Salmon in Seasonal Floodplain Wetlands of the Lower Willamette River, Oregon." *Transactions of the American Fisheries Society* 138 (1): 211–17. http://dx.doi.org/10.1577/T08-084.1.

Teel, D.J., D.L. Bottom, S.A. Hinton, D.R. Kuligowski, G.T. McCabe, R. McNatt, G. Curtis Roegner, Lia A. Stamatiou, and Charles A. Simenstad. 2014. "Genetic Identification of Chinook Salmon in the Columbia River Estuary: Stock-Specific Distributions of Juveniles in Shallow Tidal Freshwater Habitats." *North American Journal of Fisheries Management* 34 (3): 621–41. http://dx.doi.org/10.1080/02755947.2014.901258.

Telesh, I.V., S.M. Golubkov, and A.F. Alimov. 2008. "The Neva Estuary Ecosystem." In *Ecology of Baltic Coastal Waters, Ecological Studies*, vol. 197, ed. U. Schiewer, 259–84. New York: Springer. http://dx.doi.org/10.1007/978-3-540-73524-3_12.

Texas A&M University. 2014. "Department of Oceanography." http://oceanworld.tamu.edu/resources/ocng_textbook/chapter06/Images/Fig6-2.htm.

Thames Estuary Partnership. 2015. Thames Estuary Partnership. Who We Are. http://thamesestuarypartnership.org/about/who-we-are.

Thedinga, J.F., S.W. Johnson, and A.D. Neff. 2011. "Diel Differences in Fish Assemblages in Nearshore Eelgrass and Kelp Habitats in Prince Williams Sound, Alaska." *Environmental Biology of Fishes* 90 (1): 61–70. http://dx.doi.org/10.1007/s10641-010-9718-6.

Thériault, V., E.S. Dunlop, U. Dieckmann, L. Bernatchez, and J.J. Dodson. 2008. "The Impact of Fishing-Induced Mortality on the Evolution of Alternative Life-History Tactics in Brook Charr." *Evolutionary Applications* 1 (2): 409–23. http://dx.doi.org/10.1111/j.1752-4571.2008.00022.x.

Thibault, I., R.D. Hedger, J.J. Dodson, J.C. Shiao, Y. Iizuka, and W.N. Tzeng. 2010. "Anadromy and the Dispersal of an Invasive Fish Species (*Oncorhynchus mykiss*) in Eastern Québec, as Revealed by Otolith Microchemistry." *Ecology of Freshwater Fish* 19 (3): 348–60. http://dx.doi.org/10.1111/j.1600-0633.2010.00417.x.

Thom, R.M., H.L. Diefenderfer, J. Vavrinec, and A.B. Borde. 2012. "Restoring Resiliency: Case Studies from Pacific Northwest Estuarine Eelgrass (*Zostera marina* L.) Ecosystems." *Estuaries and Coasts* 35 (1): 78–91. http://dx.doi.org/10.1007/s12237-011-9430-6.

Thom, R.M., R. Zeigler, and A.B. Borde. 2002. "Floristic Development Patterns in a Restored Elk River Estuarine Marsh, Grays Harbor, Washington." *Restoration Ecology* 10 (3): 487–96. http://dx.doi.org/10.1046/j.1526-100X.2002.01038.x.

Thompson, A.M. 2007. "Amphipods Are a Strong Interactor in the Foodweb of a Brown-Water Salmon River." MS thesis, University of Montana.

Thomsen, D.S., A. Koed, C. Nielsen, and S.S. Madsen. 2007. "Overwintering of Sea Trout (*Salmo trutta*) in Freshwater: Escaping Salt and Low Temperature or an Alternate Life Strategy?" *Canadian Journal of Fisheries and Aquatic Sciences* 64 (5): 793–802. http://dx.doi.org/10.1139/f07-059.

Thomsen, M.S., T. Wernberg, J.D. Olden, J.N. Griffin, and B.R. Silliman. 2011. "A Framework to Study the Context-Dependent Impacts of Marine Invasions." *Journal of Experimental Marine Biology and Ecology* 400 (1–2): 322–27. http://dx.doi.org/10.1016/j.jembe.2011.02.033.

Thomson, R.E. 1981. *Oceanography of the British Columbia Coast*. Canadian Special Publication of Fisheries and Aquatic Sciences No. 56. Ottawa: Minister of Supply and Services Canada.

Thorpe, J.E. 1994. "Salmonid Fishes and the Estuarine Environment." *Estuaries* 17 (1): 76–93. http://dx.doi.org/10.2307/1352336.

Thorstad, E.B., I.A. Fleming, P. McGinnity, D. Soto, V. Wennevik, and F. Whoriskey. 2008. *Incidence and Impacts of Escaped Farmed Atlantic Salmon* Salmo salar *in Nature*. Report from the Technical Working Group on Escapes of the Salmon, NINA Special Report 36. Washington, DC: World Wildlife Fund.

Thorstad, E.B., F. Whoriskey, A.H. Rikardsen, and K. Aarestrup. 2011. "Aquatic Nomads: The Life and Migrations of the Atlantic Salmon." In *Atlantic Salmon*, ed. Ø. Aas, S. Einum, A. Klemetsen, and J. Skurdal, 1–32. Oxford: Wiley-Blackwell Publishing.

Thorstad, E.B., F. Whoriskey, I. Uglem, A. Moore, A.H. Rikardsen, and B. Finstad. 2012. "A Critical Life Stage of the Atlantic Salmon *Salmo salar*: Behaviour and Survival during the Smolt and Initial Post Smolt Migration." *Journal of Fish Biology* 81 (2): 500–42. http://dx.doi.org/10.1111/j.1095-8649.2012.03370.x.

Thorsteinson, F.V., J.H. Helle, and D.G. Birkholz. 1971. "Salmon Survival in Intertidal Zones of Prince William Sound Streams in Uplifted and Subsided Areas." In *The Great Alaska Earthquake of 1964*, ed. J. Nybakker, 194–219. Washington, DC: National Marine Fisheries Service.

Thurow, F. 1966. "Beitrage zur Biologie und Bestandeskunde die Atlantischen Lachses (*Salmo salar* L.) in der Ostsee." *Ber. dt. wiss. kommn. Meeresforsch.* 18 (3/4): 223–379.

Toft, J.D. 2000. "Community Effects of the Non-Indigenous Aquatic Plant Water Hyacinth (*Eichhornia crassipes*) in the Sacramento/San Joaquin Delta, California." MS thesis, University of Washington.

Toft, J.D., C.A. Simenstad, J.R. Cordell, and L.F. Grimaldo. 2003. "The Effects of Introduced Water Hyacinth on Habitat Structure, Invertebrate Assemblages, and Fish Diets." *Estuaries* 26 (3): 746–58. http://dx.doi.org/10.1007/BF02711985.

Toft, J.D., J.R. Cordell, C.A. Simenstad, and L.A. Stamatiou. 2007. "Fish Distribution, Abundance, and Behavior along City Shoreline Types, Puget Sound." *North American Journal of Fisheries Management* 27 (2): 465–80. http://dx.doi.org/10.1577/M05-158.1.

Tokranov, A.M. 1994. "The Fish Community of the Bol'shaya River Estuary (Western Kamchatka)." *Journal of Ichthyology* 34 (4): 47–57. Translated from Russian; originally published in *Voprosy Ikhtiologii* 34 (1): 5–12.

Tokranov, A.M., and V.V. Maxsimenkov. 1995. "Feeding Habits of Predatory Fishes in the Bol'shaya River Estuary (West Kamchatka)." *Journal of Ichthyology* 35 (9): 102–12. Translated from Russian; originally published in *Voprosy Ikhtiologii* 35 (5): 651–58.

Tompkins, A., and C.D. Levings. 1991. "Interspecific Interactions Affecting the Survival of Chum Salmon Fry." In *15th Pacific Pink and Chum Salmon Workshop 1991 (Symposium Conducted at the Meeting of Pacific Salmon Commission and Fisheries and Ocean Canada, Parksville, BC)*, ed. B. White and I. Guthrie, 29–36.

Tschaplinski, P.J. 1982. "Aspects of the Population Biology of Estuary-Reared and Stream-Reared Juvenile Coho Salmon in Carnation Creek: A Summary of Current Research." In *Proceedings of the Carnation Creek Workshop: Ten-Year Malaspina College Review*, ed. G.F. Hartman, 289–305. Nanaimo, BC: Malaspina College.

Tsuda, Y., R. Kawabe, H. Tanaka, Y. Mitsunaga, T. Hiraishi, K. Yamamoto, and K. Nashimoto. 2006. "Monitoring the Spawning Behaviour of Chum Salmon with an Acceleration Data Logger." *Ecology of Freshwater Fish* 15 (3): 264–74. http://dx.doi.org/10.1111/j.1600-0633.2006.00147.x.

Tully, J.P., and F.G. Barber. 1960. "An Estuarine Analogy in the Sub-Arctic Pacific Ocean." *Journal of the Fisheries Board of Canada* 17 (1): 91–112. http://dx.doi.org/10.1139/f60-007.

Tveskov, M.A., and J.M. Erlandson. 2003. "The Haynes Inlet Weirs: Estuarine Fishing and Archaeological Site Visibility on the Southern Cascadia Coast." *Journal of Archaeological Science* 30 (8): 1023–35. http://dx.doi.org/10.1016/S0305-4403(02)00291-1.

Tytler, P., J.E. Thorpe, and W.M. Shearer. 1978. "Ultrasonic Tracking of the Movements of Atlantic Salmon Smolts (*Salmo salar* L.) in the Estuaries of Two Scottish Rivers." *Journal of Fish Biology* 12 (6): 575–86. http://dx.doi.org/10.1111/j.1095-8649.1978.tb04204.x.

Umeda, K., K. Matsumura, G. Okukawa, R. Sazawa, H. Honma, M. Arauchi, et al. 1981. "Coho Salmon *Oncorhynchus kisutch* Transplanted from North America into the Ichani River, Eastern Hokkaido, Japan." *Science Reports of the Hokkaido Salmon Hatchery* 35: 9–22.

United States Environmental Protection Agency. 2000. "Aquatic Life Criteria for Dissolved Oxygen – (Saltwater) Cape Cod to Cape Hatteras." Fact sheet. http://water.epa.gov/scitech/swguidance/standards/criteria/aqlife/dissolved/dofacts.cfm.

–. 2015. "Water Quality Standards for Surface Waters." http://water.epa.gov/scitech/swguidance/standards/.

Urabe, H., M. Nakajima, M. Torao, and T. Aoyama. 2010. "Evaluation of Habitat Quality for Stream Salmonids Based on a Bioenergetics Model." *Transactions of the American Fisheries Society* 139 (6): 1665–76. http://dx.doi.org/10.1577/T09-210.1.

Urawa, S. 1993. "Effects of *Ichthyobodo necator* Infections on Seawater Survival of Juvenile Chum Salmon (*Oncorhynchus keta*)." *Aquaculture* (Amsterdam) 110 (2): 101–10. http://dx.doi.org/10.1016/0044-8486(93)90264-Y.

Urawa, S., K. Nagasawa, L. Margolis, and A. Moles. 1998. *Stock Identification of Chinook Salmon* (Oncorhynchus tshawytscha) *in the North Pacific Ocean and*

Bering Sea by Parasite Tags. North Pacific Anadromous Fish Commission Bulletin No. 1. Vancouver: North Pacific Anadromous Fish Commission.

Valle-Levinson, A. 2010. "Definition and Classification of Estuaries." In *Contemporary Issues in Estuarine Physics*, ed. A. Valle-Levinson, 1–11. Cambridge: Cambridge University Press. http://dx.doi.org/10.1017/CBO9780511676567.002.

Van den Bergh, E., S. van Damme, J. Graveland, D. De Jong, I. Baten, and P. Meire. 2005. "Ecological Rehabilitation of the Schelde Estuary (the Netherlands – Belgium; Northwest Europe), Linking Ecology, Safety against Floods, and Accessibility for Port Development." *Restoration Ecology* 13 (1): 204–14. http://dx.doi.org/10.1111/j.1526-100X.2005.00025.x.

van Hyning, J.M. 1973. "Factors Affecting the Abundance of Fall Chinook Salmon in the Columbia River." *Research Reports of the Fish Commission of Oregon* 4 (1): 1–87.

van Katwijk, M.M., A.R. Bos, V.N. de Jonge, L.S.A.M. Hanssen, D.C.R. Hermus, and D.J. de Jong. 2009. "Guidelines for Seagrass Restoration: Importance of Habitat Selection and Donor Population, Spreading of Risks, and Ecosystem Engineering Effects." *Marine Pollution Bulletin* 58 (2): 179–88. http://dx.doi.org/10.1016/j.marpolbul.2008.09.028.

van Zwol, J.A., B.D. Neff, and C.C. Wilson. 2012. "The Influence of Non-Native Salmonids on Circulating Hormone Concentrations in Juvenile Atlantic Salmon." *Animal Behaviour* 83 (1): 119–29. http://dx.doi.org/10.1016/j.anbehav.2011.10.015.

vanden Borre, J., D. Paelinckx, C.A. Mücher, L. Kooistra, B. Haest, G. de Blust, and A.M. Schmidt. 2011. "Integrating Remote Sensing in Natura 2000 Habitat Monitoring: Prospects on the Way Forward." *Journal for Nature Conservation* 19 (2): 116–25. http://dx.doi.org/10.1016/j.jnc.2010.07.003.

Varnavsky, V.S., S.V. Kalinin, N.M. Kinas, and S.S. Rostomova. 1992. "The Early Sea Life of Coho, *Oncorhynchus kisutch*, and Pink Salmon, *O. gorbuscha*, as a Period of Completion Of Smoltification." *Environmental Biology of Fishes* 34 (4): 401–8. http://dx.doi.org/10.1007/BF00004744.

Vasil'eva, E.D. 2003. "Main Alterations in Ichthyofauna of the Largest Rivers of the Northern Coast of the Black Sea in the Last 50 Years: A Review." *Folia Zoologica (Praha)* 52 (4): 337–58.

Vincent-Lang, D., M. Alexandersdottir, and D. McBride. 1993. "Mortality of Coho Salmon Caught and Released Using Sport Tackle in the Little Susitna River, Alaska." *Fisheries Research* 15 (4): 339–56. http://dx.doi.org/10.1016/0165-7836(93)90085-L.

Volk, E.C., S.L. Schroder, and J.J. Grimm. 1999. "Otolith Thermal Marking." *Fisheries Research* 43 (1–3): 205–19. http://dx.doi.org/10.1016/S0165-7836(99)00073-9.

Volk, E.C., R.C. Wissmar, C.A. Simenstad, and D.M. Eggers. 1984. "Relationship between Otolith Microstructure and the Growth of Juvenile Chum Salmon (*Oncorhynchus keta*) under Different Prey Rations." *Canadian Journal of Fisheries and Aquatic Sciences* 41 (1): 126–33. http://dx.doi.org/10.1139/f84-012.

Volk, E.C., D.L. Bottom, K.K. Jones, and C.A. Simenstad. 2010. "Reconstructing Juvenile Chinook Salmon Life History in the Salmon River Estuary, Oregon, Using Otolith Microchemistry and Microstructure." *Transactions of the American Fisheries Society* 139 (2): 535–49. http://dx.doi.org/10.1577/T08-163.1.

Volpe, J.P., E.B. Taylor, D.W. Rimmer, and B.W. Glickman. 2000. "Evidence of Natural Reproduction of Aquaculture-Escaped Atlantic Salmon in a Coastal British Columbia River." *Conservation Biology* 14 (3): 899–903. http://dx.doi.org/10.1046/j.1523-1739.2000.99194.x.

Vronskiy, B.B. 1972. "Reproductive Biology of the Kamchatka River Chinook Salmon (*Oncorhynchus tshawytscha*)." *Journal of Ichthyology* 12: 259–73. Translated from Russian; originally published in *Voprosy Ikhtiologii* 12: 259–73.

Wake, H. 2005. "Oil Refineries: A Review of Their Ecological Impacts on the Aquatic Environment." *Estuarine, Coastal and Shelf Science* 62 (1–2): 131–40. http://dx.doi.org/10.1016/j.ecss.2004.08.013.

Waldichuk, M. 1993. "Fish Habitat and the Impact of Human Activity with Particular Reference to Pacific Salmon." In *Perspective on Canadian Marine Fisheries Management* (Canadian Bulletin of Fisheries and Aquatic Sciences No. 226), ed. L.S. Parsons and W.H. Lear, 295–337. Ottawa: National Research Council of Canada.

Walters, D.M., M.A. Mills, K.M. Fritz, and D.F. Raikow. 2010. "Spider-Mediated Flux of PCBs from Contaminated Sediments to Terrestrial Ecosystems and Potential Risks to Arachnivorous Birds." *Environmental Science and Technology* 44 (8): 2849–56. http://dx.doi.org/10.1021/es9023139.

Waples, R.S., T. Beechie, and G.R. Pess. 2009. "Evolutionary History, Habitat Disturbance Regimes, and Anthropogenic Changes: What Do These Mean for Resilience of Pacific Salmon Populations?" *Ecology and Society* 14 (1): 3. http://www.ecologyandsociety.org/vol14/iss1/art3/.

Ward, B.R., and P.A. Slaney. 1990. "Returns of Pen-Reared Steelhead from Riverine, Estuarine, and Marine Releases." *Transactions of the American Fisheries Society* 119 (3): 492–99. http://dx.doi.org/10.1577/1548-8659(1990)119<0492: ROPSFR>2.3.CO;2.

Ward, D.L., A.A. Nigro, R.A. Farr, and C.J. Knutsen. 1994. "Influence of Waterway Development on Migrational Characteristics of Juvenile Salmonids in the Lower Willamette River, Oregon." *North American Journal of Fisheries Management* 14 (2): 362–71. http://dx.doi.org/10.1577/1548-8675(1994)014<0362: IOWDOM>2.3.CO;2.

Ward, D.M., and N.A. Hvidsten. 2011. "Predation: Compensation and Context Dependence." In *Atlantic Salmon Ecology*, ed. Ø. Aas, A. Klemetsen, S. Einum, and J. Skurdal, 199–220. Oxford: Wiley-Blackwell Publishing.

Warner, M.J., M. Kawase, and J.A. Newton. 2001. "Recent Studies of the Overturning Circulation in Hood Canal." In *Proceedings of Puget Sound Research 2001*, ed. T. Droscher, 9. Olympia, WA: Puget Sound Action Team.

Warren, B.A. 1983. "Why Is No Deep Water Formed in the North Pacific?" *Journal of Marine Research* 41 (2): 327–47. http://dx.doi.org/10.1357/002224083788520207.

Watanabe, K., T. Minami, H. Iizumi, and S. Imamura. 1996. "Interspecific Relationship by Composition of Stomach Contents of Fish at Akkeshi-ko, an Estuary at Eastern Hokkaido, Japan." *Bulletin of the Hokkaido National Fisheries Research Institute* 60: 239–76.

Watt, W.D. 1987. "A Summary of the Impacts of Acid Rain on Atlantic Salmon (*Salmo salar*) in Canada." *Water, Air, and Soil Pollution* 35 (1): 27–35. http://dx.doi.org/10.1007/BF00183841.

Waycott, M., C.M. Duarte, T.J.B. Carruthers, R.J. Orth, W.C. Dennison, S. Olyarnik, A. Calladine, J.W. Fourqurean, K.L. Heck, A.R. Hughes, et al. 2009. "Accelerating Loss of Seagrasses across the Globe Threatens Coastal Ecosystems." *Proceedings of the National Academy of Sciences of the United States of America* 106 (30): 12377–81. http://dx.doi.org/10.1073/pnas.0905620106.

Weatherley, A.H., and H.G. Gill. 1995. "Growth." In *Physiological Ecology of Pacific Salmon*, ed. C. Groot, L. Margolis, and W.C. Clarke, 103–58. Vancouver: UBC Press.

Webb, D.G. 1991. "Effect of Predation by Juvenile Pacific Salmon on Marine Harpacticoid Copepods. I. Comparisons of Patterns of Copepod Mortality with Patterns of Salmon Consumption." *Marine Ecology Progress Series* 72: 25–36. http://dx.doi.org/10.3354/meps072025.

Weber, E.D., and K.D. Fausch. 2003. "Interactions between Hatchery and Wild Salmonids in Streams: Differences in Biology and Evidence for Competition." *Canadian Journal of Fisheries and Aquatic Sciences* 60 (8): 1018–36. http://dx.doi.org/10.1139/f03-087.

Webster, S.J., L.M. Dill, and J.S. Korstrom. 2007. "The Effects of Depth and Salinity on Juvenile Chinook Salmon *Oncorhynchus tshawytscha* (Walbaum) Habitat Choice in an Artificial Estuary." *Journal of Fish Biology* 71 (3): 842–51. http://dx.doi.org/10.1111/j.1095-8649.2007.01553.x.

Wedemeyer, G.A., R.L. Saunders, and W.C. Clarke. 1980. "Environmental Factors Affecting Smoltification and Early Marine Survival of Anadromous Salmonids." *Marine Fisheries Review* 42: 1–14.

Weiss, G.M., G.B. McManus, and H.R. Harvey. 1996. "Development and Lipid Composition of the Harpacticoid Copepod *Nitocra spinipes* Reared on Different Diets." *Marine Ecology Progress Series* 132: 57–61. http://dx.doi.org/10.3354/meps132057.

Weitkamp, L.A. 2008. "Buoyancy Regulation by Hatchery and Wild Coho Salmon during the Transition from Freshwater to Marine Environments." *Transactions of the American Fisheries Society* 137 (3): 860–68. http://dx.doi.org/10.1577/T07-081.1.

Weitkamp, L.A., P.J. Bentley, and M.N. Litz. 2012. "Seasonal and Interannual Variation in Juvenile Salmonids and Associated Fish Assemblage in Open Waters of the Lower Columbia River Estuary." *Fisheries Bulletin* 110: 426–50.

Weitkamp, L.A., G. Goulette, J. Hawkes, M. O'Malley, and C. Lipsky. 2014. "Juvenile Salmon in Estuaries: Comparisons between North American Atlantic and Pacific Salmon Populations." *Reviews in Fish Biology and Fisheries* 24 (3): 713–36. http://dx.doi.org/10.1007/s11160-014-9345-y.

Welton, J.S. 1979. "Life History and Production of the Amphipod *Gammarus pulex* in a Dorset Chalk Stream." *Freshwater Biology* 9 (3): 263–75. http://dx.doi.org/10.1111/j.1365-2427.1979.tb01508.x.

Wendelaar Bonga, S.E. 2011. "Hormone Response to Stress." In *Encyclopedia of Fish Physiology*, ed. A.P. Farrell, 1515–23. Amsterdam: Elsevier. http://dx.doi.org/10.1016/B978-0-12-374553-8.00183-0.

Wernand, M.R., H.J. van der Woerd, and W.W. Gieskes. 2013. "Trends in Ocean Colour and Chlorophyll Concentration from 1889 to 2000, Worldwide." *PLoS One* 8 (6): e63766. http://dx.doi.org/10.1371/journal.pone.0063766.

Werner, A.E., and J. Robinson. 1978. "Acute Effects of Wood-Pulp on Sockeye Salmon (*Oncorhynchus nerka*)." *Water, Air, and Soil Pollution* 9 (1): 69–81. http://dx.doi.org/10.1007/BF00185748.

Westley, P.A., and I.A. Fleming. 2011. "Landscape Factors that Shape a Slow and Persistent Aquatic Invasion: Brown Trout in Newfoundland 1883–2010." *Diversity and Distributions* 17 (3): 566–79. http://dx.doi.org/10.1111/j.1472-4642.2011.00751.x.

Westley, P.A., D.W. Ings, and I.A. Fleming. 2011. *A Review and Annotated Bibliography of the Impacts of Invasive Brown Trout (Salmo trutta) on Native Salmonids, with an Emphasis on Newfoundland Waters*. Canadian Technical Report of Fisheries and Aquatic Sciences No. 2924. St. John's: Department of Fisheries and Oceans.

Whalen, K.G., D.L. Parrish, and S.D. McCormick. 1999. "Migration Timing of Atlantic Salmon Smolts Relative to Environmental and Physiological Factors." *Transactions of the American Fisheries Society* 128 (2): 289–301. http://dx.doi.org/10.1577/1548-8659(1999)128<0289:MTOASS>2.0.CO;2.

White, H.C. 1936. "The Food of Kingfishers and Mergansers on the Margaree River, Nova Scotia." *Journal of the Biological Board of Canada* 2 (3): 299–309. http://dx.doi.org/10.1139/f36-011.

Whoriskey, F. 2009. "Management Angels and Demons in the Conservation of the Atlantic Salmon in North America." In *Pacific Salmon: Ecology and Management of Western Alaska's Populations*, ed. Charles C. Kruger and Christian E. Zimmerman. *American Fisheries Society Symposium* 70: 1083–1101.

Wild Salmon Center. 2015. "The Wild Salmon Center, Portland, Oregon." http://www.wildsalmoncenter.org/.

World Wildlife Fund. 2001. *The Status of Wild Atlantic Salmon: A River by River Assessment*. Oslo, Copenhagen, and Washington, DC: WWF-Norway, WWF European Freshwater Program, and WWF-US. http://wwf.panda.org/about_our_earth/blue_planet/publications/?3729/TheStatus-of-Wild-Atlantic-Salmon-A-River-by-River-Assessment.

Wilkinson, B.H., and B.J. McElroy. 2007. "The Impact of Humans on Continental Erosion and Sedimentation." *Geological Society of America Bulletin* 119 (1–2): 140–56. http://dx.doi.org/10.1130/B25899.1.

Williams, D.D. 1980. "Invertebrate Drift Lost to the Sea during Low Flow Conditions in a Small Coastal Stream in Western Canada." *Hydrobiologia* 75 (3): 251–51. http://dx.doi.org/10.1007/BF00006489.

Williams, G.D., R.M. Thom, D.K. Shreffler, J.A. Southard, L.K. O'Rourke, S.L. Sargeant, et al. 2003. *Assessing Overwater Structure-Related Predation Risk on Juvenile Salmon: Field Observations and Recommended Protocols (WA-RD 573.1)*. Olympia: Washington State Department of Transportation.

Williams, G.L., and O.E. Langer. 2002. *Review of Estuary Management Plans in British Columbia*. Canadian Manuscript Report of Fisheries and Aquatic Sciences No. 2605. Ottawa: Fisheries and Oceans Canada.

Williams, J.G., R.W. Zabel, R.S. Waples, J.A. Hutchings, and W.P. Connor. 2008. "Potential for Anthropogenic Disturbances to Influence Evolutionary Change in the Life History of a Threatened Salmonid." *Evolutionary Applications* 1 (2): 271–85. http://dx.doi.org/10.1111/j.1752-4571.2008.00027.x.

Williams, R.N., ed. 2006. *Return to the River: Restoring Salmon to the Columbia River.* Amsterdam: Elsevier Academic Press. http://dx.doi.org/10.1016/B978-012088414-8/50016-3.

Wilson, C.C., and P.D. Hebert. 1998. "Phylogeography and Postglacial Dispersal of Lake Trout (*Salvelinus namaycush*) in North America." *Canadian Journal of Fisheries and Aquatic Sciences* 55 (4): 1010–24. http://dx.doi.org/10.1139/f97-286.

Wilson, G., K. Ashley, S. Mouldey Ewing, P. Slaney, and R.W. Land. 1999. *Development of a Resident Trout Fishery on the Adam River through Increased Habitat Productivity: Final Report of the 1992–1997 Project.* RD Report 68. Victoria: British Columbia Ministry of Fisheries. http://www.env.gov.bc.ca/wld/documents/fisheriesrpts/FPR68/fpr68.pdf.

Wilson, S.M., S.G. Hinch, E.J. Eliason, A.P. Farrell, and S.J. Cooke. 2013. "Calibrating Acoustic Acceleration Transmitters for Estimating Energy Use by Wild Adult Pacific Salmon." *Comparative Biochemistry and Physiology. Part A, Molecular and Integrative Physiology* 164 (3): 491–98. http://dx.doi.org/10.1016/j.cbpa.2012.12.002.

Winders, W. 2013. "Tracking the Unicorn Trout." *The Salter* (Spring 2013): 3–4. http://www.searunbrookie.org/wp-content/uploads/2013/03/Salter_Spring_2013.pdf.

Winemiller, K.O. 1989. "Patterns of Variation in Life History among South American Fishes in Seasonal Environments." *Oecologia* 81 (2): 225–41. http://dx.doi.org/10.1007/BF00379810.

Wissmar, R.C., and C.A. Simenstad. 1988. "Energetic Constraints of Juvenile Chum Salmon (*Oncorhynchus keta*) Migrating in Estuaries." *Canadian Journal of Fisheries and Aquatic Sciences* 45 (9): 1555–60. http://dx.doi.org/10.1139/f88-184.

–. 1998. "Variability of Riverine and Estuarine Ecosystem Productivity for Supporting Pacific Salmon." In *Change in Pacific Northwest Coastal Ecosystems: Proceedings of the Pacific Northwest Coastal Ecosystems Regional Study Workshop, August 13–14, 1996, Troutdale, Oregon* (NOAA Coastal Ocean Program Decision Analysis Series No. 11), ed. G.R. McMurray and R.J. Bailey, 253–94. Silver Spring, MD: National Oceanic and Atmospheric Administration Coastal Ocean Office.

Wolanski, E. 2007. *Estuarine Ecohydrology.* Queensland, Australia: Elsevier.

Wolanski, E., and D. McLusky, eds. 2011. *Treatise on Estuarine and Coastal Science.* Amsterdam: Elsevier.

Wolf, J., I.A. Walkington, J. Holt, and R. Burrows. 2009. "Environmental Impacts of Tidal Power Schemes." *Proceedings of Institution of Civil Engineers Maritime Engineering* 162 (4): 165–77.

Wonham, M.J., J.T. Carlton, G.M. Ruiz, and L.D. Smith. 2000. "Fish and Ships: Relating Dispersal Frequency to Success in Biological Invasions." *Marine Biology* 136 (6): 1111–21. http://dx.doi.org/10.1007/s002270000303.

Wood, C.C. 1987. "Predation of Juvenile Pacific Salmon by the Common Merganser (*Mergus merganser*) on Eastern Vancouver Island. 1: Predation during the Seaward Migration." *Canadian Journal of Fisheries and Aquatic Sciences* 44 (5): 941–49. http://dx.doi.org/10.1139/f87-112.

–. 2008. "Managing Biodiversity of Pacific Salmon: Lessons from the Skeena River Sockeye Salmon Fishery in British Columbia." In *Reconciling Fisheries with*

Conservation: Proceedings of the Fourth World Fisheries Congress, American Fisheries Society Symposium 49, ed. J.L. Nielsen, J.J. Dodson, K. Friedland, T.R. Hamon, J. Musick, and E. Verspoor, 349–64. Bethesda, MD: American Fisheries Society.

Wood, C.C., N.B. Hargreaves, D.T. Rutherford, and B.T. Emmett. 1993. "Downstream and Early Marine Migratory Behaviour of Sockeye Salmon (*Oncorhynchus nerka*) Smolts Entering Barkley Sound, Vancouver Island." *Canadian Journal of Fisheries and Aquatic Sciences* 50 (6): 1329–37. http://dx.doi.org/10.1139/f93-151.

Wood, C.C., J.W. Bickham, R.J. Nelson, C.J. Foote, and J.C. Patton. 2008. "Recurrent Evolution of Life History Ecotypes in Sockeye Salmon: Implications for Conservation and Future Evolution." *Evolutionary Applications* 1 (2): 207–21. http://dx.doi.org/10.1111/j.1752-4571.2008.00028.x.

Wood, P.J., and P.D. Armitage. 1997. "Biological Effects of Fine Sediment in the Lotic Environment." *Environmental Management* 21 (2): 203–17. http://dx.doi.org/10.1007/s002679900019.

Wootton, J.T., C.A. Pfister, and J.D. Forester. 2008. "Dynamic Patterns and Ecological Impacts of Declining Ocean pH in a High-Resolution Multi-Year Dataset." *Proceedings of the National Academy of Sciences of the United States of America* 105 (48): 18848–53. http://dx.doi.org/10.1073/pnas.0810079105.

Workman, M.L., and J.E. Merz. 2007. "Introduced Yellowfin Goby, *Acanthogobius flavimanus*: Diet and Habitat Use in the Lower Mokelumne River, California." *San Francisco Estuary and Watershed Science* 5 (1): 1–13.

Wright, B.E., S.D. Riemer, R.F. Brown, A.M. Ougzin, and K.A. Bucklin. 2007. "Assessment of Harbor Seal Predation on Adult Salmonid in a Pacific Northwest Estuary." *Ecological Applications* 17 (2): 338–51. http://dx.doi.org/10.1890/05-1941.

Wroblewski, J.S., L.K. Kryger-Hann, D.A. Methven, and R.L. Haedrich. 2007. "The Fish Fauna of Gilbert Bay, Labrador: A Marine Protected Area in the Canadian Subarctic Coastal Zone." *Journal of the Marine Biological Association of the United Kingdom* 87 (02): 575–87. http://dx.doi.org/10.1017/S0025315407054136.

Xie, Y., A.P. Gray, F.J. Martens, J.L. Boffey, and J.D. Cave. 2005. *Use of Dual-Frequency Identification Sonar to Verify Split-Beam Estimates of Salmon Flux and to Examine Fish Behaviour in the Fraser River*. Technical Report No. 16. Vancouver: Pacific Salmon Commission.

Yamamoto, T., and U.G. Reinhardt. 2003. "Dominance and Predator Avoidance in Domesticated and Wild Masu Salmon *Oncorhynchus masou*." *Fisheries Science* 69 (1): 88–94. http://dx.doi.org/10.1046/j.1444-2906.2003.00591.x.

Yang, Z., K.L. Sobocinski, D. Heatwole, T. Khangaonkar, R. Thom, and R. Fuller. 2010. "Hydrodynamic and Ecological Assessment of Nearshore Restoration: A Modeling Study." *Ecological Modelling* 221 (7): 1043–53. http://dx.doi.org/10.1016/j.ecolmodel.2009.07.011.

Young, K.A. 2004. "Asymmetric Competition, Habitat Selection, and Niche Overlap in Juvenile Salmonids." *Ecology* 85 (1): 134–49. http://dx.doi.org/10.1890/02-0402.

Young, P.S., C. Swanson, and J.J. Cech Jr. 2004. "Photophase and Illumination Effects on the Swimming Performance and Behavior of Five California Estuarine Fishes." *Copeia* 2004 (3): 479–87. http://dx.doi.org/10.1643/CP-03-061R1.

Yurk, H., and A.W. Trites. 2000. "Experimental Attempts to Reduce Predation by Harbor Seals on Out-Migrating Juvenile Salmonids." *Transactions of the American*

Fisheries Society 129 (6): 1360–66. http://dx.doi.org/10.1577/1548-8659 (2000)129<1360:EATRPB>2.0.CO;2.

Zacharias, M.A., D.E. Howes, J.R. Harper, and P. Wainwright. 1998. "The British Columbia Marine Ecosystem Classification: Rationale, Development, and Verification." *Coastal Management* 26 (2): 105–24. http://dx.doi.org/10.1080/08920759809362347.

Zale, A.V., D.L. Parrish, and T.M. Sutton, eds. 2012. *Fisheries Techniques*. 3rd ed. Bethesda, MD: American Fisheries Society.

Zama, A. 1987. "Biological Observations on Sea-Run Brown Trout in Fiordo Aysén, Southern Chile (Pisces: Salmonidae)." *Revista de Biologia Marina Valparaiso* 23 (2): 193–213.

Zaporozhets, O.M., and G.V. Zaporozhets. 2012. "Some Consequences of Pacific Salmon Hatchery Production in Kamchatka: Changes in Age Structure and Contributions to Natural Spawning Populations." *Environmental Biology of Fishes* 94 (1): 219–30. http://dx.doi.org/10.1007/s10641-011-9932-x.

Zaugg, W.S., E.F. Prentice, and F.W. Waknitz. 1985. "Importance of River Migration to the Development of Seawater Tolerance in Columbia River Anadromous Salmonids." *Aquaculture* (Amsterdam) 51 (1): 33–47. http://dx.doi.org/10.1016/0044-8486(85)90238-8.

Zeller, D., S. Booth, E. Pakhomov, W. Swartz, and D. Pauly. 2011. "Arctic Fisheries Catches in Russia, USA, and Canada: Baselines for Neglected Ecosystems." *Polar Biology* 34 (7): 955–73. http://dx.doi.org/10.1007/s00300-010-0952-3.

Zenkovich, V.P. 1967. *Processes of Coastal Development*. New York: Wiley-Interscience.

Zeug, S.C., and B.J. Cavallo. 2013. "Influence of Estuary Conditions on the Recovery Rate of Coded Wire Tagged Chinook Salmon (*Oncorhynchus tshawytscha*) in an Ocean Fishery." *Ecology of Freshwater Fish* 22 (1): 157–68. http://dx.doi.org/10.1111/eff.12013.

Zhou, Z., and Y. Tong. 2010. "Sediment in Rivers: Origins and Challenges." *Stockholm Water Front Magazine: A Forum for Global Water Issues* 4: 8–10. http://www.siwi.org/documents/Resources/Water_Front/WF_4-2010.pdf.

Zydlewski, J., J.R. Johnson, J. Hogle, J. Brunzell, S. Clements, M. Karnowski, and C. Shreck. 2008. "Seaward Migration of Coastal Cutthroat Trout (*Oncorhynchus clarkii clarkii*) from Four Tributaries of the Columbia River." In *The 2005 Coastal Cutthroat Trout Symposium: Status, Management, Biology, and Conservation*, ed. P.J. Connolly, T.H. Williams, and R.E. Gresswell, 65–74. Portland: Oregon Chapter, American Fisheries Society.

Index

Notes: "(f)" following a number indicates a figure; "(t)" following a number indicates a table

Å River estuary, Norway, 123
Aboriginal peoples, cultural value of salmonids, 192; enumeration of salmonid harvests, 64; subsistence fishing, 189
abundance, and fitness inferences, 144–48; as indicator of estuarine fitness for salmonids, 207–8, 210; measuring, 70
Acanthogobius flavimanus. See goby, yellowfin
acid rain, 135
Adam River estuary, BC, 60
adult salmonids, definition, 16; feeding behaviours, 78–79; fitness/survival, 139; growth, 83; holding/staging behaviour, 74; homing behaviour, 79; osmoregulation, 92–93; osmoregulation stressors, 155; as predators, 119; as prey, 109, 118; sampling and enumeration, 64–65; schooling behaviour, 79; tagging, 68; water salinity, 49; water temperatures, 48–49
agriculture, and habitat loss, 128; water quality issues, 135–36
Ainu, salmonid fishing, 3–4
Akaroa Harbour, New Zealand, 99
Alaska, estuary sizes, 219; food web, 100; habitats, 43; salmonid fishing history, 3; salmonid fitness/survival, 138, 140
Aleuts, salmonid fishing, 3
alevins, 17, 88
Allosmerus elongates. See whitebait smelt
Alosa aestivalis. See blueback herring
Alosa alosa. See shad, allis
Alosa fallax. See shad, twaite
Alosa sapidissima. See shad, American
Alosa pseudoharengus, 116
Alouette River, BC, 244
Alsea River estuary, OR, 167
amago (*Oncorhynchus masou ishikawae*), 8, 13, 90
Amu Darya River estuary, Uzbekistan, 9, 54, 134
Amur Sleeper (*Perccottus glenii*), 56
Amur River estuary, Russia, 28, 111
anadromous salmonids, conservation concerns, 8; definition, 255; evolution, 6–8; as invasive species, 57, 60, 182–84; life cycle, 5–6, 6(f)
anadromy, evolution, 8, 94
Anapka River estuary, Russia, 86, 96
aquaculture: broodstock reconditioning, 191–92; definition, 255; escaped salmonids, 59–60, 150, 173, 184, 202, 229, 233, 252; evolutionary effects, 7–8; growth of industry,

191, 229; parasite/disease spread, 194–95, 197; prohibitions, 226; siting of farms, 229. *See also* salmonids, hatchery-reared

Aral Sea, hypersalinity, 20, 134; salmonid community, 54; water levels, 24–25

archival tags, 69–70

Arctic char (*Salvelinus alpinus*), competitive interactions, 123, 126, 183; enumeration techniques, 64; estuary use, 13; evolution, 7; feeding behaviours, 95, 99, 104–5; growth, 81, 83, 85; habitats, 38–39; harvesting, 189, 225; osmoregulation, 92; as predators, 114; as prey, 118; residency behaviour, 77; sea lice on, 196; smoltification, 89; tagging, 70

Arctic estuaries, classification, 20; dearth of research, 10; fish harvesting, 189; rock and boulder beaches, 38; salmonid communities, 53; seasonal classification, 20; temperatures, 27

Argentina, invasive salmonids, 58, 60, 61(f)

artificial habitats: carrying capacity, 164; competitive interactions in, 125, 176(t); features, 37(t), 47; food web, 105, 158; and loss of natural habitats, 133, 176(t); predation in, 119, 169, 176(t); restoration from effects of, 237

Ashida River estuary, Japan, 54

Asia, habitat loss and degradation, 128; salmonids of conservation concern, 8

Asiatic smelt (*Osmerus mordax dentex*), 114–15

Atlantic cod (*Gadus morhua*), 114, 116–18, 166–67

Atlantic croaker (*Micropogonias undulatus*), 56

Atlantic estuaries, invasive salmonids, 58; subhabitats, 43; predation, 113–15, 119, 126

Atlantic herring (*Clupea harengus*), 102, 123, 180

Atlantic salmon (*Salmo salar*): abundance, 220; competitive interactions, 122–23, 126, 178–80, 182–84; conservation concerns, 8, 226, 229; distribution, 59–60, 202; estuary use, 13, 16; evolution, 7; farmed, 191; feeding behaviours, 78–79, 99, 102; fitness/survival, 138–39, 143–44, 147; food losses from habitat changes, 157; genetic structure, 139; growth, 83–84; habitats, 39, 45; harvesting, 189, 190(f), 225; as invasive species, 57, 59–60; life history types, 18; migration, 73; osmoregulation, 93; osmoregulation stressors, 151, 154–55; parasites on, 195–97; as part of estuarine ecosystem, 11; as prey, 111, 116, 118–19, 126, 166–69, 172, 174, 203; protection from fishing, 189; as reference species for biological integrity, 212–13; residency behaviour, 77, 88; smoltification, 88–89, 91; specialized adaptation to estuaries, 244; tagging, 68; territorial behaviour, 79; water temperatures, 48, 153

Atlantic salmonids, evolution, 7; habitat loss, 10; harvesting (historic), 3

Atlantic sturgeon (*Acipenser sturio*), 53

Aurland River estuary, Norway, subhabitats, 38; predation, 114

Australia, competitive interactions, 183; invasive salmonids, 58, 173

Avacha River estuary, Russia, 85

Aysén Fjord, Chile, 59, 226

baited hooks, for sampling, 65

Baltic Sea, beach habitats, 38; competitive interactions, 178–80; food web, 107; predation, 167; water quality issues, 135

bar-built estuaries. *See* lagoon estuaries

barachois estuaries. *See* lagoon estuaries

barrages, 132

beach seines, 63–64, 66, 71, 188–89

beaches, carrying capacity, 163; competitive interactions in, 123, 175, 176(t), 177, 184; definition, 35; example, 40(f); food web, 101–2; habitat loss, 129, 157, 166, 175, 176(t), 177; predation in, 114, 166, 176(t); restoration, 235; salmonid sampling techniques, 64; Svartå River estuary, Finland, 38(f); types, 36(t), 38–39

Beaufort Sea estuaries, AK, 99, 104, 123

behaviours, importance of understanding, 72; and predation risk, 112; research/data needs, 80, 250–51; stressor effects, 151(f), 155; types, 72–79

Bekanbeushi River estuary, Japan, 104, 226

Bella Coola River estuary, BC, 119

Beothuk, salmonid fishing, 3

Big Qualicum River, BC, 85, 114, 167

bioassays, definition, 255; disease testing, 193; effects of pollutants on salmonids, 211; environmental testing, 204; salmonid growth testing, 84; water quality standards, 210

biodiversity, 52, 255

biological indicators, 207–10

biomarkers, definition, 255; of water quality, 211–12

biotic interactions. *See* competition; predation

birds, as predators, 110–11, 116, 118, 121, 166, 168, 172–73; sanctuaries, 226

block nets, 65

blueback herring (*Alosa aestivalis*), 116

bluegill (*Lepomis macrochirus*), 181

Bogataya River estuary, Russia, 92, 99, 99(f)

Bol'shaya River estuary, Russia, competitive interactions, 124(f); conservation concerns, 223; estuary types, 22; food web, 103; predation, 116, 119; specialized salmonid adaptations, 243–44

boulder beaches, 36(t), 38

Britannia Creek, BC, 164

British Columbia, competitive interactions, 182; conservation strategies, 226, 230; estuary sizes, 219; fishing, 191; glacier-fed rivers, 28; habitat classification, 32; habitats, 42, 46; sea lice on salmonids, 195; seasonal habitat shifts, 32; tidal ranges, 24–25; water quality improvement, 237; water temperatures, 28

brook trout (*Salvelinus fontinalis*), competitive interactions, 182–83; distribution, 59; estuary use, 13; evolution, 7; habitat restoration, 235; habitats, 41, 45; as invasive species, 57, 59; osmoregulation stressors, 152, 155; as prey, 116, 174; sampling techniques, 63; short-term evolutionary considerations, 216–17; tagging, 69

brown bullhead (*Ictalurus nebulosus*), 55

brown trout (*Salmo trutta*), competitive interactions, 122, 126, 182; density-dependent growth, 160; distribution, 57–58; as invasive species, 57, 60, 61(f); as predator, 174. *See also* sea trout (*Salmo trutta*)

bull trout (*Salvelinus confluentus*), estuary use, 13; residency behaviour, 75

Bute Inlet, BC, 40(f), 42(f)

bycatch, 5, 188–89, 232(t), 256

California killifish (*F. parvipinnis*), 182

Campbell River estuary, BC, competitive interactions, 178, 179(f); contamination, 156; habitat restoration, 238; salmonid feeding behaviours, 78; salmonid fitness/survival, 141–43, 142(t), 145, 148; salmonid marking/tagging, 67; salmonid residency behaviour, 76, 76(f); water flow patterns, 28, 29(f); water quality, 27, 27(f)

canneries, 3–4

capelin (*Mallotus villosus*), 107; Pacific capelin (*Mallotus villosus socialis*), 116

capture techniques, for sampling, 63–64
cardinal fishes (Apogonidae), 54
Carnation Creek estuary, BC, 73, 78, 163, 219
Carneros Creek estuary, CA, 134
carp (*Cyprinus carpio*), 55
Cymatogaster aggregate. *See* shiner perch
Cyprinus carpio. *See* carp
carrying capacity, definition, 160; and density-dependent growth, 160–61; estimating for estuary, 161–63, 220; estimating for specific habitats, 163–64; evolutionary adaptations, 121; invasive species and, 164–65. *See also* competition
Caspian Sea, hypersalinity, 20; salmonid community, 54; salmonid fishing, 3; water levels, 24–25
centrarchids (Centrarchidae), 55
Channa argus. *See* northern snakehead
channels, carrying capacity, 163; competitive interactions in, 123–24, 176(t), 177; definition, 40; example, 40(f); fishing sites, 188; food web, 102–4; habitat loss, 130–32, 157–58, 166–68, 176(t), 177; predation in, 114–18, 166–68, 176(t); restoration, 235; salmonid sampling techniques, 64–65; types, 40–42
Chehalis River estuary, WA, 159, 206
chemical marking/tagging, 67–68
chemical pollutants, 135–36
Chernobyl nuclear accident, 136, 154, 256
Chesapeake Bay estuaries, US, 20
Chile, invasive salmonids, 58–60, 173, 229
Chinook salmon (*Oncorhynchus tshawytscha*), communities, 54; competitive interactions, 121–22, 124–25, 124(f), 175, 177–78, 181, 183; conservation concerns, 219–20, 220(f), 221, 223, 228; cover-seeking behaviour, 77–78; density-dependent growth, 162–63, 178–79, 179(f); distribution, 58; diversity measures, 242; estuarine fitness for, 206–7, 210; estuary use, 13; feeding behaviours, 78, 96–97, 99–100, 99(f), 102, 107, 159–60, 181; fitness/survival, 139, 141–47, 142(f), 142(t), 147–48; food losses from habitat changes, 158; growth, 82–87, 82(f), 108; habitat restoration, 235; habitats, 39, 41, 43, 45–46; holding/staging behaviour, 74; as invasive species, 57–58; and invasive species, 55; life history types, 15, 18, 221, 222(f), 243; marking/tagging methods, 67; migration, 74, 87; osmoregulation, 93; osmoregulation stressors, 150–53; parasitological studies, 70, 197; pollutant levels, 212; as prey, 111–13, 118–19, 167–70, 172, 174, 203; residency behaviour, 76–77, 76(f), 88; sampling and enumeration, 64–66; schooling behaviour, 79; short-term evolutionary considerations, 216; smoltification, 90; specialized adaptation to estuaries, 243–44; tagging, 68–69; territorial behaviour, 79; water temperatures, 47, 49, 153
chum salmon (*Oncorhynchus keta*), communities, 54; competitive interactions, 123–25, 175, 177, 180; conservation concerns, 219; density-dependent growth, 161–62; estuary use, 13, 44(f); feeding behaviours, 96–99, 101, 104; fitness/survival, 139; growth, 83, 156; habitats, 38–41, 43, 45–46; harvesting, 188; as invasive species, 57, 59; osmoregulation stressors, 153; as prey, 111–15, 118–19, 174; residency behaviour, 88; sea lice on, 195; spawning, 80; specialized adaptation to estuaries, 243; territorial behaviour, 79; water temperatures, 47, 49, 153
classification, estuaries, 20–22; habitats, 32–34; salmonids, 13–15
climate change, conservation concerns, 230, 232(t); fishing, 188; habitat loss, 131; parasite increase,

194; research/data needs, 251; sea level rise, 25; temporary estuaries, 156; water temperature, 133–34
Clowholm River estuary, BC, 38
Clupea harengus. *See* Atlantic herring
Clupea pallasi. *See* Pacific herring
clupeid (Clupeidae), 53, 56, 107, 115
Clyde River estuary, Scotland, 211
coastal estuary zone, 26(t), 27(f), 228–30
coastal plain estuaries, 21–22, 23(f)
cobble beaches, 36(t), 38–39, 64, 114
cod (Gadidae), 53, 114, 116–17, 166–67
coded wire-tagging, 67
coho salmon (*Oncorhynchus kisutch*), communities, 54; competitive interactions, 123, 179; conservation concerns, 219; cover-seeking behaviour, 78; density-dependent growth, 163; distribution, 58; domestication, 7; estuary use, 13; feeding behaviours, 78, 96, 101, 103, 159; fitness/survival, 39, 143, 145–47; food losses from habitat changes, 158; growth, 83–85; habitats, 39, 43, 45–46; holding/staging behaviour, 74; as invasive species, 57; and invasive species, 55; life history types, 18; migration, 73; osmoregulation, 93; osmoregulation stressors, 153; parasitological studies, 197; as predators, 112, 119; as prey, 111–14, 168, 173; protection from fishing, 188; residency behaviour, 76; sampling techniques, 64; smoltification, 90; specialized adaptation to estuaries, 243; water temperatures, 47, 49
cold branding, 67
Columbia River estuary, WA-OR, classifications, 22, 34; community changes, 159; competitive interactions, 124–25, 179, 181–83; conservation concerns, 219, 221, 222(f); contamination, 212, 238; food web, 96–97, 100, 102, 104, 209; habitats, 46, 206–7; hatchery-reared salmonids, 172; invasive non-salmonids, 56, 181; management planning, 228; parasitological studies, 70; predation, 111, 115–16, 167–68, 170; restoration, 10, 241–42; salmonid community, 54; salmonid evolution, 216; salmonid fitness/survival, 139, 143–44; salmonid growth, 85, 108; salmonid migration, 74; salmonid osmoregulation, 93; salmonid sampling, 64–65, 67; salmonid smoltification, 90; tidal action, 26; turbidity, 51; water quality issues, 134; wetland area, 208
communities: data limitations, 62; definition, 52–53, 256; effect of invasive nonsalmonids, 55–57; effect of invasive salmonids, 57–62; natural for salmonids, 53–54; structure and estuarine quality, 212–13
community changes, and competition, 178–84; and predation, 171–75; and salmonid feeding and growth, 158–59
competition: assessing, 122; and community change, 178–84; definition, 121–22; and estuarine fitness for salmonids, 203–4; exploitative, 122–25; and habitat loss, 175, 176(t), 177–78; interference, 125–26; research/data needs, 249–50. *See also* carrying capacity
Connecticut River estuary, CT, 133
conservation: areas of, 225–27; estuaries of concern, 8–10, 217–24; evolutionary considerations, 215–17; habitat restoration, 234–38; and landscape change, 205; management goals, 240–45; management strategies, 224–31, 233–34; and mapping, 204–5; restoration monitoring and evaluation, 238–40; salmonids of concern, 8, 15
contamination: biomarkers for, 211–12; effects on water quality, 9, 133–35; and food availability, 158; human activity–related sources,

135–36; as osmoregulatory stressor, 153–56; and predation, 169–70; treatment and control, 237–38
Cosumnes River Protected Area, CA, 241
cottids (Cottidae), 53, 99, 114, 127
cover-seeking behaviour, 77–78, 109
Cowichan River, BC, 85
cutthroat trout (*Oncorhynchus clarkii*), estuary use, 13; growth, 85; habitats, 46; as invasive species, 57; and invasive species, 55; as predators, 113, 115, 119, 122, 173; as prey, 119, 168, 170; sampling techniques, 64; tagging, 69
cyprinids (Cyprinidae), 54–55, 116, 232(t)

dace (*Rhinichthys* spp.), 54
dams, 131, 134, 216, 231
Danube River, Germany, 136
Danube River, Romania, 153, 227
data storage tags, 69–70, 257
Dee River estuary, Scotland, 73
delta estuaries, 22, 23(f)
delta size, 16
Deseado River estuary, Argentina, 203
detrital food, 98–100
Dieset River estuary, Norway, 39
diet overlap, 181
dike breaching, 236
diking, 131–32
diseases, influences on, 193; microbial, 194; parasitical, 193, 195–97; research/data needs, 197–98; threats from farmed salmonids, 195, 197, 229
disruption. *See* habitat loss and disruption
dissolved oxygen: in estuaries, changes from human activities, 134; low-level effects on salmonids, 152–53; and predation, 169; requirements, 49–50; standards, 211–12
DNA testing, advances, 243; identifying specialized adaptations, 243–44; measuring pollutant biomarkers, 212; measuring salmonid distribution and abundance, 70; measuring salmonid growth, 84
Dnieper River, Ukraine, 244
Dolly Varden (*Salvelinus malma*), competitive interactions, 123; estuary use, 14; feeding behaviours, 101; as predators, 114; as prey, 119; protection from fishing, 191
domestication of salmonids. *See* aquaculture; salmonids, hatchery-reared
Dorosoma petenense. *See* shad, threadfin
Douro River estuary, Portugal, 3, 202
dredging, 130–32, 153, 168, 224, 231
dusky flounder (*Liopsetta obscura*), 125
Duwamish River estuary, WA, 54, 239
dyes for marking, 67

ear bones. *See* otoliths
earthquakes, 128
ecological indicators, biological and habitat data, 207–10; community structure, 212–13; landscape approach, 207; mapping of salmonid habitat, 204–5; water quality, 210–12
ecological integrity, definition, 240, 257; as goal of estuary management, 240–44
Ecological Quality Ratio, 212–13
Ecopath, 160
ecosystem engineers. *See* habitat from invasive ecosystem engineers
ecotypes, 17, 18, 138, 223, 257
eelgrass, features, 36(t), 43; food web, 104, 156–57; loss of, 128, 132; restoration, 237
eels (Anguillidae), 53
elastomer dye, 67
electronic tagging, 68–70, 69(f), 80, 247, 249
endangered species, anadromous salmonids, 8; biological markers for, 213; conservation strategies, 225; protection from fishing, 188–89
England, estuary sizes, 219; fishing bycatch, 189; habitat loss, 128;

history of salmon importance, 4; salmonid fitness/survival, 139
Engraulis mordax. *See* northern anchovy
environmental flow, 230, 257
Esk River estuary, England, 77
estuaries, boundaries, 33; classification systems, 20–26, 21(f), 23(f), 26(t); as conservation areas, 225–27; conservation concerns, 8–10; conservation importance of, 217–18, 223; definition, 19, 258; ecological indicators of conditions, 204–13; global map of, 14(f); human-constructed, 128, 131, 134; key attributes for salmonids in, 199–204, 200(f); management and planning, 227–30; as portals for salmonid invasion, 61–62; productivity and species diversity, 221; research/data needs, 218, 248–49; restoration, 10; role in salmonid life cycle, 5–7, 6(f), 201; sampling from, 63–67; types used by salmonids, 16, 218; water properties, 20, 24–29, 26(t), 27(f), 28(f)
estuarine habitats, 36(t)
estuarine zones, and salmonid growth rates, 82, 85–86; and water properties, 25–26, 26(t), 27(f)
eulachon (*Thaleichthys pacificus*), 54
Europe, estuary restoration, 10; estuary types, 202; fishing, 189; habitat classifications, 32; habitat loss and degradation, 128, 135; hatchery-reared salmonid abundance, 180; invasive salmonids, 58–59, 229; river discharges, 28; salmonid fishing in history, 3–4; salmonid parasites, 195–96; salmonids of conservation concern, 8; sampling standards, 66; water quality issues, 135
European eel (*Anguilla anguilla*), 53, 241
European flounder (*Platichthys flesus*), 117, 189
European seabass (*Dicentrarchus labrax*), 189

European Union, Habitat Directive, 32, 213; Marine Strategy Framework Directive, 209; Water Framework Directive, 211–12
euryhaline salmonids, 8, 43, 258. *See also* salinity tolerance
eutrophication, definition, 258; of estuaries, 152, 213, 233; European rivers, 135
evolution: of anadromous salmonids, 6–8; importance of salmonids to understanding, 5, 7–8; research/data needs, 217; short-term considerations, 215–17
exorheic coasts, 200–2
extirpation of salmonids, 221, 239, 258

Falkland Islands, 58
fan estuaries, 22, 23(f)
Faroe Islands, 60
fat greenling (*Hexagrammos otakii*), 123–24
feeding behaviours: assessing, 78–79; food web levels, 96; laboratory studies, 156–57. *See also* food webs
fences, underwater, for fish capture, 64
fin clips, 67
Finland, brackish marsh species, 45; hatchery-reared brook trout, 59; water quality issues, 134–35
First Nations, salmonid fishing (history), 3, 149; salmonid fishing (today), 187, 189; salmonids in culture, 4, 192, 224
fish farms. *See* aquaculture
fish, nonsalmonid. *See* nonsalmonids
fish wheels, 189
fishing: bycatch, 188–89, 256; commercial, 4, 187–90, 218–19, 224, 249, 259; management strategies, 224–25, 232(t); overfishing, 9; poaching, 190, 224; as predation, 95; recreational, 3, 58, 190–91, 218, 224, 249; short-term evolutionary considerations, 216–17; subsistence, 3–4, 187–90, 192, 224–25, 263

fitness, assessing through experimentation, 141–43; assessing through observation, 143–44; assessing through statistical inference, 144–46; components of survival, 137–40; definition, 137, 258; effects of habitat loss and community change, 184–85, 185(f); and fish size, 81; modelling, 146–48. *See also* survival

fjard estuaries, 22, 23(f)

fjord estuaries, definition, 22, 23(f); depth, 22; example, 40(f); rock and boulder beaches, 38

floodplains, seasonal, 37(t), 45–46, 105

floods, 22, 37(t), 61, 131, 146, 225, 231, 232(t), 236

flounder. *See* European flounder (*Platichthys flesus*); starry flounder (*Platichthys stellatus*)

fluorescent grit, 67, 76

food availability, and carrying capacity, 162; and community changes, 158–59; and estuarine fitness for salmonids, 203, 209, 210; and growth rate, 81, 85–86; and habitat loss, 157–58; modelling, 159–60; summary of assessment methods, 165(t)

food conversion efficiency, 156, 258

food ration, 156, 157, 258

food webs: assessing and investigating, 96–97, 100–1; definition, 95, 258; effects of predation on, 110; as indicator of estuarine fitness for salmonids, 209; levels, 95, 97–99, 98(f); modelling, 159–60; nutrients in, 51; prey characteristics, 105, 106(t), 107–8; research/data needs, 249–50; variations by habitat, 101–5. *See also* competition; feeding behaviours; predation

Forth River estuary, Scotland, 45

Fowey River estuary, England, 68, 77, 92

France, brackish marshes, 45

Fraser River estuary, BC, competitive interactions, 124–25; conservation concerns, 9, 220, 220(f); fishing, 188; food web, 99–104, 103(f); habitat loss, 129, 130; habitat restoration, 235–36; habitats, 39, 41–43, 44(f), 45; invasive non-salmonids, 56, 56(f); predation, 114; salmonid community, 53–54; salmonid enumeration, 64, 66; salmonid feeding behaviours, 78; salmonid fitness, 138–39; salmonid growth, 82(f), 86; salmonid mortality, 194; salmonid osmoregulation, 93; salmonid parasites, 197; salmonid residency behaviour, 76; specialized salmonid adaptation, 244; water properties, 23, 48, 50; wetland area, 208–9

Freshwater Creek, NU, 77

freshwater habitats, competitive interactions, 121; ecological indicators, 207; invasive salmonids, 57; proxies for, 16; salmonid evolution, 6–7, 216

fry, competitive interactions, 123, 125; cover-seeking behaviour, 78; definition, 16–17; estuarine fitness for, 203; feeding behaviours, 97–99; growth, 82, 84–86; migration, 74; as prey, 103, 112–15, 117–19; residency behaviour, 76; sampling techniques, 64, 66; schooling behaviour, 79; specialized adaptation to estuaries, 243–44; tagging, 68, 70

fyke nets, 39, 43, 65, 66, 69(f), 145

Gaula River estuary, Norway, 22

genetics, advances, 247; as context for fitness, 137; and degree of estuarine dependence, 15; and homing behaviour, 79; as marker for studies, 70–71; research/data needs, 249; and run timing, 139; short-term evolutionary considerations, 215–17

geomorphology, definition, 34; and estuarine fitness for salmonids, 201–2, 205; as estuary classification basis, 20–22; and salmonid sampling, 66

Gilbert Bay, NL, 225
gillnets, 63–64, 78, 188–89
Gironde River estuary, France, as coastal plain, 23; food web, 99; geomorphic classification, 22; salmonid community, 53; water temperatures, 48
glaciation changes, and salmonid evolution, 7
globalization, definition, 258; effect on salmonid and estuarine management, 5, 228
Gobius niger. See goby, black
goby (Gobidae): black (*Gobius niger*), 56; Shimofuri (*Tridentiger bifasciatus*), 56; yellowfin (*Acanthogobius flavimanus*), 181
gravel and cobble beaches, 35, 36(t), 38–39, 123
gravel beaches, 102, 114, 129
greenling. *See* fat greenling (*Hexagrammos otakii*); whitespotted greenling (*Hexagrammos stelleri*)
grilse, 18
growth: assessing, 83–84, 86; and community change, 158–59; density-dependent, 160–65, 178, 179(f); estuarine fitness for, 203; factors affecting, 85–87, 89; and fitness/survival, 138–39, 184–85, 185(f); and food web, 97, 108; and habitat loss, 157–58; laboratory studies, 156–57; and migration, 73, 86; range of rates, 84–85; and temperature, 81; and territoriality, 160–61; and water quality, 159–60. *See also* size/weight of fish
Gyrodactylus salaris (parasite), 196–97

habitat from invasive ecosystem engineers: carrying capacity, 164; competitive interactions, 125, 177, 181; definition, 47; and food webs, 105, 158; and loss of natural habitats, 133, 176(t); predation in, 169; restoration from, 237
habitat loss and disruption: and artificial habitats, 133; beaches, 129; channels, 130–32; and competition, 175, 176(t), 177–78; definition, 129; invasive ecosystem engineers, 133; and osmoregulation, 150–56; and predation, 166–71, 176(t); research/data needs, 248; and salmonid feeding and growth, 156–58; and salmonid fitness/survival, 184–85, 185(f); short-term evolutionary considerations, 215; summary of factors affecting, 232(t); vegetation, 132
Habitat Suitability Index (HSI), 147
habitats: classification and terminology, 32–35, 36–37(t); competitive interactions in, 122–25; conservation concerns, 10, 128; definition, 31, 258; and estuarine fitness for salmonids, 203–4, 207–10; and fitness inferences, 144–48; food webs in, 95–99, 101–5; geomorphological factors affecting, 34–35, 202; predation in, 113–19; restoration, 234–37; scale of sampling, 33; types, 35–47; water properties in, 47–51. *See also* artificial habitats
haddock (*Melanogrammus aeglefinus*), 116
Haida, salmonid fishing, 3
haloclines, 21(f), 23, 26, 30, 102, 103(f)
Hals River estuary, Norway, 118–19
harvesting. *See* fishing
hatchery fish. *See* aquaculture; salmonids, hatchery-reared
Hexagrammos otakii. See fat greenling
Hexagrammos stelleri. See whitespotted greenling
Hokkaido, Japan, 4, 58, 123, 177, 180, 183
holding/staging behaviour, in aquaculture, 192; definition, 74; factors affecting, 74; food requirements, 74, 159; modelling, 159
Homathko River estuary, BC, 40(f)
homing, factors affecting, 140; and genetic differentiation, 139; purposes, 79, 139

Hood Canal, WA, 50, 115, 132, 166–67
hormonal changes in salmonids, 122, 150–52, 256
Hucho perryi. See Sakhalin taimen
human activities, changes in water quality, 50, 133–36; conservation management strategies, 224–25; dispersal of salmonids, 57–58, 60, 62, 202, 233; habitat loss, 128, 129–33, 136; social value of salmonids, 192, 224
Humber River estuary, England, 136
hydroacoustic tags, 67–68, 70, 75–77, 144, 259
hydrodynamics: and estuarine fitness, 209–10; as estuary classification basis, 24. *See also* water flow patterns
hypertidal estuaries, 24
Hypomesus japonicas. See Japanese smelt
Hypomesus pretiosus. See surf smelt

ice melt, 20, 28–29
Iceland, history of salmon importance, 4
ictalurids (Ictaluridae), invasive, 55
immersion dyes, 67
Index of Biological Integrity (IBI), 212, 241
industrial development/activity, estuarine management, 5, 9; habitat loss, 132–33, 158; sound effects, 51; water quality effects, 133–36, 153
inland seas, as estuaries, 16; temperatures, 27
International Union for Conservation of Nature (IUCN), 8
intertidal zones, connecting habitats, 31; example, 42(f); gravel and cobble subhabitats, 39; riverine channels, 40; sandflats, 39; sequences of life, 25; vegetation, 43, 44(f), 45
Inuit, salmonid fishing, 3
invasibility, definition, 259; factors affecting, 57, 60–62, 149
invasive species, carrying capacity of estuaries, 164–65; distinct from nonindigenous species, 55; factors affecting, 57, 60–62; food web changes, 158–59; nonsalmonid fishes, 55–57; parasites on, 196; predation by, 173–75, 176(t); preventing establishment of, 233–34; research/data needs, 234, 249–50; salmonids, 57–62; short-term evolutionary considerations, 215–16; upstream/downstream sources, 232(t)
Ireland, aquaculture, 191; sea lice on salmon, 195; Northern fishing, 189, 190(f)
irrigation, 131, 133
isotopic signatures, 100
iteroparous salmonids, 56(f), 120(t), 218; conservation concerns, 230; definition, 259; endangered, 190; evolution, 8; feeding behaviours, 95; growth, 81, 83–85; life cycle, 5–6, 6(f); life history types, 16–17; residency behaviour, 77; smoltification, 90
Izembek Lagoon, AK, 39

jacks, 18, 93
Japan, aquaculture, 180; invasive salmonids, 58; salmonid communities, 53–54; salmonid fishing history, 3, 4; salmonid fitness/survival, 147; salmonids in culture, 4; spawning migration of in typhoon season, 41
Japanese lamprey (*Lampetra japonica*), 111
Japanese smelt (*Hypomesus japonicus*), 125
juvenile salmonids: competitive interactions, 123; cover-seeking behaviour, 109; ecological advantages of estuarine residency, 11, 205–6; feeding behaviours, 78, 96–97, 98(f), 99–102, 104–5; fitness/survival, 140–48; growth, 82–85; holding/staging behaviour, 74; migration, 72; nomenclature, 16–17; parasitological studies, 70; pollutant effects, 211–12; as prey, 103, 111, 113, 115, 117–19, 121, 126;

research/data needs, 248–49; residency behaviour, 75–76; sampling and enumeration, 64–66; tagging, 68–69; water temperatures, 47–49

Kakanui River estuary, New Zealand, 57
Kalininka River estuary, Russia, 125
Kamchatka Peninsula, Russia, competitive interactions, 125; conservation concerns, 221, 223; estuary classifications, 32, 218; food web, 107; limans, 22; predation, 111; salmonid growth, 86; salmonid osmoregulation, 93; tides, 25
Kara River estuary, Russia, 53, 202
Karaginsky Bay, Russia, 114
Kaspi people, salmonid fishing, 3
kelp, in food web, 104; as a subhabitat, 42–43
kelts: competitive interactions, 123; definition, 16; estuarine fitness for, 203; feeding behaviours, 96, 99, 99(f), 103; osmoregulation, 92; osmoregulation stressors, 154–55; as prey, 119; reconditioning, 191, 259
Kemijoki River, Finland, 59
Kennebec River estuary, ME, 147
Keogh River watershed, BC, 143
Kerguelen Islands, 58, 60, 140, 203
Khaylyulya River estuary, Russia, 114
killifish (Fundulidae), 53, 181–82
Klamath River estuary, CA, 60, 61(f), 118, 197
kokanee, 138, 244
Koksoak River estuary, QC, 84, 99, 243
Kushiro River estuary, Japan, 48, 131
Kyronjoki River estuary, Finland, 178

La Have River estuary, NS, 68, 77
lagoon estuaries: definition, 22, 23(f); depth, 22; gravel and cobble subhabitats, 38–39; salinity, 24
Lagunitas Creek estuary, CA, 20
Lake Llanquihue, Chile, 60

lake trout (*Salvelinus namaycush*), estuary use, 14, 16; specialized adaptation to estuaries, 243
lakes, effects of invasive nonsalmonids, 55
lampreys (Petromyzontidae), 53, 213
length of fish. *See* size/weight of fish
Lepeophtherius salmonis. *See* salmon louse
Lepomis microlophus. *See* redear sunfish
life history stages, 16–18, 242
life history types, definition, 15, 17–18; and osmoregulation, 93; research/data needs, 249, 252; and smoltification, 90; terminology, 16–18
light levels in estuaries, 51, 135, 170
liman estuaries. *See* lagoon estuaries
lingcod (*Ophiodon elongatus*), 169–70
Little Susitna River estuary, AK, 78
Livar-Yakha River estuary, Russia, 53
Loch Eck, Scotland, 126
Lough Furnace, Ireland, 119
low-flow estuaries, 20, 134
lower estuary zone, 26(t), 27(f)

Mackenzie River estuary, NT, 10, 20, 27, 28
mackerel (*Trachurus symmetricus*), 167
macroalgae subhabitat, features, 35, 36(t), 39, 42–43, 42(f), 132; food web, 98(f), 104
macrotidal estuaries, 24, 39, 41
Maine, subhabitats, 45
mammals, as predators, 95, 111, 114, 116, 118, 121, 166, 173
management plans, definition, 260; estuarine, 227–30; goals, 240–44; habitat restoration, 234–38; incorporating research/data needs, 251–52
Manasquan River, NJ, 183
Manawatu River estuary, New Zealand, 226
mapping of salmonid habitat, 204–5
Margaree River estuary, NS, 111
marine mammals, predation on salmonids, 110, 168; sounds of, 51
Marine Protected Areas (MPAs), 225–26, 244

marking/tagging: chemical tags or marks, 67–68; electronic tagging, 68–70, 69(f), 80; genetic marks, 70–71; and growth measurements, 83; parasitological, 70; physical tags or marks, 67; and population estimates, 65–66; risks of, 67
marshes, brackish: cover-seeking behavior in, 78; definition, 256; features, 37(t), 43–45; subhabitat loss, 132; restoration, 236; salmonid growth, 82; salmonid sampling, 66
marshes, freshwater, 37(t), 43–45
marshes, high-salt, 37(t), 45
masu (*Oncorhynchus masou masou*), agonistic behaviour, 122; estuary use, 13; feeding behaviours, 96; as invasive species, 57; as prey, 172, 174; sea lice on, 195; smoltification, 90; water temperatures, 48
Maullín River estuary, Chile, 60
meadows, wet. *See* wet meadows
Melanogrammus aeglefinus. *See* haddock
Menidia beryllina. *See* silverside
Merlangius merlangus. *See* whiting
Mersey River estuary, England, 236, 238
mesohabitats, 34, 71, 260
mesotidal estuaries, 24
Mezen River estuary, Russia, 24
Mi'kmaq, salmonid fishing, 3
microbenthic algae, 39; definition, 260; in food web, 104–5
microconstituents, waterborne, 51
microsatellite analysis, 70–71, 242
microtidal estuaries, 24, 33
Middle East, estuary conservation concerns, 9
middle estuary zone, 26(t), 27(f)
migration, definition, 138; enhancing fitness, 138; factors affecting, 72, 86, 151–52, 155, 232(t); and life history types, 15; migratory pulse, 54, 260; for rearing and growth, 73; short-, medium-, and long-term, 73–74
minerals, and salinity, 28
mining, 128, 136, 153, 164

minnows (Cyprinidae), 53
Miramichi River estuary, NB, 53, 68, 138
mixed estuaries, 21(f), 23
modelling, estuarine restoration, 234, 239, 241, 245, 251–52; food web, 96, 101, 112–13, 160, 164, 165(t), 168; habitat changes and losses, 156–57, 159–60, 165, 174; predation, 113; salmonid evolution, 216–17; salmonid health, 195–97; salmonid survival, 144–48, 251; salmonids as invasives, 61–62; water properties, 133–34
Mohicans, salmonid fishing, 3
Morone saxatilis. *See* striped bass
mote sculpin (*Normanichthys crockeri*), 174
mRNA (messenger ribonucleic acid), 194, 260
mudflats, 36(t), 39, 40(f)
mullet (*Mugil* spp.), 189
mullets (Mugilidae), 53

Na^+-K^+–ATPase, definition, 261; in osmoregulation, 92–93; and salmonid migration, 48; in smoltification, 89–90
Nabisipi River estuary, QC, 77
Nagara River, Japan, 134
Naiba River estuary, Russia, 83
Nain Bay estuary, NL, 89
Namsen River, Norway, 86
Nanaimo River estuary, BC, carrying capacity, 161–62; competitive interactions, 175, 177–78; food web, 102; habitat loss, 129; salmonid migration, 87; salmonid sampling, 66
National Estuarine Research Reserve System, 226
National Oceanic and Atmospheric Administration (NOAA), 32
Navarro River estuary, CA, 45
negative estuaries, 20
Nehalem River estuary, OR, 167
Neiden River, Norway, 184
Netarts Bay, OR, 161–62
nets, 63–64, 188–89

Neva River estuary, Russia, food web, 102, 159; invasive nonsalmonids, 56, 159, 164; predation, 117; salmonid community, 54
New Brunswick, invasive salmonids, 58; specialized salmonid adaptations, 244
New Zealand, competitive interactions, 183; food webs, 99; habitat classification, 32; hapua, 22; invasive salmonids, 57–59
Newfoundland and Labrador, competitive interactions, 182–83
Nisqually River estuary, WA, 236
noise in estuaries: causes and ranges, 51; disruptions from human activities, 135; and predation, 170–71
nonsalmonids: as alternative prey, 116; competitive interactions, 123–26; as predators, 95, 111, 114–17, 117(f), 166–71, 173–75; as prey, 121, 123
North America, disrupted estuaries and research, 109; East Coast salmonid evolution, 7; estuary conservation concerns, 9–10; fishing, 189; habitat loss and degradation, 128; hatchery-reared salmonid abundance, 180; salmonids of conservation concern, 8; West Coast salmonid evolution, 7
North Pacific hake (*Merluccius productus*), 166–67
North Sea, sea trout growth rates, 86
northern anchovy (*Engraulis mordax*), 54
Northern Hemisphere, competitive interactions, 182–84; subhabitats, 43; research/data needs, 249; salmonid evolution, 6–7; salmonid fishing history, 3–4; salmonids of conservation concern, 18; water temperatures, 201
Northern Ireland, fishing, 189, 190(f)
northern pikeminnow (*Ptychocheilus oregonensis*), 54, 114, 172
northern snakehead (*Channa argus*), 173–74, 233

Norway, conservation strategies, 226; estuary sizes, 219; river discharges, 28; salmonid parasites, 195–97; salmonids in winter, 54
Numedals River estuary, Norway, 53
nutrients: in estuary, changes from human activities, 135; prey characteristics, 107; required, 51; sources, 96

ocean ranching, 184, 191–92, 226, 228, 232(t), 252–53
Ocean Tracking Network, 80
oceans, conservation concerns, 10, 232(t); conservation planning, 230–31, 233; research/data needs, 252; role in salmonid life cycle, 5–6, 6(f); wave action and estuaries, 25
Odontobutidae, 56
oil spills and discharges, 135–36, 154, 209
Olsen Creek estuary, AK, 39, 80
Oncorhynchus spp., conservation concerns, 8; evolution, 7; life history types, 17
Oncorhynchus clarkii. See cutthroat trout
Oncorhynchus gorbuscha. See pink salmon
Oncorhynchus keta. See chum salmon
Oncorhynchus kisutch. See coho salmon
Oncorhynchus masou ishikawae. See amago
Oncorhynchus masou masou. See masu
Oncorhynchus mykiss. See rainbow trout; steelhead
Oncorhynchus nerka. See sockeye salmon
Oncorhynchus tshawytscha. See Chinook salmon
Ophiodon elongates. See lingcod
Orkla River estuary, Norway, 22, 102, 130, 166–67
Osmerus mordax dentex. See Asiatic smelt
osmoregulation: adults, 92–93; compromised, 150–55, 151(f); definition,

261; evolution, 8; and fitness/ survival, 184–85, 185(f); kelts, 92; and migration, 72; and "reverse smolting", 155–56; salinity change as stressor, 150–52; and salmonid species composition in estuary, 202–3; specialized populations, 93, 243; and temperature, 88; water quality as stressor, 152–55. *See also* smolting/ smoltification

osmotic shock, 151–52, 155

otoliths, 67–68, 75, 83, 111, 116, 163, 244, 261

oxygen in water. *See* dissolved oxygen

Pacific North Coast Integrated Management Area, BC, 230

Pacific estuaries, habitat loss, 131; subhabitats, 43; predation, 113–15, 118–19, 126

Pacific Estuary Conservation Program, BC, 226

Pacific herring (*Clupea pallasi*), 54, 98(f), 175, 177, 210

Pacific Northwest Aquatic Monitoring Partnership, 66

Pacific lamprey (*Lampetra tridentate*), 111

Pacific redfin (*Tribolodon brandtii*), 125

Pacific salmonids, competitive interactions, 178, 182; conservation concerns, 221; evolution, 7; fitness/survival, 139, 143; oxygen in water, 49; water temperatures, 48

Pacific sand lance (*Ammodytes hexapterus*), 175

parasites, definition, 193, 261; predators as, 111; research/data needs, 197–98; as tags, 70

parr, competitive interactions, 123; definition, 16–17; estuarine fitness for, 203; fitness/survival, 138; growth, 85; residency behaviour, 77; sampling techniques, 64; schooling behaviour, 79; specialized adaptation to estuaries, 244

Parvicapsula minibicornis (parasite), 197

passive integrated transponder (PIT) tags, 69, 69(f), 70, 76, 83, 111, 144, 261

Patagonian blennie (*Eleginops maclovinus*), 174

Penobscot River estuary, ME, 73, 166

pesticides, 136, 154, 212, 261

Petitcodiac River estuary, NB, 9, 132, 235

Pettaquamscutt River estuary, RI, 136

pH: of estuarine water, 51, 134–35

pharmaceuticals, 136

phenotypes, 7, 261

photographic sampling, 65–66, 71

physical marking/tagging, 67

pigmentation, 17, 88–89

pike (*Esox lucius*), 117

pink salmon (*Oncorhynchus gorbuscha*), competitive interactions, 123, 125, 184; conservation concerns, 219; distribution, 58; estuary use, 13; and *Exxon Valdez* oil spill, 136; feeding behaviours, 96, 104; fitness/survival, 139; habitats, 39; as invasive species, 57–58, 184; osmoregulatory stressors, 156; as prey, 103, 111–17, 119; residency behaviour, 88; sea lice on, 195–96; spawning, 80; specialized adaptation to estuaries, 243; territorial behaviour, 79; water temperatures, 47

PIT. *See* passive integrated transponder (PIT) tags

Platichthys flesus. *See* European flounder

Platichthys stellatus. *See* starry flounder

poaching, 190, 224

Pollachius virens. *See* saithe

pollution. *See* contamination

ponyfishes (Leiognathidae), 54

Porcupine Creek estuary, AK, 39, 80, 101, 123

positive estuaries, 20

Potomac River, MD, 174

powan (*Coregonus clupeoides*), 126

power plants, 132–33, 136, 159, 235, 256

predation: assessing, 110–13, 120(t); avoidance, 77, 109; and community change, 171–75; definition, 110; and estuarine fitness for salmonids, 203–4; in food web, 95; and habitat loss, 166–71, 176(t); on or by hatchery-reared fish, 171–73, 176(t); intraguild, 119–21; by invasive species, 173–75, 176(t); and parasites, 193; process, 110; research/data needs, 249–50; and salmonid fitness/survival, 184–85, 185(f); and salmonid growth, 160; short-term evolutionary considerations, 216; variations by habitat, 113–19; and water quality, 169–71, 176(t)
pressure indicators, 208
prey of salmonids, characteristics, 105, 106(t), 107–8; effects of community changes, 158–59; effects of habitat loss, 156–58; effects on growth, 156–57; vertical distribution example, 103(t)
prickly sculpin (*Cottus asper*), 175, 177
Prince William Sound, AK, 136
processing plants, 3–4
Ptychocheilus oregonensis. *See* northern pikeminnow
Puelo River estuary, Chile, 28, 59
Puget Sound, WA, acidification, 164; food web, 100, 160; habitat loss, 131; mapping, 205; predation, 113, 115, 169; salmonid disease, 194; salmonid growth, 82
pumpkinseed (*Lepomis gibbosus*), 55
pulp mills, 133–35, 153, 170, 237
Puntledge River estuary, BC, 168, 170
purse seines, 64–65
Puyallup River estuary, WA, 9

Quashnet River estuary, MA, 69
Québec, competitive interactions, 183

radio tags, 68–69, 206–7, 262
radioactive contamination, 136, 154
rainbow trout (*Oncorhynchus mykiss*), competitive interactions, 122; genetics and fitness/survival, 140; as invasive species, 57, 59, 61, 229; osmoregulation stressors, 154. *See also* steelhead (*Oncorhynchus mykiss*)
rainfall, and climate change, 131, 156; and low-flow estuaries, 20; and river discharge, 28, 131; and salinity, 29; and salmonid migration, 40–41
Rakaia River, New Zealand, 99, 121–22
Ramsar Convention on Wetlands, 226, 262
rapid evolution. *See* evolution, short-term considerations
Razdol'naya River estuary, Russia, 29, 173
redd (spawning nest), 17
redear sunfish (*Lepomis microlophus*), 56
Relconcavi Fjord, Chile, 59, 174
remote sensing, 219
research/data needs, Arctic estuaries, 10; behaviours, 250–51; biotic interactions, 249–50; climate change, 251; domestication, 252–53; estuaries as alternative habitats, 218; estuary size, 248; estuary types, 248; food webs, 249–50; genetics, 249; hatchery-reared salmonids, 247; invasive species, 234, 249–50; inventory of lost, damaged and vulnerable estuaries, 248; life history types, 252; progress to date, 246–47; restoration, 249; salmonid behaviours, 80, 250–51; salmonid diseases, 197–98; salmonid evolution, 217; salmonid metrics, 248–49; stressors, 249; upstream/downstream, 252
residency: assessing, 75–77; before smolting, 88; definition, 74; estuarine fitness for, 203; and salmonid growth rates, 84
resilience, definition, 218, 240, 262; as goal of estuary management, 218, 240–44
resource polymorphisms, 244, 262

restoration, of estuaries, 10, 234–38, 262; goals, 240–44; monitoring and evaluating, 238–40; research/data needs, 249; of salmonids, 228, 262
retropinnid smelt (Retropinnidae), 99
returning veterans: definition, 16; feeding behaviours, 78–79, 103; osmoregulation stressors, 155; as prey, 109; residency behaviour, 77; sampling techniques, 64. *See also* virgin sea run
Rhine River estuary, Germany, 10
Rhine River estuary, Netherlands, 41, 86, 168, 234
Ribble River estuary, England, 155
Ribe River estuary, Denmark, 155
Rio Grande River estuary, Argentina, 61(f), 86, 126, 191
riparian vegetation, 37(t), 46
riprap, 37(t), 47, 105, 129, 207, 235, 237, 262
river discharge, along exorheic coasts, 201; fluctuations from dam spills, 29(f); and migration, 151–52; and surface salinity, 28–29
river flow. *See* water flow patterns
river lamprey (*Lampetra fluviatilis*), 213
riverine channels, example, 40(f); features, 36(t), 40–41; as fishing sites, 188
rivers, conservation concerns, 10, 232(t); conservation planning, 230–31, 233; effects of invasive non-salmonids, 55; research/data needs, 252; role in salmonid life cycle, 5–6, 6(f), 201
Rivière à la Truite, QC, 116
RNA testing, 84, 194
rock and boulder subhabitats, 36(t), 38
rockfish: copper (*Sebastes caurinus*), 169; quillback (*Sebastes maliger*), 169; yellowtail (*Sebastes flavidus*), 170
rockweed, 36(t), 42–43, 104, 169
Rosewall Creek estuary, BC, 118
Russia, habitat classification, 32; invasive salmonids, 58

Sacramento–San Joaquin River estuary, CA, classifications, 22, 34; competitive interactions, 181; conservation concerns, 9; contamination, 155; food web, 100, 102, 105, 107; habitat loss, 133; habitat restoration, 237; habitats, 41, 46; hydrodynamics, 210; invasive nonsalmonids, 55–56, 56(f), 164; predation, 174; salmonid feeding behaviours, 159–60; salmonid fitness/survival, 141, 144–45, 152; salmonid growth, 83; salmonid residency behaviour, 77; specialized salmonid adaptation, 244; water temperatures, 48, 153
saffron cod (*Eleginus gracilis*), 125
Saguenay River estuary, QC, 155
saithe (*Pollachius virens*), 116, 166
Sakhalin taimen (*Hucho perryi*), conservation concerns, 8; estuary use, 13, 15; fitness/survival, 147; osmoregulation, 92; protection, 226
salinity: changes affecting osmoregulation, 150–52, 202–3; changes from human activities, 134, 155–56; factors affecting, 28; and invasive salmonids, 60; and migration, 72; positive vs negative estuaries, 20; ranges and salmonid responses, 49; and river discharge, 28–29; and salmonid growth, 85; stratification and estuary classification, 21(f), 23–24, 26(t); units of measure, 28
salinity tolerance, 7–8. *See also* euryhaline salmonids
Salish, salmonid fishing, 3
Salmo spp., 6, 8
Salmo salar. *See* Atlantic salmon
Salmo trutta. *See* brown trout; sea trout
salmon louse (*Lepeophtherius salmonis*), 194–96, 229
Salmon River estuary, OR, competitive interactions, 177; habitats, 41; salmonid residency behaviour, 76; tagged salmonids studies, 69, 69(f), 70, 160, 177
salmonid communities. *See* communities

salmonids, biological metrics and estuary conditions, 218–21, 222(f), 223–24; conservation concerns, 8–10, 192, 218–19; diseases and parasites, 193–98; dispersal in estuaries, 199–204, 200(f); as evolutionary indicators, 5; extirpation, 221, 223; and human history, 3–5, 149; as indicator of ecosystem health, 5, 212, 242–43; as invasives, 57–62, 122, 149, 215–16; life cycle, 5–6, 6(f); life history stages and nomenclature, 16–18, 243; research/data needs, 249; sampling and enumeration, 63–67; social value, 192, 224; strongholds, 223, 262; tagging and marking, 67–71; taxa and nomenclature, 13–15

salmonids, hatchery-reared: characteristics, 15; competitive interactions, 122, 125, 149–50, 158, 176(t), 178–80, 220–21, 232(t); as conservation effort, 228; data for research, 247; disease threat, 197; evolutionary effects, 7–8; introduction and dispersal, 59, 62; marking/tagging, 68; ocean ranching, 184, 191–92, 226, 228, 232(t), 252–53; as predators, 106(t), 116, 171–73, 176(t); as prey, 167, 171–73, 176(t); production by aquaculture, 191–92; research/data needs, 252–53; salinity change as stressor, 150; schooling, 158, 173; smoltification, 90; survival, 137, 146–47, 244. *See also* aquaculture

salt marshes. *See* marshes, high-salt

salt wedge, eelgrass growth, 4; estuary classification, 21(f), 23–24; food web, 157–58; osmoregulatory success, 202–3; smolt behaviour, 156; smoltification, 91–92; turbidity, 51; water temperature, 49

Salvelinus spp., conservation concerns, 8

Salvelinus alpinus. *See* Arctic char

Salvelinus confluentus. *See* bull trout

Salvelinus fontinalis. *See* brook trout

Salvelinus leucomaenis. *See* white-spotted char

Salvelinus malma. *See* Dolly Varden

Salvelinus namaycush. *See* lake trout

sampling methods, catch and release, 63–65; international/local standards and collaboration, 66–67; marking/tagging, 66–71, 80, 83; population estimates, 66–67; visual/photographic, 65

San Antonio Creek estuary, CA, 20

San Francisco Bay, CA, food web, 100, 105, 107; hydrodynamics, 210; salmonid fitness/survival, 145

San Joaquin River estuary. *See* Sacramento–San Joaquin River estuary

sand eels (*Ammodytes* spp.), 116, 123

sandflats, example, 40(f); features, 36(t), 39; salmonid growth, 82

Santa Cruz River estuary, Argentina, 92, 140

scales, as measure of estuarine growth, 83

Scandinavia, history of salmon importance, 4

scars, as predation marker, 111

Scheldt River estuary, Belgium, 56, 130, 212, 235

schooling: alternating with territoriality, 79, 89; hatchery fish, 158, 173; and life stage, 122; and predation, 115, 118, 120(t); purposes, 79, 106(t), 118; research/data needs, 250; and smoltification, 89, 122. *See also* territoriality

sciaenid (Sciaenidae), 56

Scott Creek estuary, CA, carrying capacity, 163; classification, 22; salmonid growth, 83; salmonid residency behaviour, 77; subhabitats, 39; water flow patterns, 29

sea level rise, 25

sea lice, 194–96, 229

sea-run migrants. *See* kelts

sea trout (*Salmo trutta*): anadromy, 94; communities, 54; competitive interactions, 121–23, 126, 178, 182–

83; conservation concerns, 8, 219, 227; density-dependent growth, 162–63; distribution, 57–58, 139–40, 203; estuary use, 13; farmed, 192; feeding behaviours, 78–79, 95, 99; fitness/survival, 139–40; growth, 85–86, 88–89; habitats, 38, 41, 45; as invasive species, 58, 61(f); osmoregulation, 92; as part of estuarine ecosystem, 10–11; as predators, 119, 174; as prey, 114, 118, 168, 213; protection from fishing, 189; as reference species for biological integrity, 212–13; residency behaviour, 75, 77; sampling techniques, 63; sea lice on, 195–96; smoltification, 89; specialized adaptation to estuaries, 243–44; territorial behaviour, 79; in winter, 54. *See also* brown trout (*Salmo trutta*)
seasons, breaching of sand/gravel barriers, 201–2; estuary classification, 20; habitat, 32; residency behaviour, 75–76, 76(f); river discharge, 28–29, 133; salmonid communities, 54; salmonid growth rates, 82(f), 86; water temperatures, 26–27, 48
seawall, 130(f), 132
seawater, challenge, 150, 153, 262; conservative properties, 256
seawater hazard rating, 152
Sebastes caurinus. *See* rockfish, copper
Sebastes flavidus. *See* rockfish, yellowtail
Sebastes maliger. *See* rockfish, quillback
Sechelt Inlet, BC, 38
sediment load, conservation concerns, 232(t); correlation to delta and estuary size, 16; as estuary classification basis, 22
sediment yield, 16
Seine River estuary, France, 239
seines. *See* beach seines; purse seines
Sélune River estuary, France, 86
semelparous salmonids, definition, 262; evolution, 6–7; growth, 81, 84; life cycle, 5–6, 6(f); life history types, 17; predator avoidance, 109; smoltification, 90
Severn River estuary, UK, 50, 132, 213, 226
sewage, 136, 237–38
shad: 232(t); allis (*Alosa alosa*), 53; American (*Alosa sapidissima*), 56, 116, 210, 234; threadfin (*Dorosoma petenense*), 56; twaite (*Alosa fallax*), 53, 226
Shimanto River estuary, Japan, 43
shiner perch (*Cymatogaster aggregata*), 54, 175, 177
shipping and ports: as cause of habitat loss and degradation, 131–32, 135–36, 168, 224; and invasive species, 233–34; sound effects, 51
side channels, features, 36(t), 41; salmonid sampling techniques, 65
silverside (*Menidia beryllina*), invasive, 56
Simojoki River estuary, Finland, habitats, 45; water temperatures, 48, 167
Sixes River estuary, OR, 79, 83; carrying capacity, 163; competitive interactions, 125; food web, 102, 209
size/weight of fish: assessing over time, 81–83; and fitness/survival, 138–39; and migration, 72–73, 138; and predation, 112; and smoltification, 89. *See also* growth
Skagit River estuary, WA, 99, 131, 146, 163, 226, 239
Skeena River estuary, BC, 188
Skibotn River estuary, Norway, 77
Skjern River estuary, Denmark, 168
sloughs, 36(t), 41
smelt, 53–54, 99, 102, 114–15, 125
Smith River estuary, CA, 46
smolting/smoltification: definition, 263; process, 88–90; reverse, 92–93, 155–56; and salt wedge, 91–92; species and estuarine variation, 89; stress effects, 150–56. *See also* osmoregulation
smolts: competitive interactions, 123–24; cover-seeking behaviour, 77–78; definition, 16–17; estuarine

fitness for, 203; feeding behaviours, 99, 99(f), 102–3; growth, 83–84; holding/staging behaviour, 74; migration, 73–74; as predators, 115–17, 117(f), 119; as prey, 112–15, 118–19, 126; residency behaviour, 76–77; sampling techniques, 64, 66; specialized adaptation to estuaries, 244; tagging, 68, 70

snow melt, 28

sockeye salmon (*Oncorhynchus nerka*), conservation concerns, 219; disease mortality, 194–95; ecotypes, 18; estuary use, 13, 218; feeding behaviours, 96, 103; fitness/survival, 138, 140; growth, 86; as invasive species, 57; osmoregulation, 92; osmoregulation stressors, 153; parasites on, 197; as prey, 111, 119, 166, 174; short-term evolutionary considerations, 216; smoltification, 90; specialized adaptation to estuaries, 243–44; subhabitats, 42; water temperatures, 47–49

Somass River estuary, BC, 74, 195

Somme River estuary, France, 45, 239

sonar, definition, 263; for enumerating salmonids, 64

sonobuoys, 68

sounds. *See* noise in estuaries

Southern Hemisphere, competitive interactions, 183–84; conservation strategies, 226; subhabitats, 43; as model for estuarine ecology, 11, 199; salmonids as invasives, 57–58, 139–40, 174, 202, 204; water temperatures, 201

spear fishing, 64

species, definition, 263; diversity, 221; and growth rate, 86; smoltification variations, 90

spiny dogfish (*Squalus acanthias*), 167

splittail (*Pogonichthys macrolepidotus*), 210

Squamish River estuary, BC, food web, 97, 98(f), 102, 104; geomorphic classification, 22; habitat loss, 158; habitat restoration, 235; habitats, 46; predation, 167; salmonid growth, 86; salmonid sampling, 64

St. Lawrence River estuary, QC, 51, 130(f), 216–17

staghorn sculpin (*Leptocottus armatus*), 114–15, 169, 175, 177

starry flounder (*Platichthys stellatus*), 115, 117(f), 210

steelhead (*Oncorhynchus mykiss*), anadromy, 94; communities, 54; competitive interactions, 183; conservation concerns, 221; density-dependent growth, 163; distribution, 59; estuary use, 13; feeding behaviours, 101, 181; fitness/survival, 140, 143; growth, 81, 83; habitats, 39, 45–46; and invasive species, 55; osmoregulation, 92–93; as predators, 115–17, 117(f), 119, 173; as prey, 111, 166–67, 172; protection from fishing, 188. *See also* rainbow trout (*Oncorhynchus mykiss*)

sticklebacks (Gasterosteidae), 53, 122, 124

Stikine River estuary, AK, 10

stomach content analysis, 110–11

strand estuaries, 22, 23(f)

stratified estuaries, 21(f), 23

straying, 79

stressors, effect on osmoregulation, 150; and infection susceptibility, 194; research/data needs, 249; and "reverse smolting", 155–56; and salinity changes, 150–52; summary, 232(t); syndrome definition, 263; and water quality, 152–55, 232(t)

striped bass (*Morone saxatilis*), 174, 210

sturgeon (Acipenseridae), 53

subhabitats, 35, 36(t), 46, 71, 77, 101, 156–57, 166, 203, 206

suckers (Catostomidae), 53

surf smelt (*Hypomesus pretiosus*), 102

Surna River, Norway, 143

survival: assessing through experimentation, 141–43; assessing through observation, 143–44; assessing through statistical inference,

144–46; fitness components, 137–40; as measure of restoration, 239–40; modelling, 146–48. *See also* fitness
Svartå River estuary, Finland, 38(f)
Swinomish River estuary, WA, 75
Sylvia Grinnell River estuary, NU, 32, 83, 189
Syokanbetsu River estuary, Japan, 123
Syr Darya River estuary, Kazakhstan, 9, 134

tagging. *See* marking/tagging
Tagus River estuary, Portugal, 239
Taku River, AK, 50, 180
Tana River estuary, Norway, 116
Tasmania, invasive salmonids, 58
taxonomy of salmonids, 13–15
Tay River, Scotland, 89, 91
tectonic estuaries, 22, 263
Tees River estuary, England, 169
temperature. *See* water temperature
territoriality: and density-dependent growth, 160–61; factors affecting, 79; and life stage, 125; research/data needs, 251; and smoltification, 89; transience in estuary, 122. *See also* schooling
Thaleichthys pacificus. *See* eulachon
Thames River estuary, England, 66, 105, 132, 218, 231
thermoclines, 24, 30, 493
thinlip grey mullet (*Liza ramada*), 53
three-spined stickleback (*Gasterosteus aculeatus*), 53, 106(t), 124–25, 175, 177
tidal creeks, features, 36(t), 41–42; salmonid sampling techniques, 64–65
tidal freshwater zone, features, 26(t); temperatures, 27
tidal swamp-forested subhabitat, 37(t), 46
tides, gated, 134; range, as estuary classification basis, 24–25; role in homing/staging, 74; role in migration, 73–74; role in salmonid community, 54; role in stratification, 23

topminnows (Atherinopsidae), 53
training walls, 47, 105, 129–30
trawl sampling, 65
Tribolodon brandtii. *See* Pacific redfin
Tridentiger bifasciatus. *See* goby, Shimofuri
Tuloma River estuary, Russia, 235
turbidity: in estuaries, causes and ranges, 50–51; changes from human activities, 134; and predation, 114, 116, 170; and salmonid sampling, 65
Tyne River estuary, England, 170
typhoons, 29, 41

umbrella species, 225–26, 264
UNESCO Biosphere Reserve, 227
United Kingdom, dissolved oxygen standards, 211; habitat loss, 128; invasive salmonids, 59
United Kingdom Estuary Guide, 225
United States, dissolved oxygen criteria, 50; endangered salmonids, 15, 226; habitat classification, 32
upper estuary zone, 26(t), 27
Utka River estuary, Russia, 115–16, 117(f), 119
Utkholok River estuary, Russia, 101

Varzina River estuary, Russia, 91
Varzuga River estuary, Russia, 93
vegetation habitat, carrying capacity, 163; competitive interactions in, 124–25, 176(t), 177; definition, 42; example, 38(f); features, 36–37(t); food web, 104; habitat loss, 176(t), 177; habitat losses, 132, 158, 168–69; predation in, 118–19, 168–69, 176(t); restoration, 236–37; types, 42–46
vessel traffic. *See* shipping and ports
Vikings, salmonid fishing, 3–4
virgin sea run, 16. *See also* returning veterans
visual sampling, 65–66, 71
Volga River, Russia, 25

Waimakariri River estuary, New Zealand, 183

water balance (freshwater/ocean water), 20, 28
water exchange rate, 24
water flow patterns: effects of alteration, 9, 166, 168; and estuarine fitness for salmonids, 209–10; and osmoregulatory success, 202–3; and predation, 203; seasonal, 28–29. *See also* hydrodynamics
water properties, contaminants, 135–36; and estuary classification, 26–28; and invasive salmonids, 60–61; and migration, 72; and predation, 167; salmonid responses to, 47–51; and salmonid sampling, 65. *See also* dissolved oxygen; nutrients; pH; salinity; turbidity; water temperature
water quality, and carrying capacity, 164; and competitive interactions, 176(t), 177–78; contamination and water properties, 133–35; and food web, 158; improvement strategies, 237–38; indicators, 211–12; monitoring, 238; as osmoregulation stressor, 152–55; and predation, 169–71, 176(t); standards, 210–11
water temperature: changes from human activities, 133–34; and chemical marking, 68; and estuary classification, 24; factors affecting, 26–28; and invasive salmonids, 60; and life history types, 15; and osmoregulation, 88, 151–52; ranges and salmonid responses, 47–49, 152–53, 201; and salmonid genetics, 139; and salmonid growth, 81, 85–86; and smoltification, 88, 152–53

wave action on estuaries, 25
weight of fish. *See* size/weight of fish
Western Arm Brook estuary, NL, 39, 77, 93, 157–58
wet meadows: features, 37(t), 45–46; restoration, 236–37
wetlands, conservation, 33, 226; definition, 264; as indicator of predation refuge, 208–10; tidal swamp-forested subhabitat, 37(t), 46
White Sea, Russia, 58, 184
whitebait smelt (*Allosmerus elongatus*), 54
whitefish (Coregonidae), 52–53, 123
whitespotted char (*Salvelinus leucomaenis*), competitive interactions, 177; estuary use, 14; feeding behaviours, 99, 99(f), 101, 103–4, 107; growth, 84; life history stages, 16; osmoregulation stressors, 152; as predators, 115, 119; short-term evolutionary considerations, 216
whitespotted greenling (*Hexagrammos stelleri*), 54
whiting (*Merlangius merlangus*), 116
Wild Salmon Center, 221
Willapa River estuary, WA, 39, 43
Winchester Creek estuary, OR, 76
Wood River watershed, AK, 219
woody debris (tree trunks and stumps) subhabitat, 37(t), 46, 130(f)
wrasses (Labridae), 53

Yakoun River estuary, BC, 112
Yamato River estuary, Japan, 9, 9(f)
Yaquina River estuary, OR, 43, 179
York River estuary, QC, 73